Extreme Weather Events and Public Health Responses

W. Kirch, B. Menne and R. Bertollini (Editors)

Extreme Weather Events and Public Health Responses

With 94 Illustrations and 29 Tables

Published on behalf of
the World Health Organization Regional Office for Europe
by Springer-Verlag

Springer

Prof. Dr. med. Dr. med. dent. Wilhelm Kirch
Past President European Public Health Association (EUPHA)
Chairman Public Health Research Association Saxony
Medical Faculty, Technical University Dresden
Fiedlerstr. 27
D–01307 Dresden
Germany

Dr. Bettina Menne
Global Change and Health
WHO Regional Office for Europe
WHO European Centre for Environment and Health
Via Francesco Crispi, 10
I–00187 Rome
Italy

Dr. Roberto Bertollini
Special Programme on Health and Environment
WHO Regional Office for Europe
WHO European Centre for Environment and Health
Via Francesco Crispi, 10
I–00187 Rome
Italy

ISBN 978-3-642-06372-5 **e-ISBN 978-3-540-28862-6**

Springer is a part of Springer Science+Business Media

springeronline.com

Printed in the European Union

Editor: Thomas Mager, Heidelberg
Development Editor: Andrew Spencer, Heidelberg

Cover design: Erich Kirchner, Heidelberg

We are grateful to the Ministry of Health of Slovakia
for hosting the meeting from which this manuscript has been developed.

We are sincerely grateful to Mrs. Beatrix Hörger and Mrs. Ines Kube, Dresden,
for their excellent assistance in editing this book.

Contents

Foreword by Rudolf Zajac

Climate changes, a significant and truly global problem of mankind, represent a considerable risk factor for our environment and health. Extreme weather events are undisputed proof of climate changes. They are occuring with increasing frequency, affecting all continents of the world, with Europe being no exception. The intensity and frequency of *events* resulting from climate changes, such as floods, heatwaves and coldwaves, fires, winds and other natural disasters, have risen dramatically in recent years. The loss of homes, property, health, and human lives resulting from these disasters are a threat to people living both inland and on the coast. Therefore, it is necessary to intensify all efforts to identify effective measures to minimize the political, economic, social, environmental, and health consequences of these events.

Our present knowledge of extreme weather impact, gained by international cooperation of governmental and non-governmental institutions and organizations, has significantly contributed to the identification of factors influencing the change of climate and to the recognition of health impact assessment (WHO), but equally it shows that we are not yet adequately prepared to face threats and to overcome situations in which people are confronted with extreme weather events. Consequently, it is necessary to continue discussion on how to predict and prevent disasters, what to do once they have occured, and how to reduce the damages and the harm caused by them.

It is imperative to continue this discussion on the level of experts from various fields and professions, to inform the public, and to persuade government representatives and politicians to make reasonable decisions and to take effective measures to enable society to face the impact of climate changes on health.

Slovakia welcomed the opportunity to organize an international meeting in cooperation with the World Health Organization on the 9th and 10th of February 2004 in Bratislava and thus contribute to the discussion on the impact of extreme weather on human health. Experts from 25 countries outlined possible resources in the field of extreme climate changes. This publication is a compilation of concrete case studies and the presentations by individual countries delivered during the meeting.

I believe that this publication will be a significant asset for many countries and will serve as a knowledge base for the preparation of effective strategies, national action plans and measures, thus contributing to the minimization and the moderation of the negative consequences of global climatic changes.

March 2005

Minister of Health of Slovakia

Foreword by Karin Zaunberger

I am honoured to write a few introductory lines for the topic heat waves in the context of the book on "Extreme Weather Events & Public Health Responses". The heat wave of August 2003 in Europe was evidence that no-one is on the safe side when it comes to the impacts of climate change. Though some may argue about whether these extreme weather events are linked to global change or not, these events revealed in a rather drastic way our vulnerability and our lack of preparation.

The project cCASHh *"Climate change and adaptation strategies for human health in Europe"* (May 2001 – July 2004), co-ordinated by WHO and supported by the "Energy, Environment and Sustainable Development Programme", in the Fifth EU Framework programme for Research and Development aimed at

- identifying the vulnerability to adverse impacts of climate change on human health;
- reviewing current measures, technologies, policies and barriers to improving the adaptive capacity of populations to climate change;
- identifying for European populations the most appropriate measures, technologies and policies to successfully adapt to climate change; and
- providing estimates of the health benefits of specific strategies, or combinations of strategies, for adaptation in different climate and socio-economic scenarios.

Some of the research results are reflected in this book. Not only do these types of research activities need an interdisciplinary approach, but also prevention of and preparation for extreme weather events need cooperation at all levels and throughout disciplines. The cCASHh project was a good example and I hope that this important work will be continued.

Project Officer, European Commission, DG RTD

Preface by Wilhelm Kirch

When I was invited in November 2003 by Dr. Bettina Menne from the WHO Regional Office for Europe, European Centre for Environment and Health, Rome, to give a presentation on "Lessons to be learnt from the 2002 floods in Dresden, Germany" at a WHO conference held in February 2004 in Bratislava on "Extreme Weather Events and Public Health Responses", I was somewhat surprised since from the scientific point of view I had never had anything to do with extreme weather events. At that time I was President of the European Public Health Association (EUPHA) and in November 2002 had organized the EUPHA-conference in Dresden, which took place only three months after the severe floods of August 2002 in parts of Austria, Slovakia, the Czech Republic, Poland and in Southern Germany. The Dresden area was one of those centres of destruction caused by the floods and was really badly affected. Thus we were glad even to be able to organize the yearly congress of EUPHA after so much damage had occurred. I therefore answered Dr. Menne that my only connection to extreme weather events was that I come from Dresden, but this did not appear to disturb her, possibly in the sure knowledge of having invited enough real experts on the topic of the conference anyway.

Hence the theme of the Bratislava meeting sounded interesting to me. And because I just had edited a book about "Public Health in Europe" on the occasion of our EUPHA conference from 2002, during the preparations for my contribution on the Dresden floods the idea came up to edit another book entitled "Extreme Weather Events and Public Health Responses", to include most of the presentations of the Bratislava meeting. I suggested the idea to Dr. Menne and Dr. Bertollini from European Centre for Environment and Health of the WHO Regional Office for Europe, Rome who were the organizers of the Bratislava conference. They apparently were in favour for the book edition suggested by me and thus we started to collect manuscripts. I was surprised and appreciated very much that 25 authors out of 27 whom we asked to submit an article responded promptly and provided us with a manuscript of their contribution to the Bratislava conference. My biggest concern was to publish the book in due time, as nothing is more uninteresting than to have a publication from an event which took place a long time ago. Thus I am glad that we have managed to edit our book "Extreme Weather Events and Public Health Responses" so soon after the meeting. Furthermore, I appreciate that we have dealt in the book with several relevant aspects of the theme such as "Projected changes in extreme weather in Europe", "Heat and cold waves", "Flooding", "Public health and health care responses to extreme weather events" and to have made recommendations in this concern. The present book will be of interest not only to experts of various professions in this field, but also to people who have to deal in certain moments with extreme weather events.

Dresden, May 2005

Past President EUPHA

Preface by Wilhelm Fuchs

Preface by Jacqueline McGlade[1] and Roberto Bertollini[2]

Recent episodes of extreme weather events in Europe, including the floods of 2002 and the heat waves in the summer of 2003, have been accompanied by a significant and somewhat unexpected toll of deaths and diseases. For example, the health crisis in France caused by the 2003 heat-wave was totally unforeseen and was only detected belatedly. Health authorities were overwhelmed by the influx of patients; crematoria and cemeteries were unable to deal with the excessive number of bodies; and retirement homes were under-equipped with air-conditioning or space cooling environments and human resources. The crisis was further aggravated by the fact that many elderly people were living alone without a support system and without proper advice to protect themselves from the heat.
Because of these calamities, the linkage between extreme weather events and population health has been increasingly recognised by the scientific and decision-making communities; research and actions have been initiated to set up an efficient system for preparedness and response throughout Europe.

This book collates a number of important case studies, research and experiences on the health impacts of these recent events. They show the efforts being made by the public health and environment communities to evaluate the effectiveness of the measures taken to respond to the crises, to assess the early warning systems in place, and to use the lessons learnt to better tailor future activities. The experiences summarized in this book also underline the need to address more systematically the health system response to weather related crises as well as the knowledge gaps regarding both the effectiveness of the early warning systems in place and the interactions between different phenomena, for instance heat and air pollution.

At the Fourth Ministerial Conference on Environment and Health, held in Budapest in June 2004, it was further recognized by the European Ministers that as a consequence of our changing climate the intensity and frequency of extreme weather events may vary and probably increase in the future. Even if the extent of the association between climate change and extreme weather events is still a subject of debate in the scientific community, there is no question that there are many modulating anthropogenic influences inducing extreme weather situations and sometimes enhancing the impacts of the weather events. Changes in land use and hydrology create multiplying effects when the natural or "ecological" protection has disappeared. Examples are reduced wetland buffering areas, straightening of rivers, forestry fragmentation and logging; and in the heat wave case, the induction caused by air pollution from transport and the urban heat island effect. The complexity of the processes involved further underlines the link between ecology and human health.

Extreme weather events will continue to pose additional challenges to current and future populations, in terms of risk management and the reliability of infrastructure, including health services, power supply and others. Every effort should therefore be made by the environment and public health communities to put in place evidence based interventions and where necessary precautionary measures to limit the impacts on the environment and actively to reduce the burden of mortality and disease on human populations and ecosystems.

There is no time for complacency. Actions must be taken urgently to protect the environment of Europe and assure the health of its citizens.

[1] Executive Director, European Environment Agency
[2] Director, Special Programme on Health and Environment, WHO Regional Office for Europe

Editorial

'Si le respect de l'homme est fondé dans le cœur des hommes, les hommes finiront bien par fonder en retour le système social, politique ou économique qui consacrera ce respect'
"Lettre à un otage", Antoine de Saint-Exupéry

The global climate is changing. During the last 100 years warming has been observed in all continents with an average increase of 0.6 ± 0.2 °C (man ± SD) in the course of the 20th century. The greatest temperature changes occurred at middle and high latitudes in the northern hemispheres. The trend towards warmer average surface temperatures for the period since 1976 is roughly three times that of the past 100 years as a whole. In the last decades warming seems to be attributable to human activities (man-made environmental changes) like land-use changes, deforestation, urbanisation and the reduction of wetlands. Global climate change is likely to be accompanied by an increase in frequency and intensity of extreme weather events. Climate variability occurs at both the level of gradual change as well as the level of extreme events.

Extreme weather events are those events which society is unable to cope with. They are by definition rare stochastic events. Europe has experienced on unprecedented rate of extreme weather events in the last 30 years. Heat waves occurred in France, Italy, Portugal, the Russian Federation, Hungary and Bulgaria between 2000 and 2003. The annual number of warm extremes increased twice as fast as expected based on the corresponding decrease in the rate of cold extremes. On the other hand cold waves brought serious health problems to Northern Europe, the Russian Federation and even Bosnia and Herzegovina. In 2002 Romania suffered deleterious windstorms and Public Health responses were necessary. Last but not least, in recent years severe flooding occurred in many European countries like U.K., Poland, Czech Republic, Austria, Italy and Germany causing enormous damages, e.g. in August 2002. On the basis of current predictions on climate, more extreme weather events have to be faced in the coming years and they are likely to be more severe. Thus appropriate actions have to be undertaken in order to protect the population and the countries affected.

In the present book, articles under the following headings are published: "Climate variability and extremes in Europe", "Temperature extremes and health impact", "Response to temperature extremes", "Flooding: the impact on human health", "National case-studies on health care system responses to extreme weather events" and "Recommendations". They shed light on the mode of development and the damages caused by extreme weather events and finally give some hints of what has to be done to cope with them.

Climate Variability and Extremes in Europe

Addressing „The climate dilemma", **A. Navarra,** Bologna, comments that the concept of climate has surged to a problem of planetary relevance with an impact on several sectors of human society. The fluid envelopes of the Earth, the atmosphere and the oceans are the main components of the climate system. It is the dominant pattern of motion of the fluids that determine the climate in any given place of our planet. The distribution of land masses and of mountain ranges is also a major factor in shaping the dominant climate features. Furthermore, sea ice, the biosphere, the soil as well as land ice sheets are contributing factors. The complexity of the climate system and limitations of experimental capabilities do not appear to allow a classical scientific approach to it. This leads to a complex situation where it is sometimes difficult to differentiate between facts and assumptions. But without any doubt there is increasing evidence that two additional factors have become relevant for changes in climate over the last century, namely the steadily rising carbon dioxide and greenhouse gas concentration in the atmosphere, and the higher surface temperatures on our planet. Navarra concludes that, in the case of weather extremes which may be caused by the factors mentioned above, two levels of monitoring are necessary: short term weather forecasts (up to 8 days) and the long term view which tries to assess the frequency and characteristics of weather extremes over a period of 20 – 30 years from now. **G.R. McGregor et al**, Birmingham, state in their article "Projected changes in extreme weather and climate events in Europe" that one possible outcome of the predicted global climate change is an increase in the frequency and, possibly, intensity of extreme weather and climate events. The purpose of this chapter is to review ways in which climate change may alter the occurrence of extreme events and to consider whether certain trends predicted are reflected in the observational record of extreme events for Europe. They point out that the terms extreme weather and climate events differ from each other and refer to different phenomena. An extreme weather event like a tornado or thunder storm lasts between 1 and 6 to 10 days, whereas an extreme climate event implies a number of extreme weather events over a given time period, like hot and dry summers or wet and stormy winters. They summarize that climate change projections indicate the likelihood of substantial warming by 2100 and expect non-linear increases in extreme weather events with a change in mean climate. Trends in time series of observed extreme weather and climate indices match those suggested by climate model based projections of future climate and support the hypothesis that more frequent extreme events across Europe are associated with the climate change. **Z.W. Kundzewicz**, Poznań and Potsdam, asks in the first of his two articles in this book "Is the frequency and intensity of flooding changing in Europe?" He reports that between the 1950s and the 1990s, yearly economic losses from weather extremes have increased tenfold (in inflation-adjusted US dollars). In the last decade several destructive floods have hit Europe, of which the flood of August 2002 in Central Europe was responsible for damage costs of about 15 billion Euro. Due to global warming, precipitation has increased (2 – 4 % in the last 50 years) directly impacting on flood risk. Some recent rainfall events have exceeded all-time records. On 12 – 13 August 2002 from 6.00 a.m. to 6.00 a.m., 312 mm rain was measured in Zinnwald, Saxony, Southern Germany. Z. W. Kundzewicz concludes that in many European places flood risk is likely to have grown and a further increase of this risk is projected. **J.-C. Cohen, J.-M. Veysseire and P. Bessemoulin,** Paris, present reflections on "Bio-climatological aspects of the summer 2003 over France". During June to August 2003 there was the hottest summer period of the last 50 years in France with an extreme heat wave in the first two August weeks of 2003. In Paris, with serial data files since 1873, morning temperatures on the 11th and 12th were highest ever registered at 25.5 °C (previous record: 24 °C in 1976). The heat wave was outstanding in duration and in geographical extension (over all parts of France, including mountains and coastal regions) followed by a six month period of drought. Its tragic health impacts induced 15,000 excess deaths, probably caused by high night temperatures and high levels of pollution. Metéo-France issued a press release on 1 August 2003 announcing a progressive climb in temperature for the following period and the whole country. In response to this heat wave an early Heat Health Warning System is being established in France. Starting with the definition of New Public Health (Public Health is the science and art of preventing disease, prolonging life, and promoting health through the organized efforts of society [Committee of Inquiry into the Future Development of the Public Health Function, 1988]) **Kristie L. Ebi,** Alexandria USA, presents an article on "Improving Public Health responses to extreme weather events". Measures to reduce disease and save lives are categorized into primary, secondary and tertiary prevention. Although adverse weather and climate events cannot be prevented, primary prevention, particularly development of early warning systems, can reduce the number of adverse health outcomes that occur during and following an event. These educational programs have often been implemented in a certain region when an event has caused injuries and deaths. Few programs have been established proactively. Instead, Public Health activities have focused on surveillance and response systems (secondary prevention) to identify disease outbreaks following an event. Surveillance and response systems are ineffective for identifying and preventing many of the adverse health outcomes associated with extreme climate and weather events. The increasing ability to predict extreme events and advances in climate forecasting provide Public Health authorities with the opportunity to have early warning systems available for reducing vulnerability to extreme weather events.

Temperature Extremes and Health Impact

The only article in this book about "Cold extremes and impacts on health" is presented by **J. Hassi** from Oulu. He states that the composition of the atmosphere is changing, thereby altering the radiation balance of the earth-atmosphere system, producing the global warming and extreme conditions which were already mentioned several times. The latter include not only anomalously high but also low temperatures with extreme cold spells. Despite the fact that excess mortality related to heat is increasing, deaths from cold exposure still represent the majority of mortality excess due to extreme temperatures. Although the death rate from excessive cold has been epidemiologically quantified, less attention has been given to the Public Health actions to prevent negative impacts of cold temperature. These preventive measures should not only be related to excess cold mortality but also include actions concerning cold injuries, diseases and physiological cold stress. Furthermore, exposure to cold increases the risk of respiratory diseases, coronary heart disease and other arteriosclerotic diseases. These in particular are responsible for the excess winter mortality which varies in different European countries between 5 and 30 %, while elderly people are especially susceptible to the impact of weather changes. Countries with a high prevalence of poverty and inequity are significantly associated with winter mortality. Public Health actions for preventing cold-related health impacts include adequate weather forecast, cold wave warning systems, warm housing, protection against outdoor body-cooling and intervention programs for developing behavioural changes in cold-exposed areas. Generally people from Northern countries are more experienced and successful in handling cold exposure. **G. Havenith,** Loughborough, provides background information on "Temperature regulation, heat balance and climatic stress". He points out that in the evolutionary sense, man is considered a tropical animal. Our anatomy as well as our physiology is geared towards life in moderate and warm environments. Human body thermoregulation is discussed under certain conditions like exercise, work load or heat with regard to air humidity, wind speed, morphology and fat, gender, an underlying arterial hypertension, drug and alcohol intake or age. In good health the body can deal well with heat and cold stress, but when thermoregulation becomes impaired, as it the case with ageing, the human is at risk. Age seems to be the best predictor of the increase of mortality at high temperatures. Longer periods of hot weather, especially when little relief is given at night, have hit mainly the older population. This is consistent with the observations of J.-C. Cohen, J.-M. Veysseire and P. Bessemoulin, but also with those of other authors of this book, who found an elevated death rate during heat waves especially in the elderly population. Concerning cold exposure, G. Havenith states that the analysis of mortality and morbidity data is more complex, hence cold related problems are not always attributed to the cold in statistics. Also **Stéphanie Vandentorren and P. Empereur-Bissonnet,** Saint-Maurice, report on the "Health impact of the 2003 heat wave in France" which has already been described by T. Michelon, P. Magne and F. Simon-Delavelle as well as J.-C. Cohen, J.-M. Veysseire and P. Bessemoulin in this book. After a warm month of June in 2003, with temperatures 4–5 °C above seasonal averages and two hot last weeks in July, a heat wave struck France as a whole in August 2003. In Paris, the temperature exceeded 35 °C for as long as 10 days, a situation never observed since 1873. This led to a total mortality increase of 55 % between 1 August and 20 August compared with the expected number of deaths estimated on the basis of the mortality in 2000, 2001 and 2002 for the same period. The mortality was particularly high for elderly people, to the extent of an increase of 70 % in people >75 years of age. In order to identify etiologic factors for the increased mortality, so-called case control surveys were carried out immediately after the heat wave. The results will contribute to establishing a Heat Watch Warning System in 2004 in order to prevent excess mortality during future heat waves. Further European projects dealing with this purpose are PHEWE (Assessment and Prevention of Acute Health Effects of Weather Conditions in Europe) and the PSAS9 program. Also Portugal was hit by heat waves in June, July, August and September 2003, which is outlined by **R. Calado et al,** Lisbon, in their article "Portugal, summer 2003 mortality: the heat waves influence". The authors point out that already in 1981 and 1991 longer lasting temperature rises above 32 °C were observed. As after these periods studies had indicated that there was a strong relationship between the heat waves and excess death rates, the National Observatory of Health established a Heat Waves Vigilance and Alert System, while data from the Meteorology Institute also had to be considered. Thus since 1999 each year from 15 May to 30 September institutions like the Civil Protection and the General Directorate of Health are provided with the so called "Icaro Index" on a daily basis. The index predicts the intensity of hot weather periods, which may possibly cause deaths, three days in advance. These alerts had to be given three times during the summer 2003 and regional and local health authorities were informed. A Public Health Call Centre provided information about heat prevention measures and it answered 1400 calls during this time period. Excess deaths were averaged to 1802 cases. Finally, in Portugal a Contingency Plan for heat waves is to be established. **Anna Paldy et al,** Budapest, point out that the 3rd Ministerial Conference on Environment and Health in London 1999 recommended that national assessments of the potential health effects of climate variability should be undertaken. Thus concerning weather changes, vulnerable populations and subgroups should be identified, furthermore, interventions that could be implemented to reduce the current and further burden of corresponding diseases should be proposed. In their article on "The effects of temperature and heat waves on daily mortality in Budapest, Hungary 1970 – 2000" Anna Paldy et al. report that during these years mean daily temperature and the number of hot days increased reaching peak values in the 1990s. Concerning mortality, the authors found a considerable reduction during these 31 years (about 10 %). But with an average rise in mean temperature of 5 % during each year, the risk of mortality increased significantly. During five heat waves since 1994 mortality in the adult group did not appear to be affected. Only one heat wave in August 2000 (3 days) was associated with an excess mortality of 72 %. Analogous to what is reported by Hassi in this book for excess cold mortality, also heat wave mortality is mainly attributable to cardiovascular, cerebrovascular and respiratory diseases. Weather variability, rather than heat intensity, is often the most important factor defining human sensitivity to heat. Relative humidity had a slight, but significant effect on mortality during the winter period. The influence of air pollutants on mortality was weaker than that of temperature in the Budapest-study. **Susanna Conti et al,** Rome, report on an "Epidemiologic study of mortality during summer 2003 in Italian regional capitals: results of a rapid survey" requested by the Italian Minister of Health. The period of 1 June to 31 August 2003 was analysed and a mortality increase of 3134 deaths was found due to the unusually hot summer (compared with 2002). 92 % of the people who died were 75 years and older. The mortality rise was most pronounced in Torino (44.9 %), Trento (35.2 %), Milan (30.6 %), Genova (22.2 %), Bari (33.8 %), Potenza (25.4 %) and L'Aquila (24.7 %). Concerning the Humidex, which is a discomfort index resulting from the combined consideration of excessive humidity and high temperatures, a significant correlation was found between this parameter and mortality in cities like Turin, Milan, Genova, Rome and Bari. Calculation of the so called "lag time" allowed presentation of data on the time between exposure to heat and the occurrence of deaths. The maximum correlation was observed a few days before the fatalities: 2 days for Rome, 3 days for Bari and Genova and 4 days for Milan and Turin. The relationship between mortality on the one hand, and discomfort climate conditions (Humidex) together with the short lag time on the other, gives a clear Public Health message: preventive, social and health care actions have to be administered to elderly and frail people in order to avoid excess deaths during heat waves (see L. Abenhaim in this book). Paola Michelozzi et al, Rome, state that the relationship between weather, temperature and health has been well documented throughout the literature, both for summer and winter periods. The correlation of mortality and temperature appears graphically as a "U" or "V" shape, meaning that mortality rates are lowest when temperature ranges between 15 and 25 °C, rising progressively when it increases or decreases. In their article **Michelozzi et al** deal with "Heat waves in Italy: cause specific mortality and the role of educational level and socio-economic conditions". The authors observed excess death rates in people with a low education level e.g. by 43 % in Rome or by 18 % in Turin. Diseases of the central nervous system (CNS), of the cardiovascular, respiratory, endocrine system and psychiatric disorders were most frequently responsible for the excess mortality during heat waves in Italy in the course of summer 2003. In Rome an increase of CNS, respiratory and cardiovascular diseases causing excess death of 85 %, 39 % and 24 %, respectively, was found. For Milan corresponding values for CNS, respiratory and endocrine diseases were 118 %, 82 % and 68 %, respectively. Paola Michelozzi et al. conclude that demographic and social factors, as well as the level of urbanization, air pollution, the efficiency of social services and health care units represent relevant local determinants of the impact of heat waves on human health. Therefore prevention measures are needed which are provided in Italy by the Heat Health Watch/Warning System (HHWWS).

Response to temperature extremes

In their article on "Lessons of the 2003 heat wave in France and action taken to limit the effects of future heat waves" **T. Michelon, P. Magne and F. Simon-Delavelle,** Paris, describe the severe heat wave of August 2003 in France. As already mentioned by J.-C. Cohen, J.-M. Veysseire and P. Bessemoulin as well as Stéphanie Vandentorren and P. Empereur-Bissonnet in this book, the catastrophic health consequences of this heat wave included an estimated 15.000 excess deaths. Thus health authorities spoke of a "health crisis" in this context, which was unforeseen and which had serious repercussions in the French public. As a deficit of health information, of defined responsibilities, a work overload of health authorities (during the summer holidays), under-equipped homes (e.g. missing air-conditioning) and the lack of support systems for elderly people living alone became evident, the French government had to intervene. Several steps were undertaken to limit damages of future heat waves on public health: retrospective studies were initiated in order to identify heat wave risk factors and defining Public Health action levels determined by meteorological parameters. Furthermore, health and environmental surveillance has to be established e.g. for registration of hospital admissions and meteorological data during heat waves. Finally, action plans were made to be implemented at national and local levels before June 2004. R. Calado et al reported on the heat wave of 2003 in Portugal and pointed to the relevance of the ICARO index as a useful instrument for identifying the impact of high temperatures. **P.J. Nogueira,** Lisbon, presents additional aspects of it in his article "Examples of heat health warning systems: Lisbon's ICARO's surveillance systems, summer of 2003". Without the ICARO system intervention during the heat wave in Portugal in 2003 might not have been successful as it is a full operational heat health warning system. A higher morbidity with an increased admission of patients to hospitals as well as to healthy emergency services and excess deaths were noticed, suggesting that heat may have an "endemic aspect" which has not been referred to elsewhere. In his contribution to this book entitled "Lessons from the heat-wave epidemic in France (summer 2003)" **L. Abenhaim,** Paris, draws some Public Health conclusions from heat related events in France, but he also attempts to broaden the scope for Europe as a whole. He asks questions like "should we concentrate on epidemics or endemics of heat-related events, can epidemics of these events be predicted, can epidemics of heat related events be detected, can heat epidemics be prevented, what can be done during epidemics and what is the difference between heat related epidemics and corresponding crises? He answers and concludes that the prediction, detection and prevention of heat related epidemics is restricted by a lack of scientific knowledge and experiences concerning this topic. Air conditioning is certainly the most efficient measure to mitigate heat related symptoms and it should be available during heat waves in a continuous fashion at least for the elderly and people with health problems. This certainly contributes to the management of heat-related epidemics, by which morbidity and mortality may be reduced. **T. Kosatsky, N. King and B. Henry,** Rome, Montreal, Toronto, point out in their article that Canadian cities have initiated active heat response strategies since 1998. In this concern, they report on Montreal's and Toronto's approach of issuing public advisories for hot weather response, especially for the elderly and the homeless, in cooperation with the Canadian Meteorological Service. Research and action programs were instituted to protect residents against the effects of heat on health. Furthermore, civil defense authorities established a heat wave emergency response plan. Research activities will include the definition of a heat emergency action level, the identification of the population segments adversely affected by heat, the development of a geographic information platform, the evaluation of air conditioner use, medication practices and patient hydration in chronic care centres. The results obtained should improve our knowledge about client-specific heat health management plans.

Flooding: The Impact on Human Health

D. Meusel and W. Kirch, Dresden, present a report on the floods in the Dresden area and the "Lessons to be learned from the 2002 floods in Dresden, Germany". After unusually intense rain and thunderstorms in the second week of August 2002 catastrophic dimensions became evident in Bavaria, Austria, Slovakia, the Czech Republic and Poland. The meteorologically perfect cyclone "Ilse" with plenty of warm humidity in its lower spheres and a cold higher sphere, arrived in the mountains surrounding Dresden on the 10th of August 2002. More than 100 litres/m^2 rain during the night of 12th to 13th August caused small mountain rivers to collapse and water reservoirs to become overfilled. These masses of water and those coming from Bohemia and other parts of Saxony combined in the River Elbe causing flood damages never seen before in many cities (see figures in this article). Public Health issues of this disaster are discussed (hygiene, vaccination, problems with the decision making processes, multilevel management plans, transboundary adjustments, preventive measures). In addition to the Dresden flood experiences, **Z.W. and W.J. Kundzewicz,** Poznan and Potsdam, present a further contribution on "Mortality in flood disasters". They point out that the two most important socio-economic characteristics of disastrous floods are the number of deaths and the economic damage. Neither of these is easy to quantify in a reliable way. The term "flood related fatality" is self-explanatory and can be interpreted in a rather broad way. Certainly there is a substantial difference between the deaths of an old handicapped woman, who drowned alone in her bedroom, and that of a young, strong, and self-assured man who underestimated the danger and put himself in harm's way. In detail the authors inform about the death circumstances of 21 people all under the age of 44 who died during the July 1997 flood in Poland. As already mentioned, the damage costs of the floods of August 2002 in Central Europe are estimated at about 15 billion Euro. In the review on „The human health consequences of flooding in Europe" **S. Hajat et al,** London, Alexandria (USA), Rome, state that floods are the most common natural disaster in Europe. As already pointed out, various mechanisms may cause flooding. Flood characteristics influence the occurrence and consequences of the flood event. According to the 3rd Assessment Report of the Intergovernmental Panel on Climate Change, intense precipitation periods with floods will increase in frequency and intensity. Therefore the development and implementation of measures to prevent adverse health impacts from flooding are necessary. The health consequences of floods include drowning, injuries, anxiety and depression lasting for months after the event, whereas infectious diseases have been observed relatively seldom in Europe during and after flooding. Groups vulnerable to the health impacts of floods are the elderly, children, disabled, ethnic minorities and people with low incomes. Thus vulnerability indices have to be developed in order to establish public health interventions (risk-based emergency management programs). **E. Penning-Rowsell, Sue Tapsell and Theresa Wilson,** London, present "Key policy implications of the heath effects of floods". They point out that the impacts of floods are serious and far-reaching. Frequency and extent of flooding worldwide are expected to increase over the next 5 – 10 decades due to global warming. Despite this fact, the authors found very little information and guidance in a Europe-wide survey of emergency plans with coherent strategies for coping with health impacts of flooding or natural disasters. But there is no doubt that political measures for flood mitigation are about the priority of responses during and after flood events. In particular, a pre-planning for these activities with multi-dimensional emergency programs is needed. In this concern it has to be mentioned that early warnings of floods and the identification of those who are most vulnerable to floods and their health impacts have to be targeted. Most of the corresponding recommendations in terms of pre-event warning provision as well as post-event health care and their aftermath are straightforward. They include assistance for the elderly, for those with underlying diseases or prior-event health problems, for the poor or for dependent subjects e.g. children. In natural disasters the most striking problem is that the responsibilities for the different actions needed are split between too many organizations. **Merylyn McKenzie Hedger,** Bristol, finishes the flooding chapter with her article on "Learning from experience: evolving responses to flood events in the UK". She deals with tidal waters and spring tides (coastal floods) and also with the so-called flash floods which occurred in the UK and in Central Europe. Both Z. W. Kundzewicz and W. Kirch and D. Meusel only reported in their articles about riverine and flash floods. In the UK there were several floods from the catastrophic East coast flooding 1953 to the events in 1998 and 2000. The 1998 flood led to a management change in the responsible British Environment Agency and a new flood warning system which proved successful at the time of the 2000 floods, but has to be improved further. In 1953 a great storm surge accompanied by gale force winds swept over the North of the UK causing widespread flooding of coastal areas (more than 1000 miles). Over 300 people died, 32,000 had to be evacuated from their homes and 24,000 houses were flooded. The Easter flood 1998, however, was a flash flood caused by enormous amounts of rainfall in the preceding months affecting the Midlands and Wales leading to deaths, serious injuries and losses of homes and personal possessions. The autumn 2000 floods exceeded insured costs of >1 billion pounds. Merylyn McKenzie Hedger concludes that the policy for managing flood risk in UK is iterative and dynamic. Flood related topics such as the climate change demand further attention. As already mentioned in the case of the Dresden floods, guidance to planning authorities has to be improved. More tools and information must be delivered to local planning authorities to help them with delivery of flood risk assessment.

National case-studies on health care system responses to extreme weather events

In their article "Extreme weather events in Bulgaria from 2001 – 2003 and responses to address them" **Rajna Chakurova and L. Ivanov,** Sofia, describe the different various geographic formations of Bulgaria with consecutive climatic specifics. These led to various extreme weather events during 2001 – 2003. These included storms with hurricanes and tornados, extreme cold spells with ice-formation, warm and dry spells, torrential rains, floods, landslides or forest fires, although Bulgaria has the lowest water resources per capita of all European countries. The extreme weather events of 2001 – 2003 led to human losses and huge material damages, the most severe of which were incurred by floods. Management bodies, units of SA Civil Protection, ministries and different agencies have participated in addressing the aftermath of the disasters with their staff and equipment. In Bulgaria regulations exist for organizations for handling accidents, catastrophes and the aftermath of disasters. The authors conclude that measures for the prevention of extreme weather events should belong to the national priorities of Bulgaria. Anca Cristea, Bucarest, reports on a chain of calamities during the year 2002 in Romania. Starting with cold waves in Transylvania and the Republic of Moldova early in year, which led to a considerable decrease in the production of rape, barley, wine and fruits, a devastating drought was seen in April/May/June 2002 followed by heavy rains and floods in the centre and south of the country. The most dreadful phenomenon was a tornado in Făcăieni, South Romania, which caused severe damage. Due to unusually cold weather in the last third of the year, huge economic losses were registered once again. All of these extreme weather events had to be managed by so called Local Disaster Defence Committees which exist in every Romanian region. They care for hygienic problems, water supply, the health care of the population, its information about relevant necessary measures, but also for store houses of chemical substances. Anca Cristea assumes that the focal point in the approach of disasters is the human dimension: how prepared are the societies to cope with extreme events? She points out that any post-calamity evaluation is not able to register the real psychological impacts, the pain and the elapse of hope of every individual affected by the catastrophes. **A. Khadjibayev and Elena M. Borisova,** Tashkent, start their report on "A system of medical service to assist the population of Uzbekistan in the case of natural catastrophes" with the remark that the annual precipitation in the plain area of their country averages 120 – 200 mm, which makes Uzbekistan very vulnerable to heat waves and droughts. It has to be mentioned that in the context of the Dresden floods, 312 mm rain/m² fell within 24 hours. Thus the population of Uzbekistan is traditionally used to long, dry and hot summers and has accumulated effective measures against the heat. For cases of natural disasters Uzbekistan provides a Medical Emergency Service which functions on different levels from non-hospital medical assistance via qualified medical aid to specialized medical aid. In particular A. Khadjibayev and Elena M. Borisova point to the world's ecological tragedy, the Aral sea. Its inland waters used to provide prosperous life to the country's population. Nowadays the dried up bottom with around 700,000 tons of harmful salt damages the overall eco-system causing medical, social and economical problems ("Aral crisis"). Affected is an area of more than 100,000 km² including the Amudarya's delta. **V. Kislitsin, S. Novikov and Natalia Skvortsova,** Moscow describe a smell of burning and haze which was observed in the summer of 2002 for several days together with high concentrations of pollutants produced by forest and peatbog fires as well as industrial and vehicle emission in the Moscow region. This was preceded by a heat wave lasting several weeks leading to the pronounced smog mentioned. Smog is a well known health hazard consisting of chemical substances and suspended particulates (up to a diameter of 10 microns). Ozone, sulphur dioxide, nitrogen dioxide, carbon dioxide, benzene, formaldehyde, polychlorinated dioxins and benzofurans are among the chemical substances which were emitted. A health risk assessment methodology was used to evaluate the main adverse effects of the smog. Specifically, concentration-response functions were made for selected air pollutants. Thus a computer program was developed, the database of which contained information on the 25 most hazardous air pollutants in order to calculate different exposure outcomes. Consequently, warnings could be given to the people affected.

We have added an Annex to the book with a working paper on 'Public health responses to extreme weather events' derived from the 4th Ministerial Conference on Environment and Health which took place in Budapest from 23 – 25 June 2004. Furthermore, a description on a currently ongoing study of health effects of extreme weather events is presented. **Inge Heim,** Zagreb, describes a five year study on this topic which ends in 2004. The investigation was performed in Zagreb. So far, more than 10,000 patients with coronary disease, arterial hypertension, cardiac arrhythmias and multiple risk factors for atherosclerosis were interviewed using a questionnaire. The answers and the symptoms of the patients were correlated with meteorological parameters like air temperature, humidity or winds which were registered in defined time intervals. Furthermore, the number of patients who were daily admitted to the Zagreb hospitals due to acute myocardial infarction, unstable angina pectoris, chronic heart failure or who had a sudden death was registered. Dr. Heim expects that the study results will shed some light on the influence of weather on the course of cardiac diseases and corresponding patients. Thus Public Health measures could be developed and used for certain meteorological conditions.

Finally, recommendations are given for the prevention of health impacts of extreme weather events from **Bettina Menne** from WHO Regional Office for Europe, Rome. A corresponding working document of the **Budapest Ministerial Conference** held in June 2004 on 'Public Health Responses to Extreme Weather Events' is presented at the end of the book.

Dresden, May 2005 — W. Kirch, EUPHA

Extreme Weather Events and Health: An Ancient New Story

Bettina Menne [1]

"Two attitudes should characterize scientists: On the one hand he must honestly consider the question of the earthly future of mankind and, as a responsible person, help to prepare it, preserve it and eliminate the risk; we think that this solidarity with future generations is a form of charity. But the same time the scientist must be animated by the confidence that nature has in store secret possibilities which it is up to intelligence to discover and make use of, in order to reach the development which is in the Creator's plan".

Pope Paul VI, 19 April, 1972, address to the Pontifical Academy of Science.

Introduction

Weather is an ancient human health exposure, says Hippocrates, in "On Airs, Waters and Places", circa 400 B.C. (McMichael et al. 2003). History has shown that weather and climate variability are important determinants of health and well-being. Examples are many; like the "biblical flood" scenario purportedly 6000 B.C., the vast droughts in the MiddleAges, the severe drought in 1921 in vast areas in the former Soviet Union causing millions of deaths, the North Sea floods in 1953 causing thousands of deaths, the heatwave in 2003 causing an approximated 30,000 excess deaths. There is still considerable uncertainty about the rates of climate change that can be expected, it is now clear that these changes will be increasingly manifested in important and tangible ways, such as changes in extremes of temperature and precipitation, decreases in seasonal and perennial snow and ice extent, and sea level rise (Karl et al. 2003). Further, climate change may alter the frequency, timing, intensity, and duration of extreme weather events (Karl et al. 1995). This paper briefly summarises some of the knowledge currently available on extreme weather events and briefly introduces to the Bratislava meeting.

Extreme Weather in Europe

Human constant comparable observations of the "weather" at multiple sites are recent. Since 1861, the global surface air temperature has increased (IPCC 2001) and for most locations across Europe, increases in minimum temperature appear to be greater than in maximum temperature (Klein Tank et al. 2003) (➊ *Fig. 1*).

Several studies observed a warming tendency in winter extreme low-temperature events and summer extreme high-temperature events (Beniston 2003, Brabson et al. 2002). A lot of scientific debate is ongoing on weather the current warming trend will be also leading to increased frequency, intensity, duration and severity of extreme weather events. Several authors observed an increase (a) of the duration of heat waves

[1] with contributions from Tanja Wolf, World Health Organization

Fig. 1

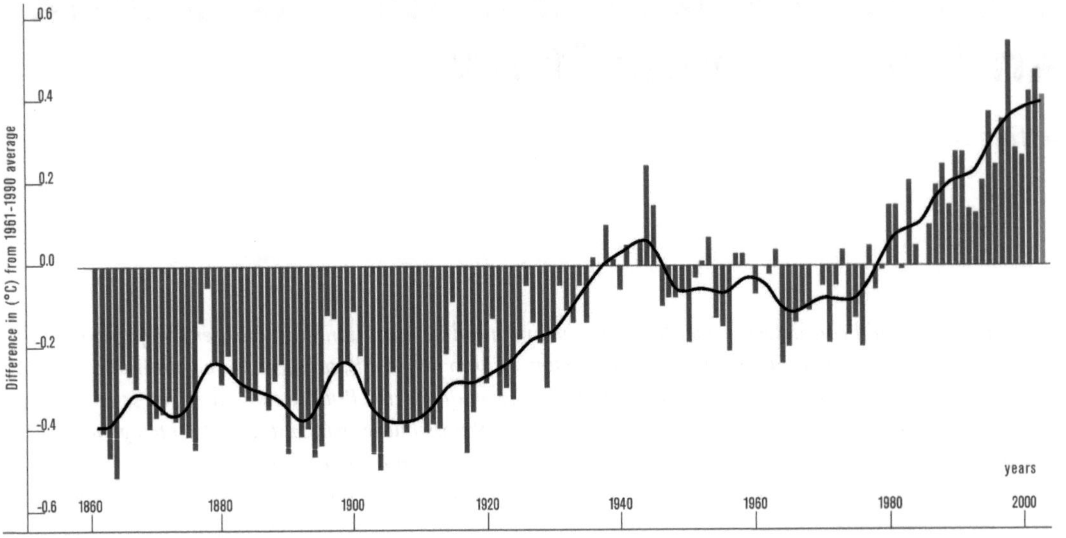

Past and future changes in global mean temperature (Hadley Center for Climate Research)

(Frich et al. 2002), (b) the summer 2003 was by far the hottest since 1500 (Luterbacher et al. 2004), (c) the 2003 heat wave bears a close resemblance to what many regional climate models are projecting for summers in the latter part of the 21st century (Beniston 2003), and (d) that the heat wave is statistically very unlikely given a shift in the mean temperature (Schar et al. 2004). An increase in variability is needed. This is also debated by the Intergovernmental Panel of Climate Change[2] (IPCC). ❯ *Figure 2* illustrates three possible scenarios of climate change with its impact on temperature: (1) an increase in mean temperature may result in less cold weather, in more hot weather and more record hot weather; (2) an increase in variance may result in more cold and hot weather as well as in more record cold and record hot weather; and (3) an increase of mean and variance might tend towards less change in cold weather, but may add to a significant increase in hot as well as record hot weather (IPCC 2001).

Using global climate models, climate change scenarios have been developed forecasting what could happen under different atmospheric concentrations of CO_2. In general, temperatures will increase over land; the exact amount is not known. Following these models, there will be more frequent extreme high temperatures and less frequent extreme low temperatures, with an associated increase (decrease) in cooling (heating) degree days; an increase in daily minimum temperatures in many regions that will exceed the increases for daytime maximum temperatures; daily temperature variability will decrease in winter but increase in summer; there will be a general drying of mid-continental areas during summer; and there will be an increase in precipitation intensity in some regions. Confidence in such projections exists because trends in observed weather and climate extremes for Europe in many ways match the expected outcomes of climate change.

The Intergovernmental Panel on Climate Change (IPCC) defines an extreme weather event 'as an

[2] The IPCC was set up in 1988, by the World Meteorological Organization (WMO) and the United Nations Environment Programme (UNEP). The role of the IPCC is to assess on a comprehensive, objective, open and transparent basis the scientific, technical and socio-economic information relevant to understanding the scientific basis of risk of human-induced climate change, its potential impacts and options for adaptation and mitigation.

Fig. 2

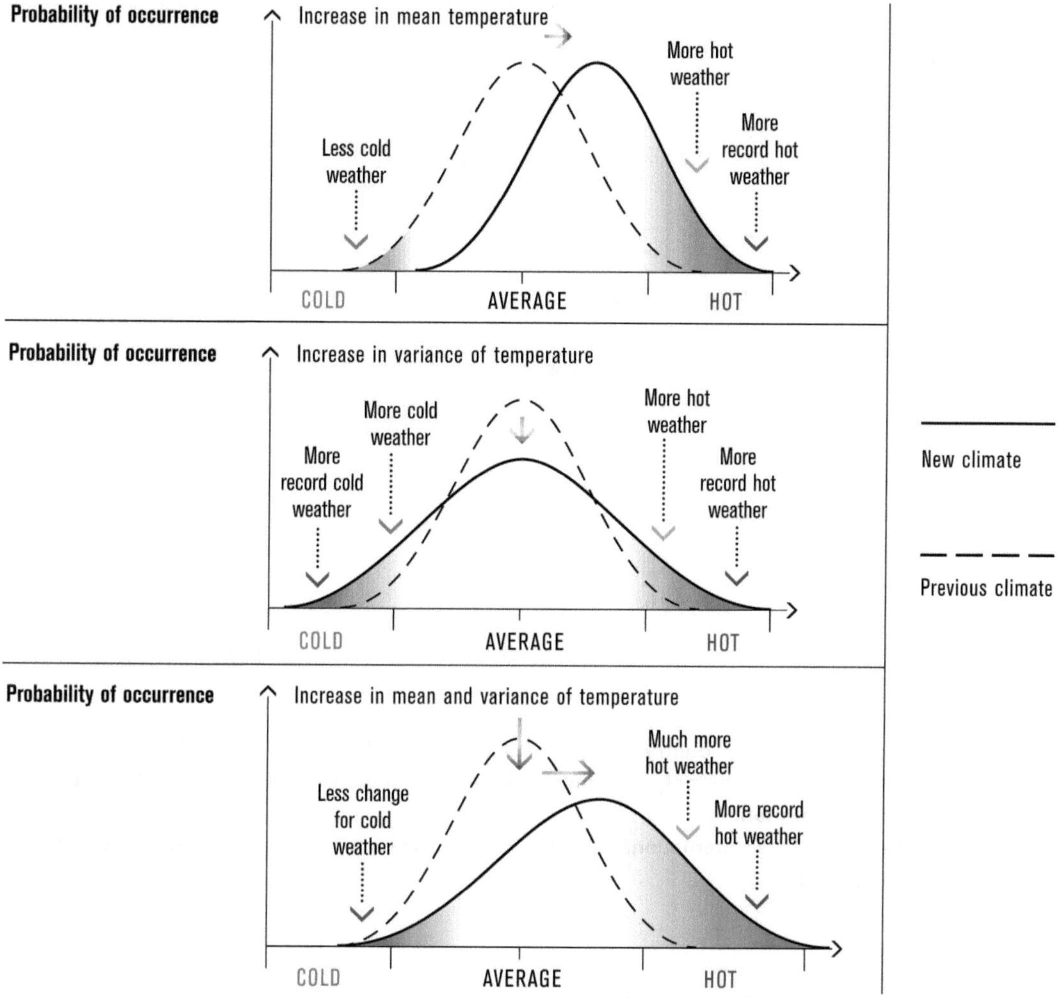

Climate change and changes in the distribution of daily temperatures (Source: Watson et al. (2001))

event that is rare within its statistical reference distribution at a particular place' and continues: 'Definitions of "rare" vary but an extreme weather event would normally be as rare or rarer than the 10th or 90th percentile' (IPCC 2001). An event may be further considered extreme merely if some of its characteristics, such as magnitude, duration, speed of onset or intensity, lie outside a particular society's experiential or coping range, whether or not the event is rare (Navarra, *Chapter 1;* McGregor, *Chapter 2*).

Figure 3, shows the distribution of natural disasters, by country and type of phenomena in Europe (1975–2001), as recorded by the EmDAT database. Although not reflected in the figure, in Europe reported extreme weather events are heatwaves, floods, windstorms, droughts and fires. The question for public health is, if extremes become more frequent and intense, will health systems and population be prepared?

Fig. 3

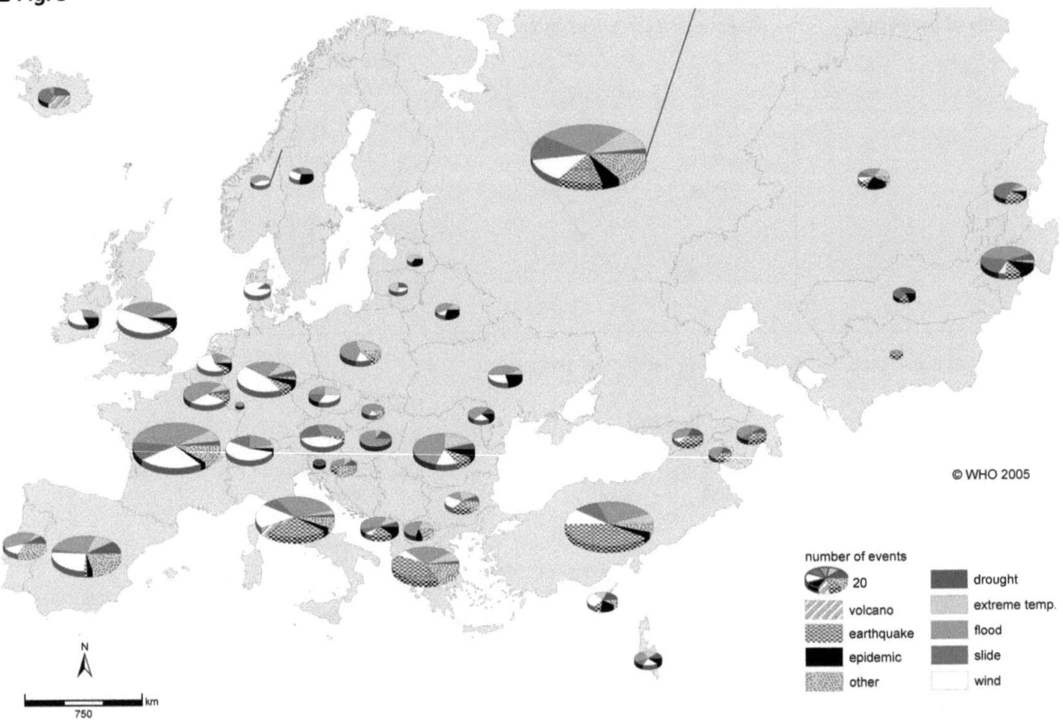

Distribution of natural disasters, by country and type of phenomena in Europe (1970–2004).
Important note: data for NIS available only since independency. Previous events have been added to the figures for the Russian Federation.
Source: EM-DAT: The OFDA/CRED International Disaster Database, www.em-dat.net – Université Catholique de Louvain, Brussels, Belgium

The health impacts of temperature extremes

Historically the relationship of temperature and mortality shows a V-like function with an optimum temperature (average temperature with lowest mortality rate), which varies with location and climate of a place (Braga et al. 2001, Huynen et al. 2001). For each degree rise above the 95th percentile of the two day mean, mortality increased by 1.9 % in London and 3.5 % in Sofia and without lag (Pattenden et al. 2003). In several studies in the United States a strong association of the temperature-mortality relation with latitude was found with warmer temperatures associated mortality in more-northern, usually cooler cities in the United States (Braga et al. 2001, Curriero et al. 2002, Keatinge et al. 2000) however this seems not to be confirmed for Europe (Michelozzi, personal communication). Several heat waves have affected the European Region during the last decades. Impacts have been elaborated in descriptive studies, mainly examining excess mortality. Excess mortality is calculated by subtracting the expected mortality from the observed mortality. The expected mortality is calculated using a variety of measures to construct averages of similar time periods of previous years. Results are difficult to compare because of the different denominators used. ❯ *Table 1* reports excess mortality rates from various sources, including country specific reports to the WHO.

Tab. 1

Heat wave events and attributed mortality in Europe (adapted from Kovats et al. 2003 and cCASHh 2005 [Climate change and adaptation strategies for human health, an EC funded research projekt]).

Place	Heat wave event	Excess mortality (all causes)	References
United Kingdom	1976 23.06. to 07.07.	9.7 % (2205) increase in England and Wales and 15.4 % (520) in Greater London	McMichael, 1998
Portugal	1981	1906 excess deaths (all cause, all ages) in Portugal, 406 in Lisbon	Garcia, 1981
Rome, Italy	1983	35 % increase in deaths in July 1983 in the over 65+ age group	Todisco, 1987
Athens, Greece	1987 21.07. to 31.07.	Estimated excess mortality > 2000	Katsouyanni, 1988
Greece (all urban areas except Athens)	1987 21.07. to 31.07.	32.5 % increase in mortality was observed in July	Katsouyanni, 1993
Portugal	1991 12.07. to 21.07.	1001 excess death (44.7 %)	Paixão and Nogeira, 2002
London, United Kingdom	1995 30.07. to 03.08.	768 (11.2 %) excess deaths occurred in in England and Wales, and 184 (23 %) in Greater London	Rooney, 1998
The Netherlands	1994 19.07. to 31.07.	1057 excess deaths (95 % CI 913, 1201) 24.4 % increase,	Huynen, 2001
Portugal	2003 30.07. to 15.08.	1854	Botelho et al, 2004
Spain (52 capitals of provinces)	2003 01.06. to31.07.	3166	Navarro et al, 2004
Italy (23 capital cities)	2003 01.06. to 15.08.	3134	Centro Nazionale de Epidemiologia, 2003
Italy (Milan)	2003 01.06. to 31.08.	559	Bisanti et al, 2004
Italy (Turin)	2003 01.06. to 31.08.	577	Bisanti et al, 2004
Italy (Rome)	2003 01.06. to 31.08.	944	Bisanti et al, 2004
France (13 cities)	2003 01.08. to 20.08.	14802	Institut de Veille Sanitaire, 2003
Germany (Baden-Wuerttemberg)	2003 01.08. to 22.08.	1415	Sozialministerium Baden-Wuerttemberg, 2003
England and Wales	2003 04.08. to 13.08.	2045	Office for National Statistics, 2003
Switzerland (Tessin)	2003 01.06. to 30.08.	No effect	Cerutti et al, 2004

Heat stress seems to most affect the aging population. A review of several epidemiological studies on heat and health underlines that persons at highest risk of death following heat-waves are over 60, or work in jobs requiring heavy labour, or live in inner cities and lower-income census tracts and thus are exposed either to low economic status or higher temperature or both (Basu et al. 2002, Keatinge et al. 2000). Prominent causes of death in studies of heat waves and elevated temperature were cardio-vascular diseases, respiratory diseases, cerebrovascular diseases and mental illness (Basu et al. 2002). People with pre-existing illness, especially heart and lung diseases, are at higher risk of dying in heat waves. In fact, cardiovascular and respiratory causes of deaths are most strongly linked with changes in temperature and this makes elderly people and people with impaired health but also those suffering from poor social conditions most susceptible to impact of weather changes (Ballester et al. 2003, O'Neill et al. 2003). Also mental illness shows a positive association with heat related death (Kaiser et al. 2001).

The adverse health effects of heat are more evident and more often studied in urban areas. The Urban Heat Island, as the fact that within cities the ambient air temperature is higher than in the surroundings, poses to risk the urban inhabitants' health (Montavez et al. 2000). Many of today's large cities tend to amplify extremes of temperature. The heat island in summer is because of the expanse of brick and asphalt heat-retaining structures, the treeless expanses of inner cities and the physical obstruction of cooling breezes (McMichael 2001). The health of urban inhabitants may in addition be impaired due to urban air pollution from industry and traffic (Sartor et al. 1995, Smoyer et al. 2000).

Beneath the demographical risk factors there are behavioural risk factors like living alone, being confined to bed, not being able to care for oneself, having no access to transportation or not leaving home daily and social isolation (Semenza et al. 1996). Similar finding have been drawn also from the 2003 heat wave in France (Empereur-Bissonnet, ➤ *Chapter* 8). A case-control study (INVS 2004) highlights the significant correlation with death during the heat wave of socio-professional categories (workers at risk), the degree of autonomy (people confined to bed at risk or not autonomous in getting washed and dressed), the health status (at risk patients with cardiovascular, psychiatric or neurological diseases) and the quality of thermo-isolation of the home. To wear fewer clothes and the use of a "refreshment measure" have shown some protective effect (INVS 2004). However, many more efforts are needed to understand how best to predict, detect and prevent the heat waves associated health impacts and how best to target intervention strategies.

Health impacts from cold spells can be classified as being derived from acute exposure (hours, days) as well as chronic exposure to cold (weeks, month, years). Exposure to cold temperatures can result in several negative health consequences, including death, disease, injury, other health complaints, degradation of performance, and degradation of motivation. Accidental cold exposure occurs mainly outdoors, in socially deprived people, workers, alcoholics, the homeless, the elderly in temperate cold climates. A simple lack of awareness combined with a lack of protective clothing, for instance, may carry a risk of death from hypothermia even during outdoor temperatures as mild as 0 °C. The onset of air-related frostbite appears at an environmental temperature of −11 °C. Wind, high altitude and wet clothing lead to onset of injury at higher environmental temperatures. The incidence of more serious frostbite requiring hospital treatment increases at temperatures of −15 °C and below. Mortality with respect to chronic exposure to cold is subject to seasonality. In many temperate countries 'all-cause mortality' as well as cardiovascular and respiratory mortality is higher during the winter months. Some epidemiologists use the term excess winter mortality to describe this seasonal phenomenon. Most European countries suffer from 5–30 % excess winter mortality. Ironically, increases in mortality because of cold temperatures occur more often in the warmer regions of Europe compared the colder regions. By means of protective clothing and a better infrastructure, North Europeans seem to be better adapted to extreme cold conditions (Hassi, ➤ *Chapter* 6). However there is significant scientific debate and uncertainty on whether the warming occurring has been or will be beneficial in reducing winter seasonal mortality.

Health impacts of floods

Europe experiences three types of floods: flash, riverine, and storm surges. *Figure 4* illustrates a European Map on sites of floods that occurred since the 19th century. Events as registered by the EmDAT database were used for the compilation of this map.

Fig. 4

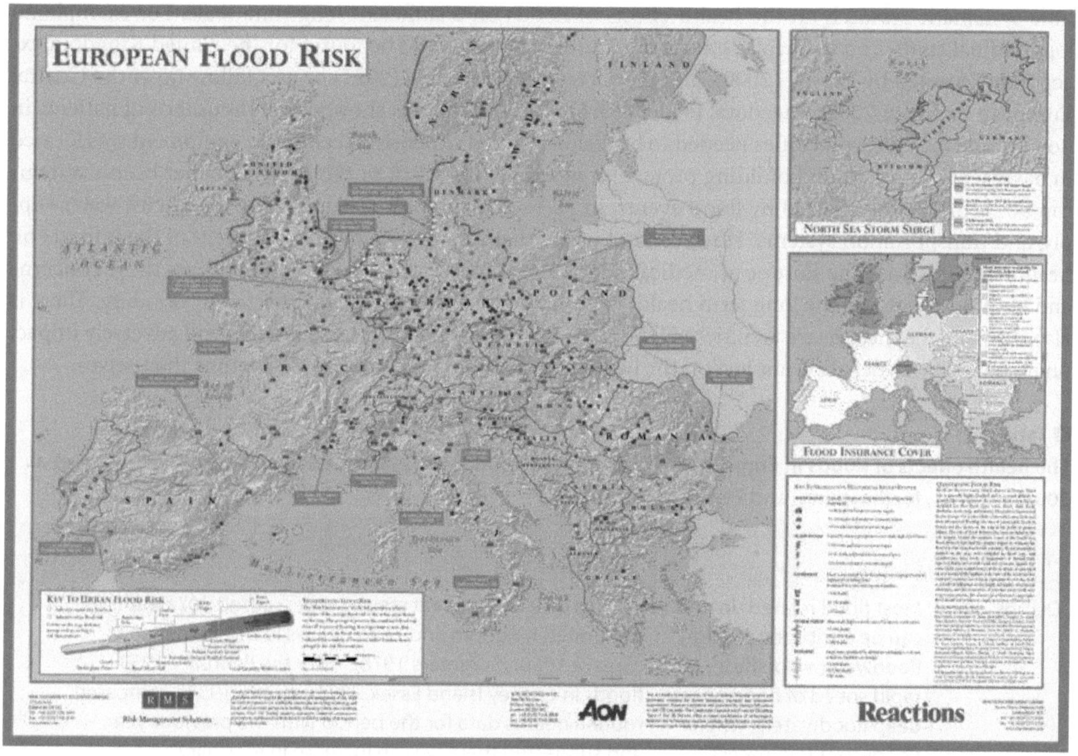

Flood risk map for Europe. Source http://www.rms.com/Publications/UK_Flood.asp accessed on 30.01.2005

The adverse human health consequences of flooding can be complex, far-reaching, and difficult to attribute to the flood event itself (Hajat et al. 2003; *Chapter 16*). Floods can cause major infrastructure damage, including disruption to roads, rail lines, airports, electricity supply systems, water supplies, and sewage disposal systems. The economic consequences are often greater than indicated by the physical effects of floodwater coming into contact with buildings and their contents. Economic damage may reach beyond the flooded area and last longer beyond the event (Kundzewicz, *Chapter 16;* Ebi, *Chapter 5*).

Adverse health impacts of flooding can arise from a combination of some or all of the following factors: characteristics of the flood event itself (depth, velocity, duration, timing, etc.); amount and type of property damage and loss; whether flood warnings were received and acted upon; the victims' previous flood experience and awareness of risk; whether or not flood victims need to relocate to temporary housing; the clean-up and recovery process, and associated household disruption; degree of difficulty in dealing with builders, insurance companies, etc.; pre-existing health conditions and susceptibility to the physical and mental health consequences of a flooding event; degree of concern over a flood recurrence; degree of financial concern; degree of loss of security in the home; and degree of disruption of community life.

The physical health effects can be further categorized into direct effects caused by the floodwaters (such as drowning and injuries) and indirect effects caused by other systems damaged by the flood (such as water- and vector-borne diseases, acute or chronic effects of exposure to chemical pollutants released into floodwaters, food shortages, and others) (*Tab. 2*) (Menne et al. 2000). There is a common perception that the problems associated with a flooding event end once the floodwaters have receded. However, for many victims, this is when most of their problems begin.

From several international assessments and literature reviews carried out it is apparent there is very little systematic research on the health effects of floods on a sufficient long time scale. For example no longitudinal studies on the health effects of natural disasters could be found for the United Kingdom except that reported by Tapsell (2000), and more recently by Hepple (2001) or as a follow up of the Central European floods in 2002. Anecdotal evidence from the 2002 floods showed that thousands of patients in flood prone health care facilities needed to be dislocated and expensive health care equipment was located in basements without flood building protection measures (*Kirch in this book*). Further health systems can be further badly affected by flood events, in particular through disruptions to electricity, water supply, and transportation systems. However, in Europe there is no systematic assessment of the impacts on health care systems. The European medical communities need to be prepared to address these concerns and both the short and the long term health needs of people who have been affected by flooding. There is also the issue that healthcare facilities will be stretched at times of disasters, and this will adversely impact on normal service delivery, not just on the healthcare provision for the disaster victims themselves.

Tab. 2
The health effects of floods in Europe, with examples of flood events [adapted from Hajat, Ebi, Kovats, et al. 2003; Ebi et al., forthcoming; and Few et al. 2004]

Health outcome	Comment	Example
Deaths	Most flood related deaths can be attributed to: high floodwater velocities; rapid speed of flood onset; deep floodwaters, where floodwater is in excess of 1 metre depth; long duration floods; debris load of floodwaters; characteristics of accompanying weather and clean up activities in the aftermath of floods	February 1953, the great storm surge, caused 307 deaths in the United Kingdom and 1795 deaths in the Netherlands (Greave 1956) (Summers 1978). After the February 1953 floods in Canvey Island Essex, UK, Lorraine (1954), compared routine deaths data for the period with the previous year, and suggested there was an increase in mortality. In the UK, Bennet (1970) conducted a retrospective study of the 1968 Bristol floods, and found a 50 % increase in the number of deaths among those whose homes had been flooded, and the most pronounced rise was in the 45–64 age group. In October 1988, a flash flood occurred in the Nimes region of France, 9 deaths occurred (Duclos, Vidonne, Beuf et al. 1991). In 1996, 86 people died in the town of Biescas in Spain as a consequence of the water and mud that suddenly covered a campsite located near a canalized river. In 1997, river floods in central Europe caused more that a 100 fatalities (Kriz, Benes, Castkova et al. 1998).

Tab. 2 (Continued)

Health outcome	Comment	Example
Deaths (Continued)		In the 1998 flood in Sarno, Italy, 147 people were killed by a river of mud that rapidly destroyed an urban area (Thonissen 1998). Between 1980 and 1999, an annual rate of 1.3 deaths and 5.7 injuries occurred per 10,000,000 population due to inland floods and landslides in Western Europe. (McMichael, Campbell-Lendrum, Kovats et al. 2002).
Injuries	Surveillance of morbidity following floods is limited. Little reliable information on injuries found in relation to European floods.	Duclos et al. (1991) report that in their community survey (108/181 households completed a questionnaire) of the 1988 floods in Nîmes, France, 6 % of households surveyed reported mild injuries (contusions, cuts, and sprains) related to the flood.
Infectious disease	Small risk of communicable disease following flooding, although severe occurrences are rare due to the public health infrastructure in place prior to and following a flood event such as water treatment and effective sewage pumping.	Leptospirosis outbreak occurred after the flooding in the Czech Republic in 1997. (Kriz et al. 1998). No increase in infectious disease was observed following the flash flood in Nimes in 1988, or 1995 river floods in eastern Norway, or in floods in UK in 2002. (Duclos et al 1991, (Aavitsland, Iversen, Krogh et al. 1996). Finland reported 13 waterborne disease outbreaks with an estimated 7300 cases during 1998 – 1999, associated with untreated groundwater from mostly flooded areas (Miettinen, Zacheus, von Bonsdorff et al. 2001).
Respiratory disease	Very little information available, mainly anecdotal	Following the floods in the north-eastern Republic of Sakha (Yakutia) in July 1998, a high incidence of respiratory diseases was observed by the International Federation of Red Cross (IFRC) (IFRC, personal communication).
Mental health (anxiety depression)	Few well conducted studies – but clearly an area for further investigation.	Higher levels of depression among flooded households compared to controls following floods in South east of UK. (Tapsell, Tunstall 2001) (Green, Emery, Penning-Rowsell et al. 1985). In a retrospective case-control study of the 1968 floods in Bristol, UK, Bennet (1970) found a significant increase (18 % versus 6 %; $\times$ 2 7.57; $p < 0.01$) in the number of new psychiatric symptoms (considered to comprise anxiety, depression, irritability, and sleeplessness) reported by flooded female respondents compared with the non-flooded group. In the Netherlands, Becht et al. (1998) interviewed children (n=64) and their parents (n=30) 6 months post-flood, and found 15 – 20 % of children having moderate to severe stress symptoms. After the 1997 floods in Opole, Poland (Bokszczanin 2000, 2002) studied children aged 11 – 14 years, and 11 – 20 year olds. Results confirm long-term negative effects on wellbeing of children, with resultant PTSD, depression and dissatisfaction with ongoing life.

Conclusions

Extreme weather events impact negatively on public health at many dimensions. Increased rates of mortality and morbidity are among the most important (Meusel et al. 2004). The meeting in Bratislava had the objective of exchanging information and discussing and developing recommendations on public health responses. These recommendations are discussed in ➲ *Chapter 26*. There is growing recognition that climate variability and change are causing serious risks to human health. How much climate variability may increase over the next decades is highly uncertain. Changes in extreme events may be experienced as changes in the rate or frequency of events and/or as changes in their intensity or magnitude. Spatial and temporal clustering of events may become more common. For example, heat-waves may increase in both frequency and duration in coming decades. Projections for cold spells are more uncertain. What is certain is that increasing climate variability will challenge public health systems. These possible changes require policy makers at all levels to take a proactive, anticipatory approach to designing strategies, policies and measures to reduce current and future burdens of climate-sensitive diseases. There is a need to increase collaboration and coordination between the health and meteorological communities, including the use of meteorological indicators by the health community.

Heat-waves are associated with an increase in all causes of death. Many knowledge gaps exist: in characterizing the relationship between heat exposure and a range of health outcomes, in understanding interactions between harmful air pollutants and extreme weather and climate events as well as on analysis of the health-threatening characteristics of heat wave episodes as opposed to the more general assessment of the overall relationship between temperature and health. Research is also needed on what information is necessary and how that information should be communicated, to motivate appropriate changes in behaviour during heat-waves. People perceive risks differently and have different responses to perceived risks. More information is needed on how to effect appropriate behavioural changes in vulnerable populations. Finally, criteria need to be developed for how to identify regions with more vulnerable populations. The assessment of the environmental and health consequences of heat-waves highlighted a number of knowledge gaps and problems in public health responses. Until 2003, heat-waves have not been considered a serious risk to human health with "epidemic" potential in the European Region. In order to reduce the health impacts of future heat-waves, fundamental questions need to be addressed, such as can a heat-wave be predicted, can it be detected, can it be prevented and what can be done.

The risk of floods will probably increase during the coming decades. Two trends point to this. Firstly, the magnitude and frequency of floods are likely to increase in the future as a result of climate change, i.e. higher intensity of rainfall as well as rising sea levels. Secondly, the impact of flood events may increase, because more people live in areas at risk of flooding and also more economic assets (business and industry) are located in such areas. Moreover, human activities such as the clearing of forests, the straightening of rivers, the suppression of natural flood plains and poor land planning, have contributed significantly to increasing the risk of floods. In 2002, the flooding in central Europe was of unprecedented proportions, with dozens of people losing their lives, extensive damage to the socioeconomic infrastructure, and destruction of the natural and cultural heritage. Germany, the Czech Republic, and Austria were the three countries most severely affected. Estimates of the economic and insured losses were € 11.0 billion in Germany, € 3.9 billion in the Czech Republic, and € 3.4 billion in Austria (Munich Re Group, 2001). With regards to floods there is no European comprehensive data base and there is no common understanding on what best health targeted measures are needed. The health sector should be more pro-active in planning for and providing pre and post-flood event assistance. With better information, the emphasis in disaster management could shift from post-disaster improvisation to pre-disaster planning. A comprehensive, risk-based emergency management programme of preparedness, response and recovery has the potential to reduce the adverse health effects of floods.

This meeting contributed to the preparatory process for the Fourth Ministerial Conference on Environment and Health (Budapest, June 2004), by submitting a working document (➔ *Annex 1)* to the 4th Ministerial Conference in Budapest. This was endorsed in paragraph 7b of the declaration of the 4th Ministerial Conference. (◄) Chapter 26 of this book describes the ongoing preventive activities and the potentials for additional cooperation to further prevent health effects.

◄ "We (Ministries of Health and Environment) recognize that climate is already changing and that the intensity and frequency of extreme weather events, such as floods, heat-waves and cold spells, may change in the future. Recent extreme weather events caused serious health and social problems in Europe, particularly in urban areas. These events will continue to pose additional challenges to health risk management and to the reliability of the power supply and other infrastructure. This demands a proactive and multidisciplinary approach by governments, agencies and international organizations and improved interaction on all levels from local to international. Based on the working paper Public health responses to extreme weather and climate events (➔ *Annex 1)*, we decide to take action to reduce the current burden of disease due to extreme weather and climate events. We invite WHO, through its European Centre for Environment and Health, in collaboration with the World Meteorological Organization, the European Environment Agency (EEA) and other relevant organizations, to support these commitments and to coordinate international activities to this end. We agree to report on progress achieved at the intergovernmental meeting to be held by the end of 2007".

References

Aavitsland P, Iversen BG, Krogh T, Fonahn W, Lystad A (1996) Infections during the 1995 flood in Ostlandet. Prevention and incidence. Tidsskr Nor, 116 2038 – 2043

Baden-Wuerttemberg (2004) Gesundheitliche Auswirkungen der Hitzewelle im August 2003. Sozialministerium Baden-Wuerttemberg, Stuttgart

Ballester F, Michelozzi P, Iniguez C (2003) Weather, climate, and public health. J Epidemiol Community Health 57(10):759 – 760

Basu R, Samet JM (2002) Relation between elevated ambient temperature and mortality: a review of the epidemiologic evidence. Epidemiological Reviews 24(2):190 – 202

Becht MC, van Tilburg MAL, Vingerhoets AJJM, Nyklicek I, de Vries J, Kirschbaum C, Antoni MH, van Heck GL (1998) Watersnood. Eenverkennend onderzoek naar de gevolgen voor het welbevinden en de gezondheidvan volwassenen en kinderen. Flood: A pilot study on the consequences for wellbeing and health of adults and children. Tijdschrift voor Psychiatrie 40:277 – 289

Beniston M (2003) The 2003 heat wave in Europe: A shape of things to come? An analysis based on Swiss climatological data and model simulations. Geophys Res Lett 31(2)

Bennet G (1970) Bristol floods 1968. Controlled survey of effects on health of local community disaster. BMJ 3 454 – 458

Bisanti L, Russo A, Cadum E, Costa G, Michelozzi P, Perrucci C (2004) 2003 Heat Waves and Mortality in Italy. ISEE presentation

Bokszczanin A (2000) Psychologiczne konsekwencje powodzi u dzieci i mlodziezy. Psychological consequences of floods in children and youth. Psychologia 43:172 – 181

Bokszczanin A (2002) Long-term negative psychological effects of a flood on adolescents. Polish Psychological Bulletin 33:55 – 61

Botelho J, Catarino J, Carreira M, Calado R, Nogueira PJ, Paixao E, Falcao JM (2004) Onda de calor de Agosto de 2003. Os seus efeitos sobre a mortalidade da populacao portuguesa. Instituto Nacional de Saude Dr. Ricardo Jorge, Losboa

Braga AL, Zanobetti A, Schwartz J (2001) The time course of weather-related deaths. Epidemiology 12(6):662 – 667

Campbell-Lendrum D, Pruss-Ustun A, Corvalan C (2003) How much disease could climate change cause? In: McMichael A, Campbell-Lendrum D, Corvalan C, Ebi K, Githeko A, Scheraga J, Woodward A (eds) Climate Change and Health: Risks and Responses. WHO/WMO/UNEP, Geneva

Centro Nazionale de Epidemiologia (2003) Sorveglianza e Promozione della Salute, Ufficio di Statistica. Indagine Epidemiologica sulla Mortalità Estiva. Presentazione dei dati finali (Susanna Conti). Rome: Istituto Superiore di Sanità; (http://www.epicentro.iss.it/mortalita/presentazione%20mortalità%20estiva2.pdf)

Cerutti B, Tereanu C, Domenighetti G, Gaia M, Bolgani I, Lazzaro M, Cassis I (2004) La mortalita estiva in Ticino nel 2003. Ufficio del medico cantonale

Curriero F, Heiner KS, Samet J, Zeger S, Strug L, Patz JA (2002) Temperature and mortality in 11 cities of the Eastern United States. American Journal of Epidemiology 155(1):80–87

Duclos P, Vidonne O, Beuf P, Perray P, Stoebner A (1991) Flash flood disaster – Nimes, France. European Journal of Epidemiology 7(4):365–371

Frich P, Alexander LV, Della-Marta P, Gleason B, Haylock M, Klein Tank AMG, Peterson TC (2002) Observed coherent changes in climatic extremes during the second half of the twentieth century. Climate Research 19:193–212

Garcia AC, Nogueira PJ, Falcao JM (1999) Onda de calor de Junho de 1981 em Portugal: efeitos na mortalidade. Revista Portuguesa de Saude Publica 1:67–77

Greave H (1956) The great flood. Report to Essex County Council. Essex County Council, Essex

Green C, Emery P, Penning-Rowsell E et al (1985) The health effects of flooding: a case study of Uphill. Middlesex University Flood Hazard Research Centre, Enfield

Hajat S, Ebi KL, Kovats RS, Menne B, Edwards S, Haines A (2003) The human health consequences of flooding in Europe and the implications for public health: a review of the evidence. Applied Environmental Science and Public Health 1(1):13–21

Hepple P (2001) Research into the long-term health effects of the flooding event in Lewes, Sussex of October 12th 2000: a case-control study. Unpublished Masters thesis. London School of Hygiene and Tropical Medicine, London.

Huynen MM, Martens P, Schram D, Weijenberg MP, Kunst AE (2001) The impact of heat waves and cold spells on mortality rates in the Dutch population. Environ Health Perspect 109(5):463–470

Huynen MM, Martens P, Schram D, Weijenberg MP, Kunst AE (2001) The impact of heat waves and cold spells on mortality rates in the Dutch population. Environ Health Perspect, 109(5):463–470

Institut de Veille Sanitaire (2003) Impact sanitaire de la vague de chaleur en France survenue en août 2003. Progress report, 29 August 2003. (http://www.invs.sante.fr/publications/2003/chaleur_aout_2003/index.html)

INVS (2004) Etude des facteurs de deces des personnes agees residant a domicile durent la vague de chaleur d'aout 2003. Institut de veille sanitaire, Paris

IPCC (2001) Climate Change 2001: The Scientific Basis. Contribution of Working Group I to the Second Assessment Report of the Intergovernmental Panel on Climate Change. Cambridge University Press, New York

Kaiser R, Rubin CH, Henderson AK, Wolfe MI, Kieszak S, Parrott CL, Adcock M (2001) Heat-related death and mental illness during the 1999 Cincinnati heat wave. Am J Forensic Med Pathol 22(3):303–307

Karl TR, Knight RW, Plummer N (1995) Trends in high-frequency climate variability in the twentieth century. Nature 377:217–220

Karl TR, Trenberth KE (2003) Modern global climate change. Science 302(5651):1719–1723

Katsouyanni K, Pantazopoulou A, Touloumi G, Tselepidaki I, Moustris K, Asimakopoulos D, Poulopoulou G, Trichopoulos D (1993) Evidence for interaction between air pollution and high temperature in the causation of excess mortality. Arch Environ Health 48(4):235–242

Katsouyanni K, Trichopoulos D, Zavitsanos X, Touloumi G (1988) The 1987 Athens Heatwave. Lancet 332(8610):573

Keatinge WR, Donaldson GC, Cordioli E, Martinelli M, Kunst AE, Mackenbach JP, Nayha S, Vuori I (2000) Heat related mortality in warm and cold regions of Europe: observational study. BMJ 321(7262):670–673

Klein Tank AMG, Wijngaard J, van Engelen A (2003) Climate of Europe: Assessment of observed daily temperature and precipitation extremes. European Climate Assessment 2002. KNMI

Kriz B, Benes C, Castkova J, Helcl J (1998) Monitorov n¡ Epidemiologick' Situace V Zaplavenych Oblastech V Eesk' Republice V Roce 1997. [Monitoring the Epidemiological situation in flooded areas of the Czech Republic in 1997.] In: Proceedings of the Conference DDD'98, 11–12th May 1998, Prodebrady, Czech Republic

Luterbacher J, Dietrich D, Xoplaki E, Grosjean M, Wanner H (2004) European seasonal and annual temperature variability, trends, and extremes since 1500. Science 303(5663):1499–1503

McMichael AJ, DH C-L, Corvalan C, F., Ebi KL, Githeko AK, Scheraga JS, Woodward A (eds) (2003) Climate Change and Human Healh. Risk and Responses. World Health Organization, Geneva

McMichael AJ, Kovats RS (1998) Assessment of the Impact on Mortality in England and Wales of the Heatwave and Associated Air Pollution Episode of 1976. LSHTM, London

McMichael T (2001) Human frontiers, environments and disease. Past patterns, uncertain futures. Cambridge University Press, Cambridge

Menne B (2000) Floods and public health consequences,

prevention and control measures. UN 2000 (MP.WAT/SEM.2/1999/22)

Meusel D, Menne B, Bertollini R, Kirch W (2004) Extreme Weather Events and Public Health Responses – A review of the WHO-Meeting on this topic in Bratislava on 9–10 February 2004. J Public Health 12(6):371–381

Miettinen IT, Zacheus O, von Bonsdorff CH, Vartiainen T (2001) Waterborne epidemics in Finland in 1998–1999. Water Science and Technology 43(12):67–71

Montavez J, Rodriguez A, Jimenez J (2000) A study of the urban heat island of Granada. International Journal of Climatology 20:899–911

Munich Re Group (2001) Topics: Natural Catastrophes 2000. Munich Re, Munich

Navarro F, Somin-Soria, Lopez-Abente G (2004) Valoracion del impacto de la ola de calor del verano de 2003 sobre la mortalidad. Gaceta Sanitaria, 18 (Suplemento sespas).

Office for National Statistics [homepage on the internet] (2004) Summer mortality – deaths up in August heatwave. [Posted 3 October 2003; cited 8 March 2004]. Available from: http://www.statistics.gov.uk/cci/nugget.asp?id=480

O'Neill MS, Zanobetti A, Schwartz J (2003) Modifiers of the temperature and mortality association in seven US cities. Am J Epidemiol 157(12):1074–1082

Paixao E, Nogueira PJ (2002) Estudio da Onda de Calor de Julho de 1991 em Portugal: Efeitos na Mortalidade. Insitituti Nacional de Saude Dr. Ricardo Jorge, Projecto ICARO

Pattenden S, Nikiforov B, Armstrong BG (2003) Mortality and temperature in Sofia and London. J Epidemiol Community Health 57(8):628–633

Rooney C, McMichael A, Kovats R, Coleman M (1998) Excess mortality in England and Wales, and in Greater London, during the 1995 heatwave. J Epidemiol Community Health 52(8):482–486

Sartor F, Snacken R, Demuth C, Walckiers D (1995) Temperature, Ambient Ozone Levels, and Mortality during Summer, 1994, in Belgium. Environmental Research 70(2):105–113

Schar C, Jendritzky G (2004) Climate change: Hot news from summer 2003. Nature 432(7017):559–560

Semenza JC, Rubin CH, Falter KH, Selanikio JD, Flanders WD, Howe HL, Wilhelm JL (1996) Heat-related deaths during the July 1995 heat wave in Chicago. N Engl J Med 335(2):84–90

Smoyer KE, Kalkstein LS, Greene JS, Ye H (2000) The impacts of weather and pollution on human mortality in Birmingham, Alabama, and Philadelphia. Pennsylvania. International Journal of Climatology 20:881–897

Summers D (1978) The East Coast Floods. David & Charles, London

Tapsell S, Penning-Rowsell E, Tunstall S, Wilson T (2002) Vulnerability to flooding: health and social dimensions. Phil Trans R Soc Lond A 360:1511–1525

Tapsell SM (2000) Follow-up study of the health effects of the 1998 Easter flooding in Banbury and Kidlington. Final report to the Environment Agency Thames Region. Middlesex University Flood Hazard Research Centre, Enfield, London

Thonissen C (1999) Water management and flood prevention in the framework of land-use management [presentation]. In: Workshop held by DGXVI of the European Commission; 1998 Jul 2–3, Thessaloniki, Greece

Todisco G (1987) Indagine biometeorologica sui colpi di calore verificatisi a Roma nell'estate del 1983 [Biometeorological study of heat stroke in Rome during summer of 1983]. Rivista di Meteorologica Aeronautica XLVII (3–4):189–197

List of Authors

Prof. Dr. L. Abenhaim, MD, PhD
Professor of Epidemiology and Public Health
CHU Cochin Port Royal
124 Rue Saint Jacques
F-75014 Paris, France

Dr. Carmen Beltrano
Bureau of Statistics,
National Centre of Epidemiology
Surveillance and Promotion of Health
Italian National Institute of Health
Viale Regina Elena, 299
I-00161 Rome, Italy

Dr. Roberto Bertollini
Special Programme on Health and Environment
WHO Regional Office for Europe
WHO European Centre for Environment and Health
Via Francesco Crispi, 10
I – 00187 Rome, Italy

Dr. Pierre Bessemoulin
Director of Climatology
Météo France
2 Avenue RAPP
F-75007 Paris, France

Dr. Luigi A. Bisanti
ASL Città di Milano, Servizio di Epidemiologia
Corso Italia, 19
I-20122 Milano, Italy

Dr. J. Bobvos
Fodor József National Center for Public Health
National Institute of Environmental Health
Gyali 6-8
H-1097 Budapest, Hungary

Dr. Elena M. Borisova
Deputy Director of Republican Research
Centre of Emergency Medicine
National Scientific
2, Farkhadskaya Street
UZ-700107 Tashkent, Uzbekistan

Dr. J. S. Botelho
General Directorate of Health
Information and Analysis Department
Alameda D. Afonso Henriques 45-7°
P-1049-005 Lisbon, Portugal

Dr. Ennio Cadum
Agenzia Regionale per la Protezione Ambientale
del Piemonte,
Area Funzionale Tecnica di Epidemiologia Ambientale
Via Sabaudia, 164
I-10095 Grugliasco, Italy

R. M. D. Calado
Director; Department of Epidemiology
General Directorate of Health
Information and Analysis Department
Alameda D. Afonso Henriques 45-7°
P-1049-005 Lisbon, Portugal

Dr. M. Carreira
General Directorate of Health
Information and Analysis Department
Alameda D. Afonso Henriques 45-7°
P-1049-005 Lisbon, Portugal

Dr. Judite Catarino
General Directorate of Health
Information and Analysis Department
Alameda D. Afonso Henriques 45-7°
P-1049-005 Lisbon, Portugal

Prof. Dr. Rayna Chakurova
Social Medicine and Health Management Department
Disaster Medicine Sector
National Center of Public Health
8, Byalo More St.
BG-1000 Sofia, Bulgaria

Dr. J.-C. Cohen
Coordinator in Biometeorology
Météo France
2 Avenue RAPP
F-75007 Paris, France

Dr. Susanna Conti
Chief of the Statistical Bureau
National Centre for Epidemiology
Surveillance and Promotion of Health
Italian National Institute of Health
Viale Regina Elena, 299
I-00161 Rome, Italy

Dr. G. Costa
Department of Epidemiology of Rome
Via di S. Costanza 53
I-00198 Rome, Italy

Dr. Anca Cristea
Head of Environmental Health Department
Directorate of Public Health
21, Nicolae Balcescu Street
RO-700117 Iasi, Romania

Francesca de'Donato
Department of Epidemiology of Rome
Via di S. Costanza 53
I-00198 Rome, Italy

M. DeMaria
Department of Epidemiology of Rome
Via di S. Costanza 53
I-00198 Rome, Italy

Mariangela D'Ovidio
Department of Epidemiology of Rome
Via di S. Costanza 53
I-00198 Rome, Italy

Dr. Kristie L. Ebi
Senior Managing Scientist Exponent
1800 Diagonal road, Suite 355
Alexandria, VA 22314, USA

Dr. Sally Edwards
London School of Hygiene and Tropical Medicine
Keppel Street, WCIE 7HT
UK-London, United Kingdom

Dr. P. Empereur-Bissonnet
Institut de Veille Sanitaire
Département Santé-Environnement
12, Rue du Val D'Osne
F-94415 Saint Maurice Cedex, France

C. A.T. Ferro
Department of Meteorology
University of Reading
Earley Gate, RG6 6BB
UK-Reading, United Kingdom

Prof. A. Haines
Dean
London School of Hygiene and Tropical Medicine
Keppel Street, WCIE 7HT
UK-London, United Kingdom

Dr. S. Hajat
Public & Environmental Health Research Unit
London School of Hygiene & Tropical Medicine
Keppel Street, WCIE 7HT
UK-London, United Kingdom

Prof. Dr. J. Hassi
Research Professor, Director
Thule Institute, Centre for Arctic Medicine
University of Oulu
PO Box 5000
FIN-90014 Oulu, Finland

Prof. Dr. G. Havenith
Reader Environmental Physiology and Ergonomics
Dept. Human Sciences
Human Thermal Environments Laboratory
Loughborough University
UK-LE11 3TU Loughborough, United Kingdom

Dr. Inge Heim
Head of the Department of Epidemiology
Institute for Cardiovascular Prevention and Rehabilitation
Draskoviceva 13
HR-10 000 Zagreb, Croatia

Prof. Bonnie Henry
Department of Public Health Sciences
Faculty of Medicine
University of Toronto
12 Queen's Park Crescent West
Toronto, Ontario M5S 1A8, Canada

Prof. L. Ivanov
Director
National Center of Public Health
Center of Hygiene and Ecology
15, Dimitar Nestorov Blvd.
BG-1000 Sofia, Bulgaria

Prof. A.A. Khadjibayev
General Director of Emergency Medicine Service
2, Farkhadskaya Street,
Chilanzar
UZ-700107 Tashkent, Uzbekistan

Dr. N. King
Direction de Santé publique Montréal
1301 Rue Sherbrooke Est
Montreal H2L IM3, Canada

Prof. Dr. Dr. W. Kirch
European Public Health Association
Past President
Faculty of Medicine
Dresden University of Technology
Fiedlerstr. 33
D-01307 Dresden, Germany

Dr. V. Kislitsin
Leading Researcher
Research Institute of Human Ecology & Environment Health
of Russian Academy of Medical Sciences
Pogodinskya St. 10/15 Bld. 1
RUS-Moscow, Russian Federation

Dr. T. Kosatsky
WHO European Centre for Environment and Health
Via Francesco Crispi, 10
I-00187 Rome, Italy

Dr. Sari Kovats
Lecturer
Public & Environmental Health Research Unit
London School of Hygiene & Tropical Medicine
Keppel Street, WCIE 7HT
UK-London, United Kingdom

W.J. Kundzewicz
Research Centre for Agricultural and Forest Environment
Polish Academy of Sciences
Bukowska 19
PL-60-809 Poznań, Poland

Prof. Dr. Z.W. Kundzewicz
Research Centre for Agricultural and Forest Environment
Polish Academy of Sciences
Bukowska 19
PL-60-809 Poznań, Poland
and
Potsdam Institute for Climate Impact Research
Postbox 60 12 03
D-14412 Potsdam, Germany

Dr. P. Magne
General Directorate of Health
Ministry of Health
8, Avenue de Ségur
F-75007 Paris, France

Prof. Jacqueline McGlade
Executive Director
European Environment Agency
International Cooperation
Kongens Nytorv
DK-1050 Copenhagen K, Denmark

Dr. G.R. McGregor
Reader in Synoptic Climatology
The University of Birmingham
Edgbaston Park Road, B15 2TT
UK-Birmingham, United Kingdom

Dr. Merylyn McKenzie Hedger
Climate Change Policy Manager – Environment Agency
Rio House
Aztec West, BS32 4UD
UK-Bristol, United Kingdom

Dr. Paola Meli
Bureau of Statistics, National Centre of Epidemiology
Surveillance and Promotion of Health
Italian National Institute of Health
Viale Regina Elena, 299
I-00161 Rome, Italy

Dr. Bettina Menne
Global Change and Health
WHO Regional Office for Europe
WHO European Centre for Environment and Health
Via Francesco Crispi, 10
I-00187 Rome, Italy

Dr. D. Meusel
Research Association Public Health Saxony
Faculty of Medicine
Dresden University of Technology
Fiedlerstr. 33
D-01307 Dresden, Germany

Dr. T. Michelon
Head of Department – Ministry of Health
General Director of Health
8, Avenue de Ségur
F-75007 Paris, France

Paola Michelozzi
Department of Epidemiology of Rome
Via di S. Costanza 53
I-00198 Rome, Italy

Dr. Giada Minelli
Bureau of Statistics, National Centre of Epidemiology
Surveillance and Promotion of Health
Italian National Institute of Health
Viale Regina Elena, 299
I-00161 Rome, Italy

Dr. A. Navarra
Dirigente di Ricerca
Istituto Nazionale di Geofisica e Vulcanologia
Via Donato Creti, 12
I-40128 Bologna, Italy

Dr. P. J. Nogueira
Estatista, Onsa – Observatório Nacional de Saúde
Instituto Nacional de Saúde
Avenida Padre Cruz
P-1649-016 Lisboa, Portugal

Prof. Dr. S. Novikov
Research Institute of Human Ecology & Environment Health
of Russian Academy of Medical Sciences
Pogodinskya St. 10/15 Bld. 1
RUS-Moscow, Russian Federation

Dr. B. Nyenzi
Chief
World Climate Applications and CLIPS Division
World Meteorological Organization
7 Avenue de la Paix, C.P. 2300
CH-1211 Geneva 2, Switzerland

Dr. Anna Páldy
Deputy Director
Fodor József National Center for Public Health
National Institute of Environmental Health
Gyali 6–8
H-1097 Budapest, Hungary

Prof. Dr. E. Penning-Rowsell
Director, Middlesex Research and Head
Flood Hazard Research Centre
Middlesex University
Queensway Enfield, EN 34SF
UK-Middlesex, United Kingdom

Dr. L. Perini
Bureau of Statistics, National Centre of Epidemiology
Surveillance and Promotion of Health
Italian National Institute of Health
Viale Regina Elena, 299
I-00161 Rome, Italy

Dr. C.A. Perucci
Department of Epidemiology of Rome
Via di S. Costanza 53
I-00198 Rome, Italy

Dr. A. Russo
Department of Epidemiology of Rome
Via di S. Costanza 53
I-00198 Rome, Italy

F. Simon-Delavelle
General Directorate of Health
Ministry of Health
8, Avenue de Ségur
F-75007 Paris, France

Natalia Skvortsova
Research Institute of Human Ecology & Environment Health
of Russian Academy of Medical Sciences
Pogodinskya St. 10/15 Bld. 1
RUS-Moscow, Russian Federation

Dr. Renata Solimini
Bureau of Statistics, National Centre of Epidemiology
Surveillance and Promotion of Health
Italian National Institute of Health
Viale Regina Elena, 299
I-00161 Rome, Italy

Dr. D. B. Stephenson
Department of Meteorology
University of Reading
Earley Gate, RG6 6BB
UK-Reading, United Kingdom

Sue Tapsell
Middlesex Research and Head
Flood Hazard Research Centre
Middlesex University
Queensway Enfield, EN 34SF
UK-Middlesex, United Kingdom

Dr. Virgilia Toccaceli
Bureau of Statistics, National Centre of Epidemiology
Surveillance and Promotion of Health
Italian National Institute of Health Rome
Viale Regina Elena, 299
I-00161 Rome, Italy

Dr. A. Vámos
Fodor József National Center for Public Health
National Institute of Environmental Health
Gyali 6-8
H-1097 Budapest, Hungary

Dr. Stéphanie Vandentorren
Institut de Veille Sanitaire
Department Santé-Environment
12, rue du Val D'Osne
F-94415 Saint Maurice Cedex, France

Dr. J.-M. Veysseire
Deputy Director of Climatology
Météo France
2 Avenue RAPP
F-75007 Paris, France

Dr. Monica Vichi
Bureau of Statistics, National Centre of Epidemiology
Surveillance and Promotion of Health
Italian National Institute of Health
Viale Regina Elena, 299
I-00161 Rome, Italy

Theresa Wilson
Middlesex Research and Head
Flood Hazard Research Centre
Middlesex University
Queensway Enfield, EN 34SF
UK-Middlesex, United Kingdom

Climate Variability and Extremes in Europe

1

The Climate Dilemma

Antonio Navarra

Abstract

Climate has become one of the most topical issues over the last two to three decades. It has graduated from the status of obscure scientific debate to that of a global geopolitical issue. Climate itself is a sophisticated concept that is somewhat different to that used or discussed in everyday life. The basic nature of the globe's climate is regulated by the global energy balance, as are the principal climate mechanisms. These in turn are modulated via the complex nonlinear interactions between the components that comprise the global climate system. These non-linear interactions generate an intense variability in climate that makes detection of small, secular trends in climate very difficult. The increase of carbon dioxide and surface temperatures is now being established as a fact, but the attribution of the temperature increase to carbon dioxide increases is a complex challenge. Due to constraints imposed by the current level of climate modeling technology we cannot perform crucial experiments in climate science. Accordingly we have to rely on a combination of numerical experiments, often with a considerable degree of parameterization of key climate processes and consensus among experts to reach provisional explanations concerning the causes, magnitude and intensity of climate change. Although the scientific research procedure is incremental in nature, the process of data collection, experimentation and verification of modeling outcomes, results in the steady accumulation of knowledge. It is this knowledge on which we rely for drawing conclusions about the state of the globe's climate and that policy makers use in drawing up recommendations related to mitigation of and adaptation to climatic variability and change.

Introduction

Over the last two to three decades climate has become a word used frequently beyond the small circle of climate specialists who concern themselves with the workings of the global climate system. Climate has graduated to become a problem of planetary relevance as respectively climatic variability and change have clear and potentially strong impacts on several sectors of human society. One of the reasons for the wide awareness of climate is that on a daily basis we use climate information in a variety of ways, for example to plan a vacation or activities in the workplace or outdoors.

For most of us climate is "the prevailing environmental conditions in a specific location". However, this simple definition becomes problematic if applied as a basis for the construction of a scientific theory of the earth's climate system, as well as to understanding its workings, reconstructing its history and ultimately to making predictions. For example, "the prevailing conditions" depend on time, as conditions may change, sometimes drastically on a seasonal basis. Therefore a definition that implies climate is a static phenomenon is inappropriate. Furthermore, the "prevailing conditions" are very elusive. Even if we resolve to use longer and longer time averages to define "prevailing conditions" climate still shows an amazing amount of variability. For example, a particular winter may be quite anomalous or successive winters may be very different, the same applies for other time periods such as months, years, decades and millennia. As climate is non-stationary (its mean and variability characteristics are not constant), to successfully model its characteristics at a variety of spatial and temporal scales, we will need to strive to understand and describe its evolution, statistics and dynamics. These requirements are very different to those expounded during the days of the geographic exploration of climate.

If a dynamical definition of climate is adopted then we can then apply theoretical ideas in physics, mathematics and chemistry to its description (Oort and Peixoto 1992) and modelling. Moreover, the only way to test hypotheses and verify theories about climate is to apply an experimental approach using numerical (climate) models, that is, mathematical representations of the processes and interactions that are believed to force the global climate system. Such models contain, in distilled form, the present body of knowledge on the workings of the climate system. They are the best means that we have at our disposal for making predictions about the complex world around us. Climate models are verified by comparing their output with observations of the real atmosphere. The comparison is as quantitative as possible, but the final evaluation, even if is supported by objective indices and statistical methods, always possesses a certain amount of subjective interpretation. The climate dilemma sits at this crucial juncture. Ultimately, the most sophisticated models must, along with expert opinion, assist in finding a general consensus about what is the best estimate of the truth. Such "organized skepticism" is the process through which science progresses, providing temporary solutions to problems that are modified and/or discarded as new observational evidence becomes available. In the following, the constituent parts of this dilemma will be discussed.

The climate machine

The sole energy source for the climate machine or global climate system is the sun. The quantity of energy received at the top of the atmosphere is referred to as the solar constant, namely the amount of energy per second per square meter. The constant has a nominal value of 1380 W/m^2. This energy is apportioned and transferred between the components of the climate system. Our understanding of the components of the climate system has expanded considerably over the years. The fluid envelopes of the Earth, the atmosphere and the oceans, are the main components of the climate system. It is the dominant pattern of motion within these that determines the climate at any given place on our planet. The distribution of landmasses and mountain ranges, in addition to the characteristics of the land surface, the cryosphere and biosphere are also major factors in shaping climate. The fact that the east coast of continents exhibits in general a harsher climate than the west coast is due to the presence of the continents. It is now becoming increasingly clear that consideration of the atmosphere and the oceans is not sufficient to give a proper description of the climate. Other components of the climate system must be added, like the sea-ice, the evolution of the atmosphere's chemical composition, the biosphere, the soil, and for treatment of climate over geological times, land based ice sheets. The common threads that link these seemingly disparate components of the climate systems are water and energy. It is the unique capability of water to change phase between the three states of liquid (water), gas (water vapour), and solid (ice) that weaves these separate components into an interlocked system. Further every time water changes state there is energy release or absorption.

Energy input to the earth's surface results in water evaporating from the land surface, land surface water bodies and the ocean. It is transported horizontally and vertically as vapour by the wind. Following the process of condensation within the upper atmosphere, the water falls back to the surface as rain or snow. The water that is not absorbed by the soil, the run-off, flows across the earth's surface in river systems eventually finding its way to the oceans. Understanding the path of water through the climate system, or in other words the hydrological cycle, facilitates an insight into its workings. This is because water is an extremely effective energy vector. Water vapor contains the energy (referred to as latent heat) that was necessary to evaporate it from the ocean or land surface. The atmosphere can carry this energy for thousands of kilometers as it transports water vapor, to release it where condensation takes place. In short it is not possible to describe or understand the climate system, without taking great care in understanding water and its transformations.

A further challenge to understanding the workings of the climate system is that its components interact strongly in a non-linear manner. Moreover, they have different times of reaction; the atmosphere is fastest, the ocean is slower. For example, events occurring in the atmosphere, will impact through their statistics on processes occurring in the ocean on slower time scales. Compared to the ocean, reactions times of sea-ice and the biosphere are even slower, and then there are the geological agents, like the land ice sheets and the main mountain ranges, which have the slowest response times. Climate models attempt to describe these strongly nonlinear interactions but in doing so do not ignore fast variations even if it is the slow variations of climate over centuries or so that are of interest.

As noted earlier, the chief energy source for the global climate system is the sun. However the energy received at the earth's surface cannot be stored. Instead the same amount as that received must be re-emitted to space so as to preserve the global energy balance. The equilibrium between incoming and outgoing energy can be measured by a macroscopic parameter, the temperature. The equilibrium temperature of the Earth is about −18 °C. However average global temperature is around 15 °C. What accounts for this difference is the fact that the Earth has an atmosphere. The atmosphere is a crucial component of the climate system because it is essentially transparent to incoming solar radiation, but relatively opaque to the outgoing longwave radiation emitted from the earth's surface. The outgoing radiation absorbed by the atmosphere is re-emitted in all directions. Some will eventually escape to outer space, but a considerable portion will be re-emitted back towards the surface thus maintaining the global surface temperatures above that for the situation of no atmosphere. This effect is referred to as the natural Greenhouse Effect. To preserve the surface equilibrium temperature of the earth radiation must be emitted to offset the warming caused by the natural Greenhouse Effect, otherwise there will be a shift of the equilibrium point and the surface temperature will increase.

➤ *Figure 1*

Although the natural Greenhouse Effect is a benign effect it is ultimately responsible for the maintenance of life on earth, a rather different role to that perceived by the public over the last two to three decades. Of current concern is the extent to which human society is enhancing the natural greenhouse effect thus altering its intensity and characteristics. Human usage of fossil fuels has been changing the composition of the atmosphere, increasing its opacity and therefore making more efficient the atmospheric trapping of the outgoing radiation emitted from the earth's surface. The current reason for concern is not the existence of the greenhouse effect per se, but the inadvertent modification of our environment that may result from an anthropogenically enhanced Greenhouse Effect.

The complexity of the climate system and the limitations to our experimental capability prevent a classical scientific approach to modeling the impacts of changes in the nature of the Greenhouse Effect on climate. Sometimes it is difficult to tell fact from opinion. However, there are two facts that are based on observational evidence that have become established beyond a reasonable level of doubt. These are the concomitant rise in atmospheric concentrations of carbon dioxide (CO_2) and temperature.

The concentration of carbon dioxide has been increasing steadily over the past century (➤ *Fig. 2*). It has increased from 315 parts per million (in volume) in the late 1940s to about 370 ppmv at the end of the 20th century. While these are very small numbers, because of the great efficiency of carbon dioxide as a Greenhouse gas such increases in CO_2 concentrations can have a large impact on the surface temperature. To understand the significance of the increase in CO_2 concentrations, which is mainly due to the burning of fossil fuels, it is best to look back at the past history of atmospheric CO_2 levels. Fortunately there are very good records of the past history of carbon dioxide concentrations, because trapped within the layers

Fig. 1

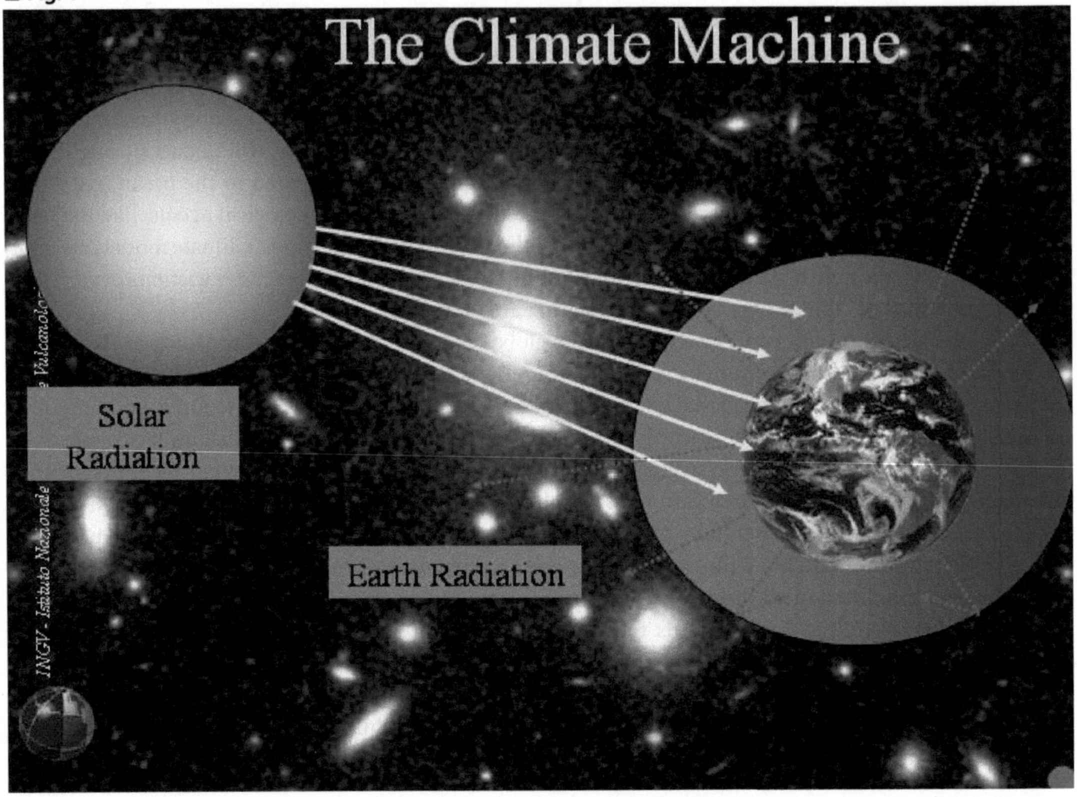

The basic working of the climate machine. The solar radiation arrives on the Earth (white rays) and it is balanced by surface Earth radiation (straight grey arrows), but the atmosphere traps some of the Earth radiation, reradiating it towards the space and again toward the surface.

of ice that make up the Antarctica and Greenland ice sheets are small bubbles of air containing CO_2 that date back to the era of formation of the ice. By recovering long ice cores from the ice sheets and analyzing the composition of the bubbles of air, it is possible to reconstruct a record of the earth's history of CO_2 concentrations. Such reconstructions reveal considerable changes over the last 420,000 years. CO_2 concentrations have fluctuated between 200 and 280 ppmv often following the behavior of temperature as indicated by proxy measurements (lower blue line). Often slow rises in concentration have been followed by rather sharp decreases, on timescales of a few thousand years. However, the period in which we are living today is unique in the context of the last 420,000 years. The Earth has never seen such high CO_2 concentrations over the period of human history and it has never witnessed changes of such magnitude and intensity (about 40 units in 40 years). Clearly the past century has been extraordinary.

The second fact concerns the historical instrumental record of temperature, which stretches for more than over a hundred years. The challenge in reconstructing this record has been to derive reliable estimates of global temperatures from measurements that are sparse in time and space. The meteorological and climate community has been concerned with this kind problem since the very early days of climate research as a regularly spaced climate observing network does not exist across the earth's land and ocean surfaces. The different types of observation systems, the variety of platforms and techniques and the range of parameters

Fig. 2

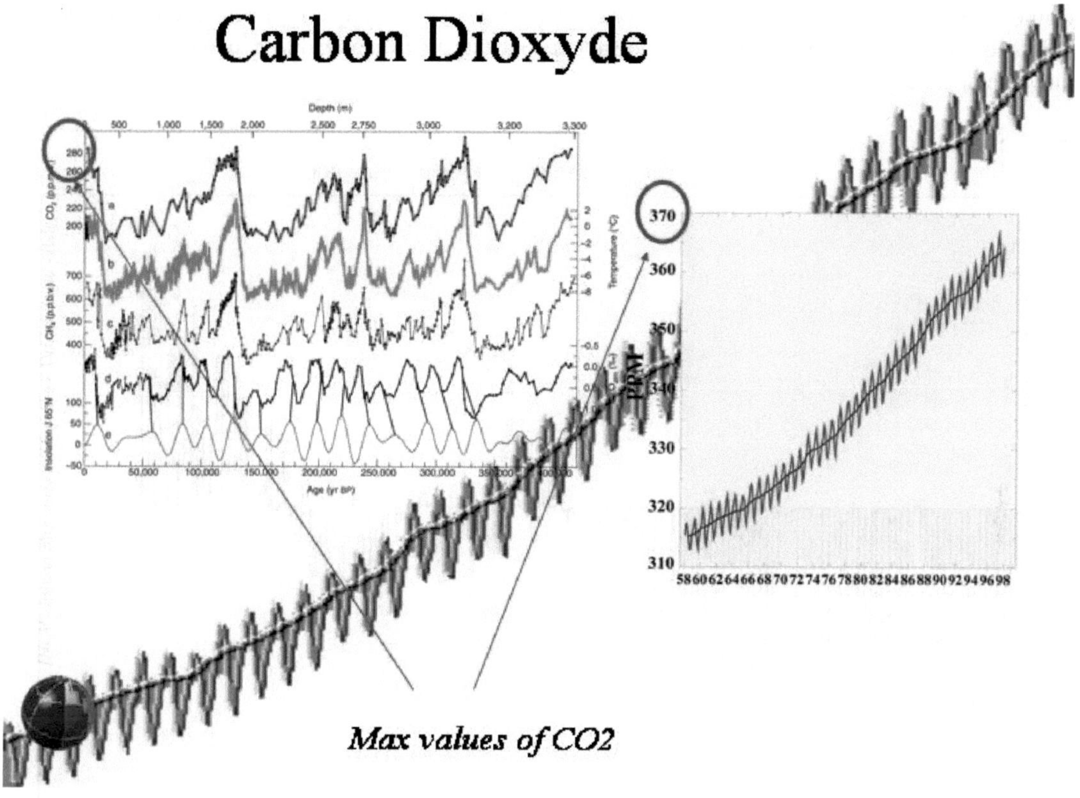

The increase of concentration of carbon dioxide in the last century (right, from Bell and Halpert 1998) compared to the concentration of CO_2 (upper curve) in the last 420,000 years from the geological ice record (left, from Petit et al 1999). In the past the CO_2 content has oscillated between 200 and 280 ppmv, whereas it started at 310ppm in 1948 and it has now reached about 370 ppmv around the year 2000.

being measured, has prompted the development of a strong branch of climate research that deals with making space-time estimates of fields that are dynamically and physically consistent. These methods have produced estimates of the global surface temperatures that are quite reliable.

Figure 3 shows the trend of the globally averaged surface temperatures for the period 1861 – 2003, expressed as anomalies or deviations from the 30 year reference period 1961 – 1990. A marked increase in hemispheric and global surface temperature is clearly visible. For both hemispheres the rate of increase has accelerated since the 1960s although it is somewhat weaker for the southern hemisphere. Although the increases in temperature are impressive they still refer only to the Earth's surface. Moreover the records on which such constructions of the global temperature curve are based are plagued by the usual problems associated with climatic data: they are sparse in space and time and there are quality control issues. Furthermore, for sometime the satellite-based measurements of mean tropospheric temperatures did not show any indication of warming as recorded by the surface record. However, a recent evaluation of the measurement errors associated with the satellite-based records of temperature has revealed a tropospheric warming trend that is consistent with the observed trend at the surface (Vinnikov and Grody 2003). Such convergence of evidence provides strong support for the debate that indeed the Earth is warming at the rates indicated by the surface record for the past decades. Of course the big question is whether the warming can be attributed to the increase in carbon dioxide. Although there is no direct evidence of this casual connection, alternative explanations are becoming less likely plus all modeling and observational evidence points to the fact that increases in Greenhouse gas concentrations can lead to concomitant increases in global temperature.

Fig. 3

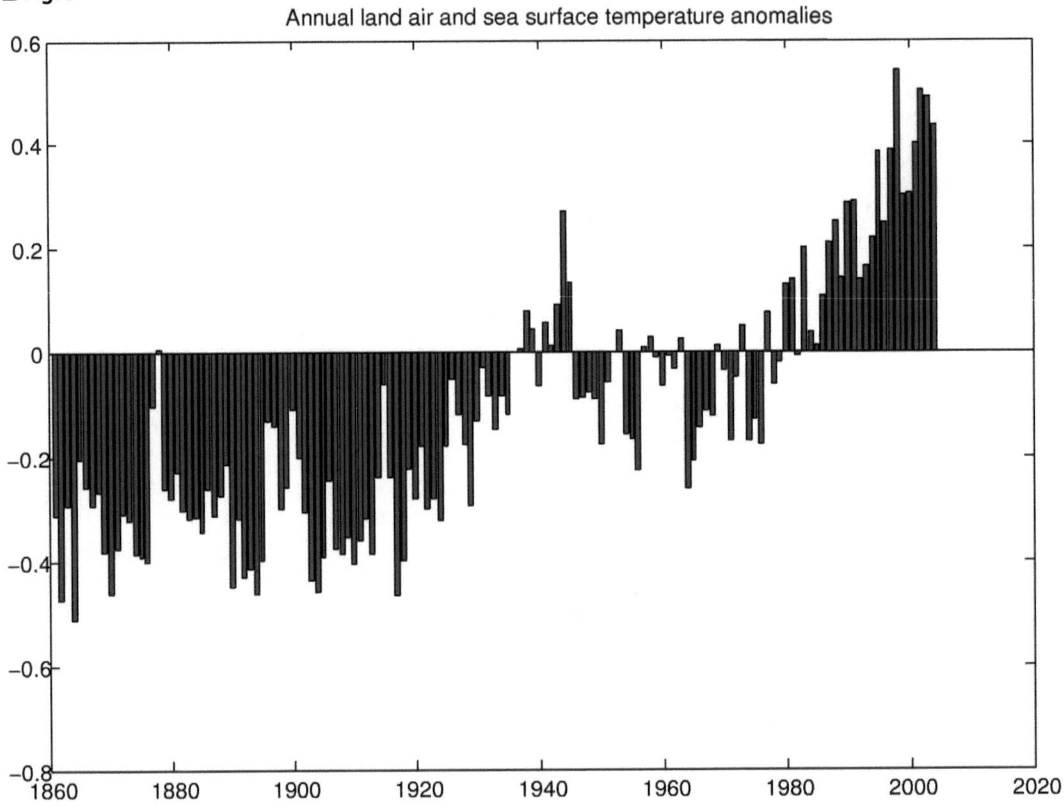

The increase of global surface temperature in the last century (from NOAA, http://www.ncdc.noaa.gov/oa/climate/research/anomalies/anomalies.html). The values are deviations from the average for the period 1880–2003

The interannual variability of climate

While the statistically smoothed curve of mean global temperature presented in *Figure 3* might create the impression that there has been a steady incremental increase in temperature from year to year, a closer examination of the actual anomalies for individual years indicates a considerable degree of interannual variation in climate. In short the climate system departs from its climatological average almost all the time. The averaged blue line is a convenient way to express in a synthetic way, the overall effect of the variations that contribute to the mean itself. From *Figure 4* it is clear that there were some anomalously hot years in the late 90s, but there were also some relatively cool years. A warming trend therefore does not imply that every year will be a record-breaking year, but that something about the characteristics of the climate system's variations is changing to reflect a small change of the mean. It then becomes imperative to understand the nature of the variations and to try to predict how their statistics will modify under changing conditions. *Figure 5* shows the temperature anomalies over France in the summer of 2003. The anomaly is defined as the deviation of the actual temperature from the climatological normal. The summer of 2003 was clearly extraordinary but emphasizes the fact that summer temperatures can exhibit large variations from one year to another. The presence of these large variations makes the detection of any trend with some sort of reliable statistical confidence difficult.

Fig. 4

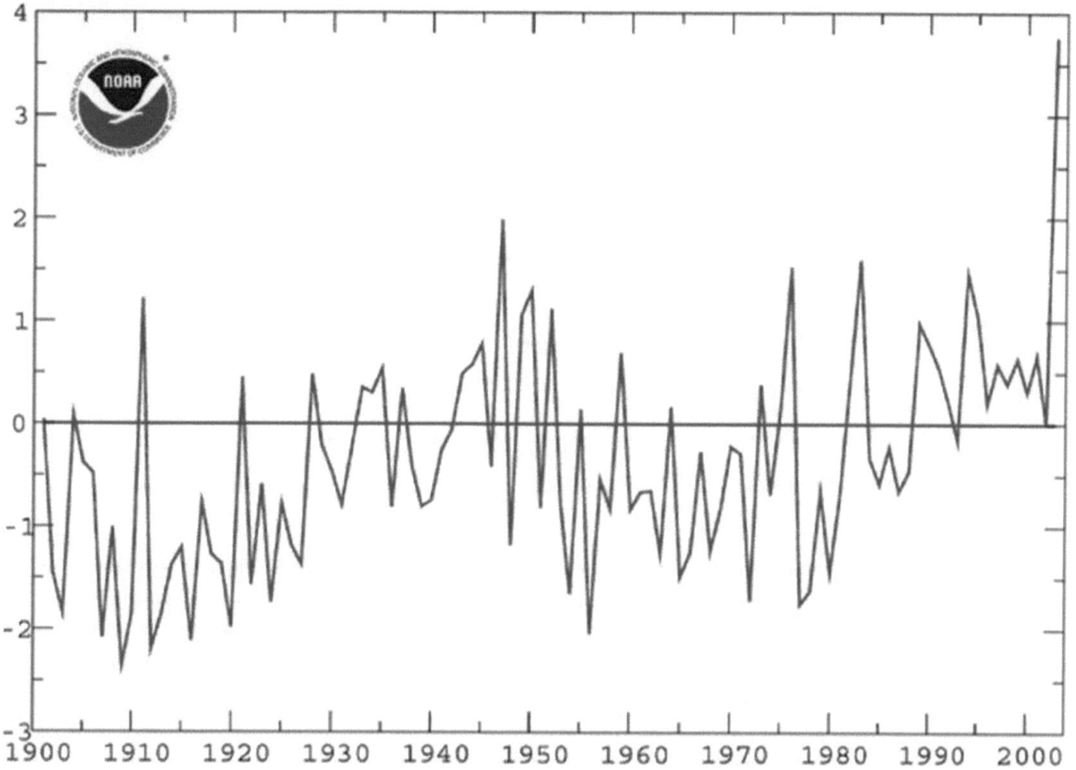

Summer anomaly temperature in France (from the NOAA, Climate Report for 2003, http://www.ncdc.noaa.gov/oa/climate/research/2003/aug) showing the variations of the summer temperature in France for the XX century. The anomaly is defined as deviations from the climatology. The critical summer of 2003 is well above the rest of the century. Note the large deviations from one year to next that make very difficult the identification of any trend before 2003. Obviously, 2003 alone is not sufficient to establish such a trend, but it has a remarkable large value.

The anomalous temperatures of summer 2003 were not limited to France, but covered most of Western Europe and extended into northern Europe and Scandinavia (*Fig. 6*). France, Italy and some regions of Spain, experienced an exceptional heat wave. The Mediterranean Sea was also anomalously warm with sea surface temperatures being as high as 4 degrees above normal.

The anomalous summer of 2003 did not occur in isolation as other parts of the Northern Hemisphere also experienced departures from normal conditions such as over the east coast of the United States where summer was cooler than normal. This is not a surprise as climate variations are not independent, but they

Fig. 5

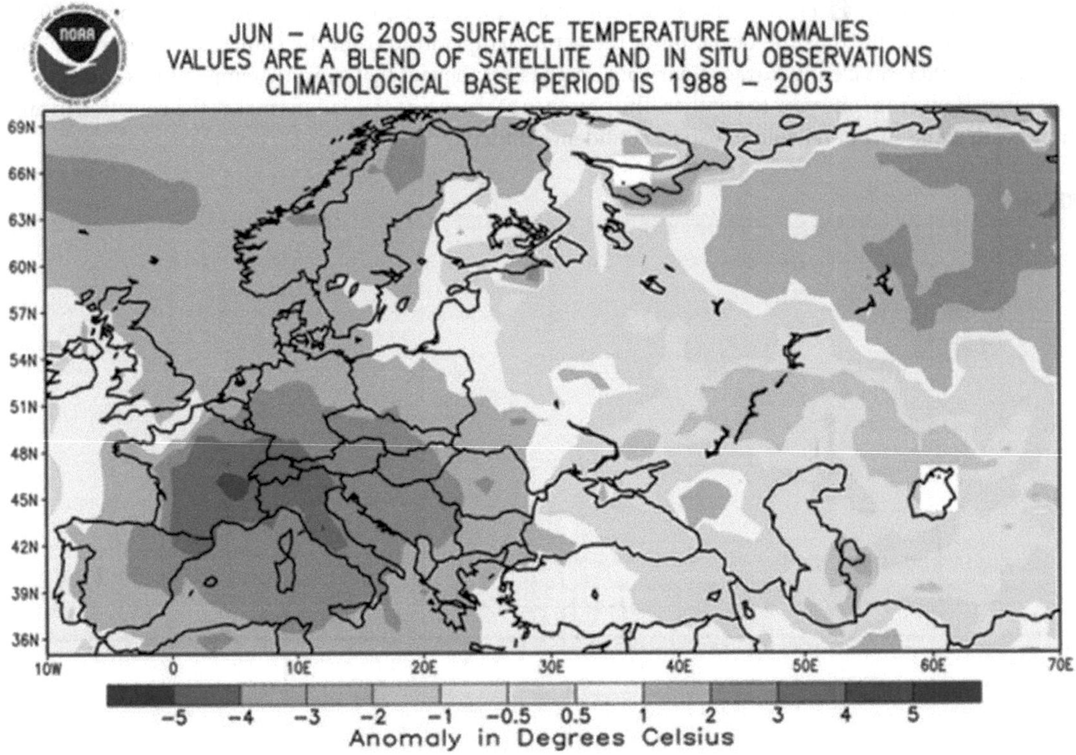

Summer anomaly temperature over Europe for the summer 2003 (from the NOAA, Climate Report for 2003, http://www.ncdc.noaa.gov/oa/climate/research/2003/aug) showing the exceptional warm summer season. The anomaly is defined as deviations from the period 1988-2003. Large areas of France and Italy show anomaly temperature in excess of 4 degrees. The warming is part of a larger regional pattern that extends from the Mediterranean to Scandinavia.

are connected, so that the climatic variations in distant regions may be correlated. Such patterns are called teleconnections. These often express themselves in the form of atmospheric and temperature and occasionally precipitation variations of the opposite sign so that in one region the atmospheric pressure may be anomalously high while in another region the pressure will be abnormally low. One of the most often referred to teleconenctions is the Southern Oscillation (SO), which is often equated with the El-Nino Southern Oscillation (ENSO) phenomenon. In addition to this are a number of other major teleconnections, including those such as the Pacific- North American pattern (PNA) and the North Atlantic Oscillation (NAO) both of which extend into the mid-latitudes. The precise origin of the teleconnections is not completely known, but they certainly arise from a complex web of interactions between the ocean, via the Sea Surface Temperature (SST), and the overlying atmosphere. While atmosphere-ocean interactions in the Pacific are somewhat better understood (Philander 1990), the situation in the Atlantic is more obscure.

The Mediterranean and Southern Europe is only affected marginally by the aforementioned large teleconnection patterns, except for the influence of the North Atlantic Oscillation. However, this region may be affected by climate change through a different mechanism. The Mediterranean region is a border region. During the summer it is under the influence of the descending branch of the tropical Hadley Cell and is characterised by a climate that is dry and hot. During the winter it is within the mid-latitude cir-

Fig. 6

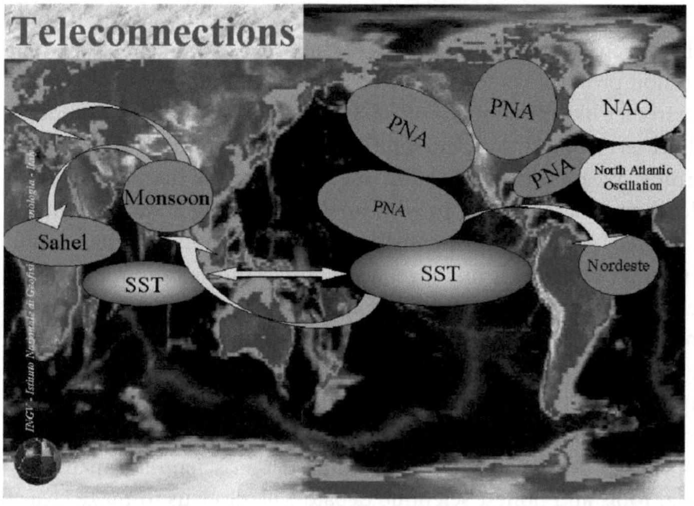

The interactions between atmosphere and oceans in the tropics dominate the variability at interannual scales. The main player is the variability in the equatorial Pacific. Wave trains of anomaly stem from the region into the mid-latitudes, as the Pacific North American Pattern (PNA). The tropics are connected through the Pacific SST influence on the Indian Ocean SST and the monsoon, Sahel and Nordeste precipitation. It has been proposed that in certain years the circle is closed and a full chain of teleconnections goes all around the tropics. Also shown is the North Atlantic Oscillation a major mode of variability in the Euro-Atlantic sector whose coupled nature is still under investigation.

culation regime, and it is traversed by a fair number of low pressure systems of Atlantic origin that advect large amounts of moisture and relatively cool air from the west and north resulting in relatively fresh temperatures. The equitable Mediterranean climate owes its origin to this alternation of hot summers and wet winters. A change in the position of the Hadley Cell and its associated circulation or the northward or southward shift of the winter-time Atlantic storm track, possibly as a result of climate change, could have very drastic consequences. A northward shift of the Hadley Cell would put the Mediterranean permanently under the influence of the tropical Hadley circulation. This would mean a reduction in the winter rains and a shift of the Mediterranean climate toward what is today the climate of North Africa. This kind of change would be visible only after several years and most likely manifest itself through an increase in the frequency of events like the summer of 2003, which would eventually have an affect on the mean climate. The question that will have to be answered in the next few years is "was the summer of 2003 an isolated anomalous event or the first sign of an impending change?" It will therefore become important to this end to monitor closely the behavior of climate for any sign of a significant change in the statistics of precipitation and drought.

Conclusions

We have seen that in order to detect climate changes we need to use more than just our intuition. We have to use powerful analysis methods that use advanced statistical techniques to distill from the complex behaviour of climate an underlining trend. Extensive usage of statistics is required, but we have to be very careful to always accompany the presentation of the results of statistical analysis with a good physical explanation of any findings. (Von Storch and Navarra 1999). Numerical modelling is another powerful weapon that we have at our disposal to unravel the workings of the climate system and also to address the much more challenging issue of climate change prediction. However, in order to validate predictions and to address the climate change detection problem we will need to continue to monitor the state of the climate system. In the case of climate extremes like the summer of 2003, there must be two levels of monitoring. A watch must be kept on the short term, and this type of monitoring will be connected to weather services for short-term alarm and warning. Weather forecast skill is approaching eight days at the global level and two-three days at regional level and therefore can provide an early warning about impending heat waves and other extreme weather events. However importantly the limiting factor in the organization of an effective and proper response to weather extremes is no longer a scientific issue, but more an institutional, social and cultural one. The second level of monitoring will need to take a long-term view with the aim of assessing how the frequency and characteristics of weather and climate extremes may change over the next 20 – 30 years. We are far from a definitive answer in this area, but numerical models are one of the best tools at our disposal. It is to be expected that the next generation of models currently under development, will be able to offer more accurate and detailed information on this issue. Although unraveling the complexities of the climate system appears to be a challenge beyond our current scientific capabilities, a wise usage of numerical and statistical tools and honest scientific debate will allow us to make steady progress towards solving the climate dilemma.

References

Bell GD, Halpert MS (1998) Assessment for 1997, 1998, Bull Amer Meteor Soc 32:302 – 350

Navarra A (ed) (2001) Beyond El Niño: Decadal and Interdecadal Climate Variability. Springer, Berlin Heidelberg New York, pp 333ff

Petit JR, Jouzel J, Raynaud D, Barkov NI, Barnola J-M, Basile I, Bender M, Chappellaz J, Davisk M, Delaygue G, Delmotte M, Kotlyakov VM, Legrand M, Lipenkov VY, Lorius C, Pepin L, Ritz C, Saltzmank E, Stievenard M (1999) Climate and atmospheric history of the past 420,000 years from the Vostok ice core, Antarctica. Science 429:436

Oort AH, Peixoto JP (1992) Physics of Climate. Springer, Berlin Heidelberg New York pp 560ff

Philander SGH (1990) El Niño, La Niña, and the Southern Oscillation. Academic Press, Inc., San Diego, pp 293ff

Vinnikov YK, Grody NC (2003) Global warming trend of mean tropospheric temperature observed by satellite. Science 302:269 – 272

Von Storch H, Navarra A (1999) Analysis of Climate Variability. Applications of Statistical Techniques. Springer, Berlin Heidelberg New York, pp 330ff

Projected Changes in Extreme Weather and Climate Events in Europe

Glenn R. McGregor · Christopher A. T. Ferro · David B. Stephenson

Abstract

Extreme weather and climate events have wide ranging impacts on society as well as on biophysical systems. That society, on occasions, is unable to cope with extreme weather and climate events is concerning, especially as increases in the frequency and intensity of certain events are predicted by some global climate change projections. Extreme events come in many different shapes and sizes. The multitude of extreme event types has also led to a proliferation of definitions appropriate for different applications at different times and places.

Conceptually, climate change may lead to an alteration of extreme weather and climate events across Europe. Trends in time series of observed extreme weather and climate indices are suggestive of changes in the climatology of extreme events over Europe. This paper reviews theoretical approaches associated with the scientific assessment of extreme weather events.

Introduction

Extreme weather and climate events have wide ranging impacts on society as well as on biophysical systems. In 2003 for example there were several noteworthy events across Europe, including annual temperature anomalies of 1 – 2 °C over Central and Western Europe, prolonged summer drought, a major heat wave, and severe wildfires in Portugal, France and the Mediterranean. The consequences were significant economic losses as well as human fatalities. The economic losses have been estimated at US$ 18,619 million with drought accounting for one third of the total (Munich Re, 2004); the majority of the estimated 30,000 natural hazard related fatalities are most likely attributable to the heat wave that gripped Europe for almost two weeks in early August. Clearly 2003 is a pertinent example of how the weather and climate can produce conditions that are outside the coping range of society.

That society, on occasions, is unable to cope with extreme weather and climate events is concerning, especially as increases in the frequency and intensity of certain events are predicted by some global climate change projections. The purpose of this chapter is to illustrate ways in which climate change may affect the occurrence of extreme events, consider the observational record of climate extremes for Europe and present the results of one projection of the impact of climate change on extreme events over Europe. Some of the issues associated with the analysis of extremes are presented first.

Types of Extreme Events

Extreme events come in many different shapes and sizes. Events occurring over short time scales, between less than 1 day and 6 – 10 days, are often referred to as extreme weather events. For example, tornadoes and thunderstorms usually have durations shorter than a day. The latter period, which is often referred to as the synoptic time scale, is typical for the weather associated with low- or high-pressure systems that usually bring unsettled (wet and windy) or settled (hot/cold and dry) weather respectively. Beyond the synoptic time scale lie extreme climate events. Examples include hot and dry summers or wet and stormy winters, which may be the product of the cumulative effect of a number of hot dry periods or a high frequency of intense rain-bearing cyclonic systems. Extreme events also occur on a range of spatial scales, from concentrated events such as tornadoes to diffuse events such as droughts. Extreme events can further be classified as simple or complex. Simple extremes are characterised by a single variable such as temperature, whereas complex extremes might involve a critical combination of variables associated with a particular weather or climate phenomenon such as a cyclone (rainfall and wind) or drought (little precipitation and high temperatures).

The different forms of extreme events have implications for monitoring. Generally it is easier to monitor acute events, such as floods or windstorms, because of their high magnitude and short duration. In contrast, the onset of chronic events such as droughts is difficult to detect because of the long lead-time and imperceptible transformation of the environment. The multitude of extreme event types has also led to a proliferation of definitions appropriate for different applications at different times and places. For example, the Intergovernmental Panel on Climate Change (IPCC) defines an extreme weather event as 'an event that is rare within its statistical reference distribution at a particular place' and continues: 'Definitions of "rare" vary but an extreme weather event would normally be as rare or rarer than the 10th or 90th percentile' (IPCC 2001, p 790). Other definitions might be more meaningful in other contexts however: an event may be considered extreme merely if some of its characteristics, such as magnitude, duration, speed of onset or intensity, lie outside a particular society's experiential or coping range, whether or not the event is rare. Applying definitions out of context can therefore be misleading and causes confusion, and attempting to formulate a universal definition of 'extreme' is misguided. Studies should simply adopt the most apposite definition and state it precisely.

Climate Change and Changes in Extreme Events

In order to gauge how extreme a particular event is, the typical values of climate variables and the relative frequencies of those values must be known. These statistical properties are captured by the probability distribution of a climate variable. For example, the distribution of temperature is often approximately Gaussian, identified with a symmetric bell-shaped curve, while precipitation approximately follows the asymmetric gamma distribution that is truncated at zero and has a longer upper tail of high values. Useful summaries of probability distributions include measures of the location, scale (or spread) and shape (or skewness). The mean value is a common measure of location, while the standard deviation (the square root of the variance) measures the scale. Another measure of location is the median, the level below which values from the distribution fall 50 % of the time. This is also known as the 50th percentile. Other percentiles are defined similarly and a small set of them can provide an informative summary of a distribution: for example, the 90th, 95th and 99th percentiles are sometimes used to summarise the upper tail of a distribution. Percentiles can also be used to define measures of scale such as the inter-quartile range, which is the difference between the 75th and 25th percentiles, and measures of shape such as the Yule-Kendall skewness statistic, positive (negative) values of which indicate that the upper (lower) tail of the distribution is the longer.

The statistical characteristics of weather and climate events could be affected by climate change in a number of ways (*Fig. 1*). For example, in the case of temperature there simply could be a shift in location towards higher (warmer) values (*Fig. 1a*). This would result in an increase in the number of extreme events at the hot end and a decrease at the cold end of the distribution. Consequently, there would be not only more hot weather but also more record hot and less cold weather. As the probability of exceeding a fixed threshold changes non-linearly with shifts in location, a small change in the location can result in a large relative change in the probability of extremes (Mearns et al. 1984; Meehl et al. 2000; IPCC 2001).

Fig. 1

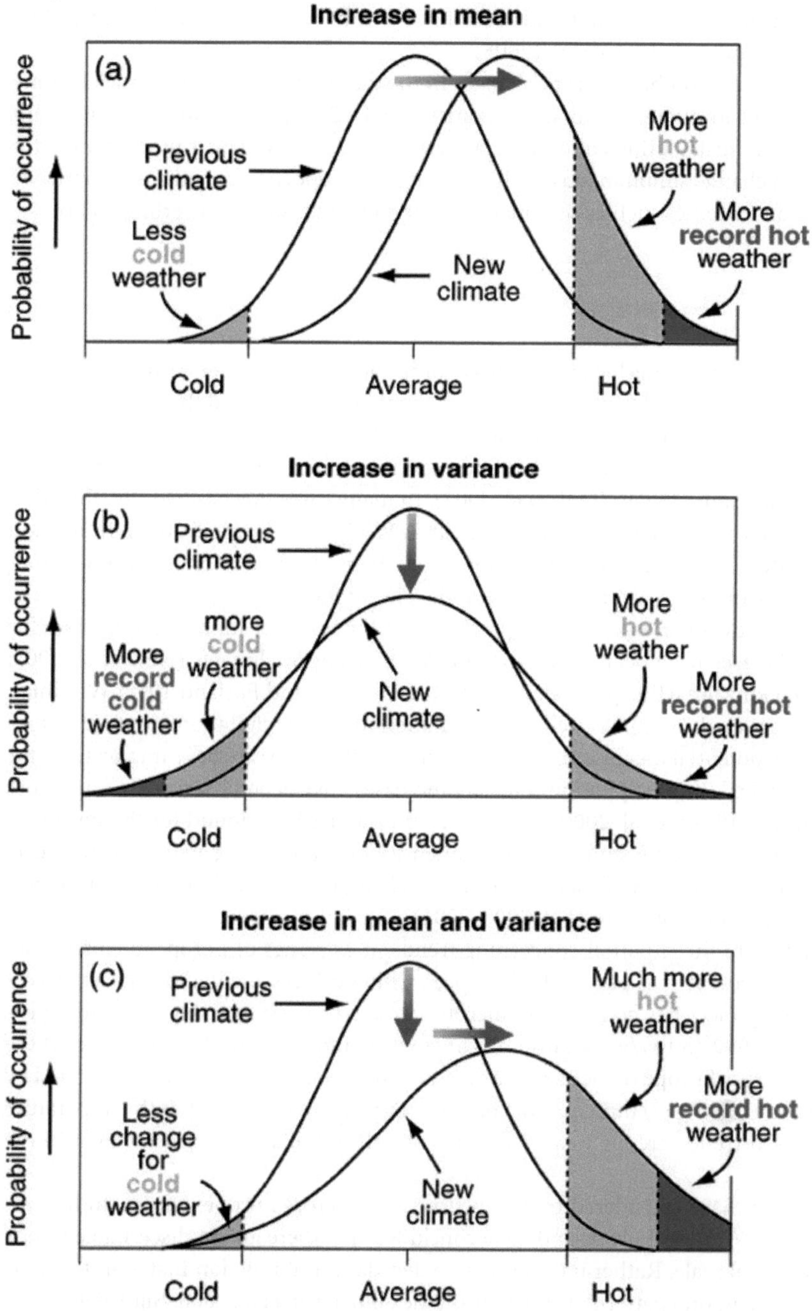

Theoretical changes in the distribution of climate variable values. (a) change in mean (location); (b) change in variance (scale); (c) change in mean, variance and skewness (shape)

In addition to a change in location, there may be a change in the scale of the distribution (➤ *Fig. 1b*). This would produce changes in the occurrence of extreme events at either end of the distribution and potentially have a greater effect on the frequency of extreme events than a simple change in location (Katz and Brown, 1992). In addition to changes in the location and scale, changes in the shape of a distribution can be envisaged such that the distribution, of temperature for example, becomes skewed to the right (➤ *Fig. 1c*). This would result in much more hot and record hot weather and fewer cold events. More discussion of possible changes is given in Ferro et al. (2005).

Many other types of change and combinations of changes in the distributions of climate variables are of course possible and will result in different climate outcomes (IPCC, 2001). For example, a pure shift in location is unlikely to occur in distributions of non-negative valued variables such as precipitation, for which location and scale often change simultaneously. This can alter disproportionately the occurrence of various aspects of precipitation extremes such as seasonal totals or daily intensities (Easterling 2000).

Trends in Observed Climate Extremes

Studies concerning climate trends at various locations across Europe abound. However, results from these are not directly comparable because of the contrasts in data set length and quality and the different methods used for data processing and trend analysis (Wijngaard et al., 2003). Nevertheless, common patterns appear to be emerging. For most locations across Europe, increases in minimum temperature appear to be greater than in maximum temperature (Klein-Tank et al. 2002). In many cases this has been attributed to increasing nocturnal cloud cover (Brazdil et al., 1996; Huth, 2001; Wibig and Glowicki 2002). In relation to human thermal comfort, McGregor et al. (2002) have noted for Athens, Greece, a tendency towards an increase in the length of the discomfort season over the period 1966 – 1995.

Precipitation studies have shown increases in total precipitation for some locations but decreases for others. Generally, rainfall increases have been noted for non-Mediterranean climates (New et al. 2001; González-Rouco et al. 1999; González-Hidalgo et al. 2001; Hanseen-Bauer and Forland 1998; Windman and Schar 1998; Garcia-Herrara et al. 2003). As well as precipitation totals, precipitation intensity has received some attention because potential increases in this precipitation characteristic have implications for flooding and soil erosion. For the UK, precipitation intensity increases have been observed and are more marked for the winter months (Osborn et al. 2000). This matches what has been found for the European Alpine region (Frei and Schar 2001). Some studies point to intensity increases being associated with certain types of weather systems (Windman and Schar 1998) and the changing relationships between wet day occurrence and wet day rainfall totals (Brunetti et al. 2000; 2001).

Perhaps the clearest picture of the situation concerning trends in extremes of European climate can be garnered from the analyses presented as part of the European Climate Assessment and Dataset project (Klein-Tank et al. 2002). This project has compiled numerous climate time series using consistent procedures for the period 1946 – 1999 (http://www.knmi.nl/samenw/eca/index.html). An analysis of these reveals that observed trends across Europe demonstrate a greater consistency with the trends predicted by climate models for temperature (➤ *Fig. 2*) than for precipitation. Further, in the case of both temperature and precipitation, the observed trends are far weaker in the first half of the analysis period than in the second half (1975 – 1999).

A limited number of studies have considered trends in storminess across Europe. Of the studies undertaken most have focused on Western Europe and have concluded that there are no discernable trends in storminess at this geographical scale. Rather inter-annual to decadal variability dominates and there is geographical variability in the temporal pattern of storminess (Alexandersson et al. 2000; Bijl 1999; Flocas et al. 2001; Maheras et al. 2001; Pryor and Bathelmie 2003).

Fig. 2 a and b

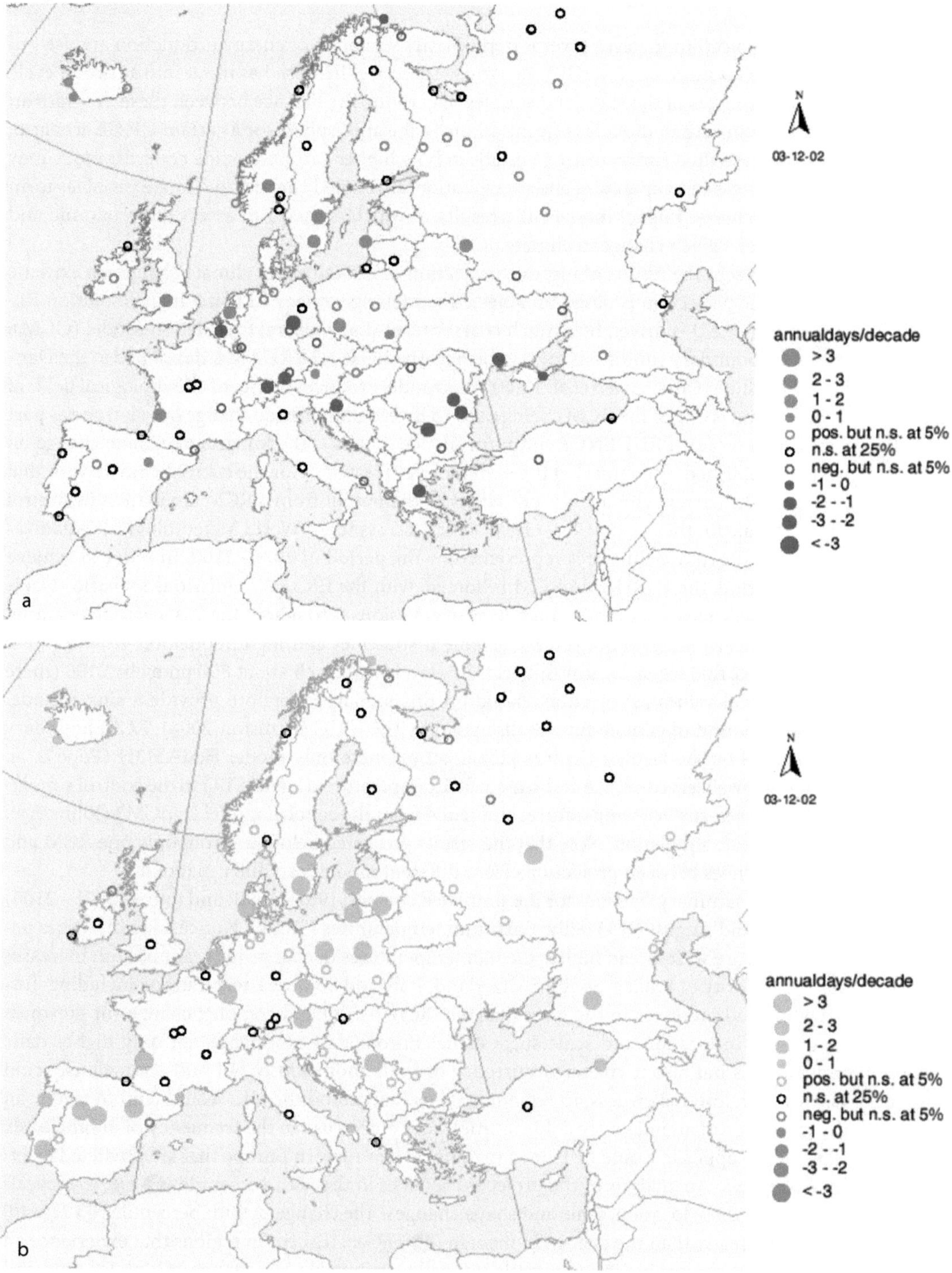

Observed changes in heat wave duration index, (a) 1946–1999; (b) 1976–1999

Projected Changes of Extreme Events

Arriving at a climate projection involves several steps. Firstly scenarios of energy production are used to construct Greenhouse Gas Emission (GHGE) scenarios. These are then used as input into a carbon cycle model that provides estimates of the sinks and sources of carbon. The balance between these provides an estimate of the increase in carbon dioxide concentrations in the atmosphere for a certain GHGE scenario. Global climate models are then run in order to establish how higher carbon dioxide concentrations may affect, for example, changes in temperature and precipitation. Estimated changes in climate variables form the input into climate change impact models, the results of which are used to assess the economic and societal consequences of a given change in climate.

This section will present the results of just one projection of the impact of climate change on extreme events over Europe. The projection is based on a nested modelling strategy in that a high resolution Regional Climate Model (RCM) is driven by a much coarser resolution General Circulation Model (GCM); the latter provides the boundary conditions for the former. The RCM is HIRHAM4, developed at the Danish Meteorological Institute (Christensen et al. 1998). This model produces output of climatological fields at a resolution of 50 km and is one of the RCMs being used to investigate climatic change over Europe as part of the European Union project PRUDENCE (http://prudence.dmi.dk). The impacts of climate change on extreme events are established by comparing the extreme event statistics for the current climate with that of the future climate. The current climate is represented by the output from a RCM simulation (a control simulation) of the climate for the period 1961 – 1990, which is a standard World Meteorological Organization reference period. The future climate is represented by the period of 2071 – 2100. In order to achieve predictions for this period, the HIRHAM4 RCM is 'forced' with the IPCC A2 emissions scenario (Nakicenovic et al. 2000). This scenario assumes a high level of emissions throughout the 21st century, resulting from low priorities concerning greenhouse-gas abatement strategies and high population growth in the developing world. Under this scenario, atmospheric CO_2 levels will reach about 800 ppmv by 2100 (three times their pre-industrial values). Projections based on this scenario therefore provide a single-model estimate of the upper bound of climate futures discussed by the IPCC (Beniston 2004). RCM boundary conditions are supplied by the Hadley Centre's global, atmosphere-only model HadAM3H (Pope et al. 2000), which is driven by observed sea ice and sea-surface temperatures (HadISST1) in the control simulation, and by sea ice and sea-surface temperatures simulated from the coupled model HadCM3 (Johns et al. 2003) in the future climate simulation. Note that the results presented below are from only one RCM and that quantitative differences between projections from different models can differ markedly.

➍ *Figure 3* displays summary statistics for the simulated control (1961 – 1990) and future (2071 – 2100) climates for June, July and August (JJA) daily maximum temperatures (Tmax). Noticeable differences between the two climates are widespread higher median temperatures (a change in location) with increases of 4 – 8 °C over the majority of France, Spain, Switzerland, Italy and South-eastern Europe including Turkey, similarly widespread increases in the inter-quartile range (scale) and complex changes in skewness (shape). The increases in location and scale suggest that Europe will experience not only higher daily maximum temperatures but also a greater occurrence of both anomalously hot and anomalously cold summer days. However, lower skewness, in regions such as France and the UK, would tend to favour an increase in the frequency of anomalously cold summers and a decrease in the frequency of anomalously hot summers, while the opposite would be true in regions such as Eastern Europe that are predicted to experience higher skewness. An analysis of the projected changes in the 90th percentile (➍ *Fig. 4a)* reveals the combined effects of these location, scale and shape changes. The change in 90th percentile (➍ *Fig. 4b)* is almost everywhere greater than the change in median (➍ *Fig. 4c)*. Except in regions that experience an increase in skewness, the change in 90th percentile is mostly attributable to a change in both location and scale (➍ *Fig. 4d)*, not just location alone.

Fig. 3

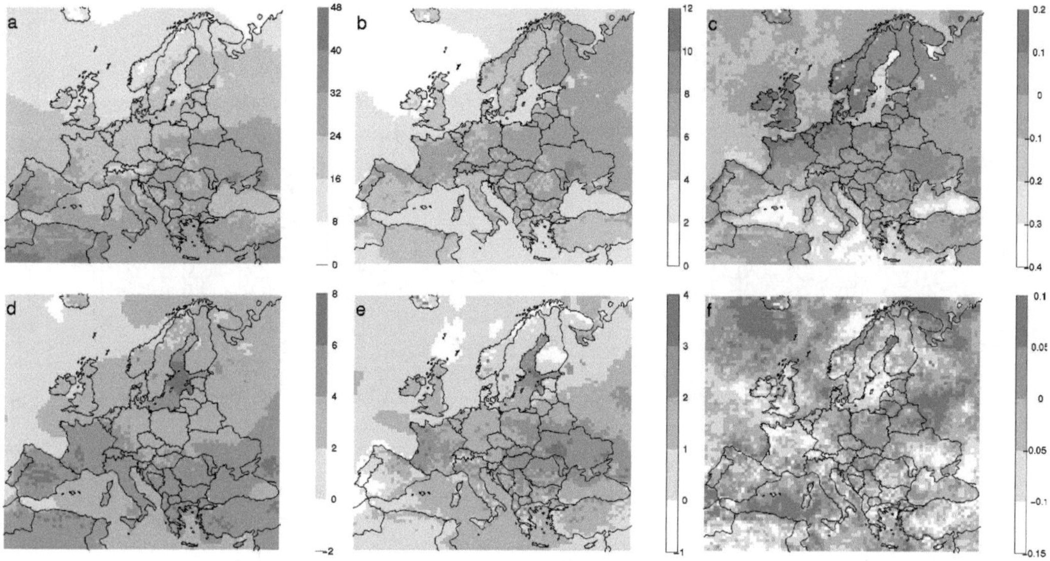

a) Median (°C), b) inter-quartile range (°C) and c) skewness measure for JJA daily maximum temperatures in the control; d), e) and f) the differences in the summary statistics between the scenario and control

If these projected changes in summer temperatures were to materialize then the implications for human health and availability of water resources in summer could be severe. With regards to human health, an analysis of the number of days with maximum temperature above 30 °C provides a crude approximation of the way in which heat-wave incidence and duration might change across Europe. According to the HIRH-MAM RCM control simulation, the majority of Western Europe currently experiences about 5 – 10 days per summer with maximum temperatures in excess of 30 °C. However, for the period 2071 – 2100 the situation could be quite different with the model simulation predicting increases of up to 60 days per summer in Mediterranean countries (*Fig. 5*). By 2100, countries such as France may experience temperatures above 30 °C as often as Spain and Sicily currently experience such events. Consequently, this climate model predicts increases in heat-wave frequency and duration across most of Europe, along with prolonged dry periods and increased probability of summer drought. Using a simple definition of a heat wave, based on three successive days above 30 °C, a three and ten fold increase in the duration and frequency of heat waves might be expected for many places across Europe by the end of the current century (Beniston 2004). The extent to which such projected changes in the heat wave climate of Europe would bring about increases in heat-related mortality and morbidity depends on the level of societal adaptation to such events. With respect to this, the summer of 2003 in Europe may be a harbinger of the future, as in statistical terms the maximum temperature climate of 2003 resembles far more that predicted for 2071 – 2100 than the current climate (Beniston 2004).

In addition to changes in temperature, climate change is likely to bring about changes in precipitation amount and spatial distribution. Climate models show the precipitation response to be far less certain than temperature (Deque 2003). Generally, winter precipitation increases are predicted for most of Europe apart from the far south. In summer, Europe-wide precipitation decreases are predicted apart from the far north where summer wetness may increase. Although summer precipitation amounts are predicted to reduce substantially over large parts of Central and Southern Europe, the opposite may hold true for

Fig. 4

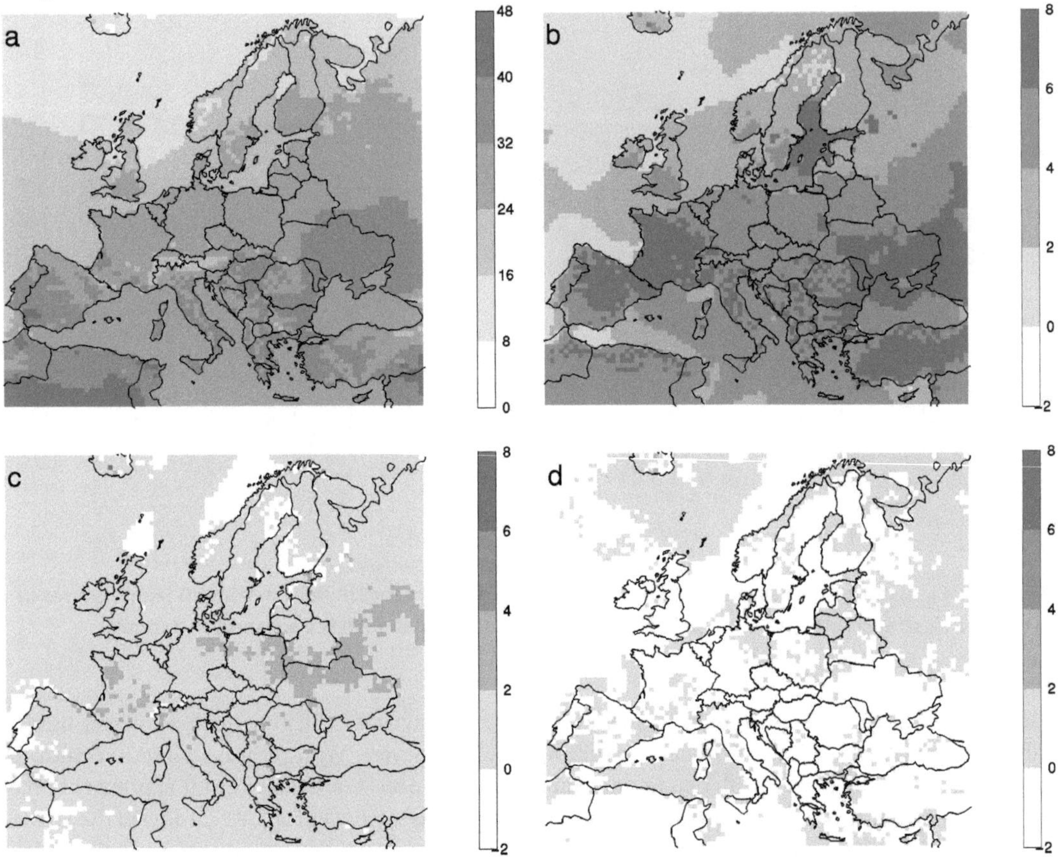

The 90 % quantile (°C) of JJA daily maximum temperatures: a) in the control, b) the difference between the scenario and control, and the differences after adjusting for c) location, and d) both location and scale.

trends in heavy precipitation amounts. This is because climate change projections point to heavier/more intense summertime precipitation events over large parts of Europe despite overall decreases in summer precipitation amounts. This has clear implications for flash flooding in summer (Christensen and Christensen 2003) such that intense precipitation events that lead to flooding in the Elbe, Donau, Moldau and the Rhone in 2003 are likely to become more frequent. For winter, increases in precipitation, relative to the current climate, have been simulated by both GCMs and RCMs for northern Europe (Johns et al. 2003; Ferro et al. 2005). This is mainly because in a warmer climate the atmosphere will contain more water. As for summer, increases in winter precipitation have implications for flooding especially as increased precipitation will bring soil moisture capacities closer to their maximum. Thus wet antecedent conditions are likely to increase the probability of wintertime flooding.

More humidity may also provide an additional source of energy via latent heat release during cyclogenesis and could lead to the intensification of low-pressure systems and make more water available for precipitation (Frei et al. 1998). Increases in winter precipitation over Northern Europe may also be due to changes in the position of winter storm tracks (McCabe et al. 1999), which may partly account for the slight decreases in simulated precipitation for Southern Europe. The situation regarding projections of future storminess is far less clear than that for temperature and precipitation. Although there are a growing

Fig. 5

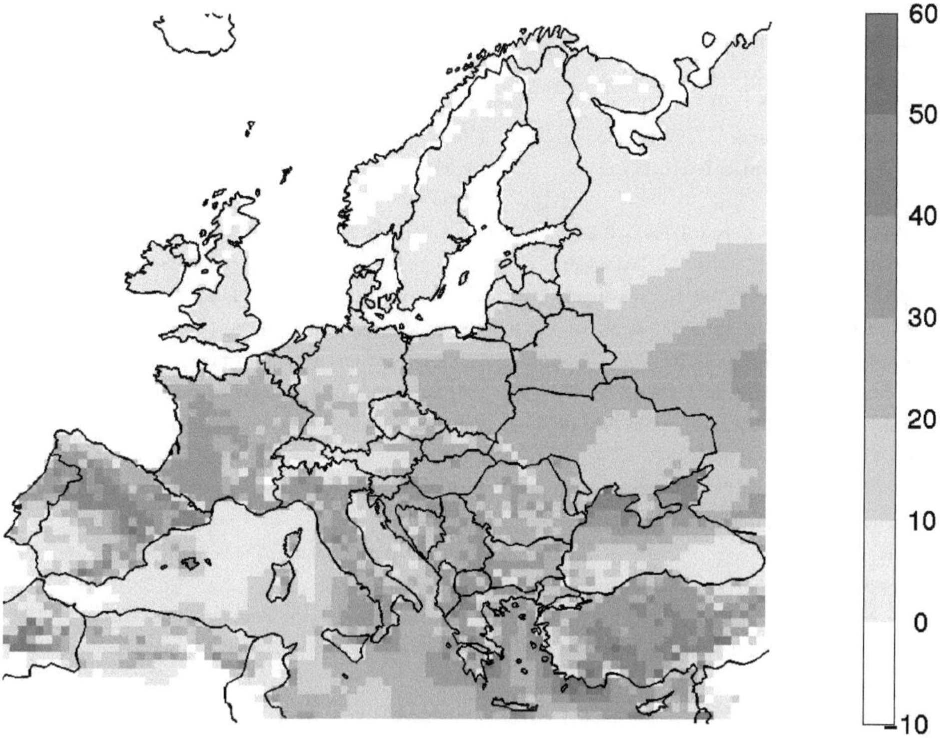

Difference between the mean number of JJA days with maximum temperature exceeding 30 °C in the scenario (2071 – 2100) and the control (1961 – 1990)

number of studies addressing changes in storm activity as a result of climate change there is little consensus yet (Meehl et al. 2000b; IPCC 2001).

Conclusions

Conceptually, climate change may lead to an alteration of extreme weather and climate events across Europe. Trends in time series of observed extreme weather and climate indices are suggestive of changes in the climatology of extreme events over Europe. Climate change projections indicate not only the likelihood of substantial warming by 2100 but also non-linear changes in the probability of extreme events relative to the mean climate. This finding lends weight to the hypothesis that increases in the probability of extreme events across Europe are likely with climate change. Changes in the climatology of extreme events holds pressing implications for European society and economy, especially in terms of the development of effective adaptation measures to reduce the vulnerability of people, property, livelihoods and infrastructure to the fatal and otherwise damaging effects of extreme weather and climate events.

References

Alexandersson A, Tuomenvirta H, Schmith T, Iden K (2000) Trends of storms in NW Europe derived from an updated pressure data set. Climate Research 14:71–73

Bijl W, Flather R, de Ronde JG, Schmith T (1999) Changing storminess? An analysis of long-term sea level data sets. Climate Research 11:161–172

Brázdil R, Budíková M, Auer I, Böhm B, Cegnar T, FašKo P, Lapin M, Gajič-Čapka M, Zaninovič K, Koleva E, Niedzwiedz T, Ustrnul Z, Szalai , Weber RO (1996) Trends of maximum and minimum daily temperatures in central and southeastern Europe. International J Climatology 16:765–782

Brunetti M, Colacino M, Maugeri M, Nanni T (2001) Trends in the daily intensity of precipitation in Italy from 1951 to 1996. International J Climatology 21:299–316

Brunetti M, Buffoni L, Maugeri M, Nanni, T (2000) Precipitation intensity trends in northern Italy. International J Climatology 20:1017–1031

Christensen JH, Christensen OB (2003) Severe summer-time flooding in Europe. Nature 421:805–806

Christensen OB, Christensen JH, Machenhauer B, Botzet M (1998) Very high-resolution regional climate simulations over Scandinavia: Present climate. J Clim 11 3204–3229

De Luís M, Raventós J, González-Hidalgo JC, Sánchez J, Cortina J (2000) Spatial analysis of rainfall trends in the region of Valencia (east Spain). International J Climatology 20:1451–1469

Deque M (2003) Uncertainties in the temperature and precipitation response of PRUDENCE runs over Europe. Abstract from the European Science Foundation and PRUDENCE 3rd Annual Conference on "Regional Climate Change in Europe", Wengen, Switzerland, September 29–October 3, 2003

Ferro CAT, Hannachi A, Stephenson DB (2005) Simple nonparametric techniques for exploring changing probability distributions of weather and climate (in press)

Flocas HA, Maheras P, Karacostas TS, Patrikas I, Anagnostopoulou C (2001) A 40-year climatological study of relative vorticity distribution over the Mediterranean. International J of Climatology 21:1759–1778

Frei C, Schär C (2001) Detection Probability of Trends in Rare Events: Theory and Application to Heavy Precipitation in the Alpine Region Journal of Climate 14:1568–1584

Frei C, Schär C, Lüthi D, Davies HC (1998) Heavy Precipitation Processes in a Warmer Climate. Geoph Res Lett 25:1431–1434

García-Herrera R, Gallego D, Hernández E, Gimeno L, Ribera P, Calvo N (2003) Precipitation trends in the Canary Islands. International J Climatology 23:235–241

González-Hidalgo JC, De Luis M, Raventós J, Sánchez JR (2001) Spatial distribution of seasonal rainfall trends in a western Mediterranean area. International J Climatology 21:943–860

Gregory J, McCabe M, Clark P, Serreze MC (2001) Trends in Northern Hemisphere surface cyclone frequency and intensity. Journal of Climate 14:2763–2768

Hanssen-Bauer I, Førland EJ (1998) Long-term trends in precipitation and temperature in the Norwegian Arctic: can they be explained by changes in atmospheric circulation patterns? Climate Research 10:143–153

Huth R (2001) Disaggregating climatic trends by classification of circulation patterns. International J Climatology 21:135–153

IPCC (2001) Climate Change. The Scientific Basis. Cambridge Univ. Press, pp 881ff

Johns TC et al. (2003) Anthropogenic climate change for 1860 to 2100 simulated with the HadCM3 model under updated emission scenarios. Clim Dyn 20:583–612

Klein Tank AMG et al. (2002) Daily dataset of 20th-century surface air temperature and precipitation series for the European Climate Assessment. International J Climatology 22:1441–1453

Maheras P, Flocas HA, Patrikas I, Anagnostopoulou C (2001) A 40 year objective climatology of surface cyclones in the Mediterranean region: spatial and temporal distribution. International J Climatology 21:109–130

McGregor GR, Markou MT, Bartzokas A, Katsoulis BD (2002) An evaluation of the nature and timing of summer human thermal discomfort in Athens, Greece. Climate Research 20:83–94

Nakicenovic N et al. (2000) IPCC Special Report on Emission Scenarios, Cambridge Univ. Press, Cambridge New York, pp 599ff

New M, Todd M, Hulme M, Jones P (2001) Precipitation measurements and trends in the twentieth century. International J Climatology 21:1899–1922

Osborn TJ, Hulme M, Jones PD, Basnett T (2000) Observed trends in the daily intensity of United Kingdom precipitation. International J Climatology 20:347–364

Pope DV, Gallani M, Rowntree R, Stratton A (2000) The impact of new physical parameterizations in the Hadley Centre climate model HadAM3. Clim Dyn 16:123–146

Pryor SC, Barthelmie RJ (2003) Long-term trends in near-surface flow over the Baltic. International J Climatology 23:271 – 289

Türke M, Sümer UM, Demir I (2002) Re-evaluation of trends and changes in mean, maximum and minimum temperatures of Turkey for the period 1929 – 1999. International J Climatology 22:947 – 977

Wibig J, Glowicki B (2002) Trends of minimum and maximum temperature in Poland. Climate Research 20:123 – 133

Widmann M, Schär C (1997) A principal component and long-term trend analysis of daily precipitation in Switzerland. International J Climatology 17:1333 – 1356

Wijngaard JB, Klein Tank AMG, Können GP (2003) Homogeneity of 20th century European daily temperature and precipitation series. International J Climatology 23:679 – 692

Is the Frequency and Intensity of Flooding Changing in Europe?

Z. W. Kundzewicz

Introduction

Floods have recently become more destructive and projections show that this tendency may become even more pronounced. The Intergovernmental Panel on Climate Change (IPCC, 2001a) reports that the costs of extreme weather events have exhibited a rapid upward trend. From the 1950s to the 1990s, yearly economic losses from weather extremes increased tenfold (in inflation-adjusted dollars). A part of this trend is linked to socio-economic factors, such as population increase and accumulation of wealth in vulnerable areas, another part is probably linked to increased reporting. However, these factors alone cannot explain the whole observed growth, and possibly a portion of it is linked to climate.

Hydrological variables, such as precipitation, river flow, soil moisture and groundwater levels, display strong spatial and temporal variability. From time to time they take on extremely high values, exerting considerable impacts on ecosystems and human society. Floods have been a major concern since the dawn of human civilization and continue to hit every generation of human beings. In fact, flood losses worldwide continue to rise. According to the global data compiled by the Red Cross for 1971 – 1995, in an average year floods killed nearly 13 thousand people worldwide and made over three million homeless, affecting more than 60 million people, and the mean annual flood damage exceeded 40 billion USD. From the global data compiled by Berz (2001), one may conclude that the number of great flood disasters in the nine years 1990 – 1998 was higher than in earlier three-and-half decades, 1950 – 1985, together. Most flood fatalities, 140 thousand, resulted from a storm surge in Bangladesh in April 1991 (Munich Re 1997), while the highest material damage (exceeding 30 billion dollars) has been caused by floods in China in summer of 1998 (Kundzewicz & Takeuchi 1999). Since 1990, there have been over 30 floods worldwide in each of which the material losses exceeded one billion USD and/or the number of fatalities was greater than one thousand. The majority of recent large floods have occurred in Asia, but indeed only few countries worldwide are free of the flood danger. Destructive floods have also occurred in arid and semiarid regions.

Increase in the total national flood damage, adjusted for inflation, with the rate of 2.92 % per year was found by Pielke & Downton (2000) for the data from the USA, covering the period 1932 – 1997. This rise has been related to both climate factors (increasing precipitation) and socio-economic factors (increasing population and wealth).

Several destructive floods have been experienced in the last decade in Europe. It is estimated that the material flood damage recorded in the European continent in 2002 has been higher than in any single year before. The floods in Central Europe in August 2002 alone (on the rivers Danube, Labe/Elbe and their tributaries) caused damage exceeding 15 billion Euro, therein 9.2 in Germany, 3 each in Austria and in the Czech Republic. A little later, during severe storms and floods on 8 – 9 September 2002, 23 people were killed in southern France (Rhone valley), while the total losses went up to 1.2 billion USD.

Has there been a climate track in the apparent rise of flood and drought hazard? Is it likely to manifest itself in the future? The present paper reviews various aspects of these issues.

Observed variability of intense precipitation and projections for the 21st century

The mean global land precipitation has increased in the warming world. Moreover, according to the Clausius-Clapeyron law, the atmosphere´s water holding capacity, and hence the potential for intense precipitation, increases with temperature. Since evaporation grows with temperature rise, this means that the actual atmospheric moisture increases, favouring stronger rainfall events and increasing the risk of flooding (Trenberth 1998, 1999). Trenberth (1999) presented a physically-based conceptual framework for changes in hydrological extremes with climate change.

Higher and more intense precipitation has been already observed. As stated in IPCC (2001), it is very likely "that in regions where total precipitation has increased … there have been even more pronounced increases in heavy and extreme precipitation events". Moreover, increases in intense precipitation have also been documented in some regions where the total precipitation has remained constant or even decreased. That is, the number of days with precipitation may have decreased more strongly than the total precipitation volume.

Over the latter half of the 20th century it is likely that there has been a 2 to 4 % increase in the frequency of heavy precipitation events reported by the available observing stations in the mid- and high latitudes of the Northern Hemisphere. The area affected by most intense daily rainfall is increasing and significant increases have been observed in both the proportion of mean annual total precipitation in the upper five percentiles and in the annual maximum consecutive 5-day precipitation total. The latter statistic has increased for the global data in the period 1961 – 1990 by 4 % (IPCC, 2001).

Some recent rainfall events in Europe have exceeded all-time records. On 12 – 13 August 2002, a new German record of one-day precipitation (from 6.00 a.m. 6.00 a.m.) of 312 mm was measured at the gauge Zinnwald-Georgenfeld, while in the category of 24-hour precipitation (from 5.00 a.m. to 5.00 a.m.) the record went up to 352 mm. The list of all-time extreme precipitation totals observed in Germany contains several entries from the last ten years (in several rainfall duration classes), cf. Rudolf & Rapp (2003).

However, projections of extreme events for future climate are highly uncertain. There are large quantitative differences between scenarios and models. Yet, based on global model simulations and for a wide range of scenarios, global average water vapour concentration and precipitation are expected to increase further during the 21st century, while precipitation extremes are projected to increase more than the mean, with consequence for the flood risk. The frequency of extreme precipitation events is projected to increase almost everywhere (IPCC 2001).

According to the material compiled in (IPCC 2001a), wetter winters are projected throughout Europe, with the two regions of highest increase being the Northeast of the continent and the northwestern Mediterranean coast. This is of direct importance for flood hazard.

Palmer & Räisänen (2002) analyzed the modelled differences between the control run with 20th century levels of carbon dioxide and an ensemble with transient increase to doubled CO_2 concentration (61 – 80 years from present). They found a considerable increase of the risk of a very wet winter in Europe. The modelling results indicate that the probability of total boreal winter precipitation exceeding two standard deviations above normal will considerably increase over large areas of Europe. For example, an over five-fold increase is projected over Scotland, Ireland and much of the Baltic Sea basin, and even over seven-fold increase for parts of the Russian Federation.

The Modeling the Impact of Climate Extremes (MICE) project within the 5th Framework Programme of the European Union examines changes in precipitation between the control period, 1961 – 1990 and the 2070 – 2099 period, using climate models, HadCM3 and HadRM3. Based on the results of the climate models, it is projected that in most of Europe intense precipitation will increase, even over vast areas where decrease of mean precipitation is expected. This is shown in analyses performed within the MICE project for extremes simulated by HadRM3, between the control period in the 20th century, 1961-1990, and the period of interest in the 21st century, 2070 – 2099 (cf. Kundzewicz et al., 2005). These results agree qualitatively with the findings of Christensen & Christensen (2003), based on a different model.

Increase in intense precipitation over many areas in the future (IPCC; 2001a) will have multiple adverse consequences, such as: increased risk of such damaging events as flood, landslide, avalanche, and mudslide; increased soil erosion; increased pressure on government and private flood insurance systems and on disaster relief.

Floods

Observed changes in extreme river flow

Where data are available, changes in annual streamflow usually relate well to changes in total precipitation (IPCC 2001). However, published results of change detection in flood flows show no uniform greenhouse signature. The statement that severe floods are becoming significantly more frequent and intense is supported by several studies, while other publications do not report such evidence. This deserves a closer look, and possibly consideration of different flood-generation mechanisms.

In a comprehensive study of US river flow records by Lins & Slack (1999), all, but the highest quantiles of discharge were found to increase across broad areas. However, no general dominating tendency of increase or decrease in the 90th and 100th percentiles of daily streamflow could be detected, despite the documented increase in extreme precipitation events (e.g. Easterling et al. 2000).

Regional changes in timing of floods have been observed in many areas, with increasing late autumn and winter floods (caused by rain) and less snowmelt floods, e.g. in Europe. Also the number of inundations caused by ice jams goes down in effect of warming (more rivers do not freeze at all) and human capacity to cope with ice-based obstructions of flow. This has been a robust result. Looking at the data assembled by Mudelsee et al. (2003), in the last 150-year time series of maximum daily flow on the Elbe (gauge Dresden), the level of 3000 m^3/s was exceeded 11 times: eight times in winter (1862, 1865, 1876, 1881, 1895, 1900, 1920, and 1940), and only three times in summer (May 1896, September 1890, and August 2002); the last discharge being highest. This illustrates that severe winter floods, frequent in the days of yore, have not occurred for 63 years now.

On the other hand, intensive and long-lasting summer precipitation episodes (in particular, related to the Vb cyclone track after van Bebber, cf. Kundzewicz et al. 2005) have led to disastrous recent flooding in Europe, e.g., the Odra / Oder deluge in 1997 (cf. Kundzewicz et al. 1999), and the 2002 flooding on the Elbe and its tributaries, the Danube, and other rivers. Not only historic records of material losses were exceeded, but also records of hydrological variables, such as stage or discharge. The maximum stage of the river Elbe at Dresden (940 cm in August 2002) was far above the former record (877 cm in 1845) and the peak discharge at Raciborz-Miedonia on the Odra in July 1997 was twice as high as the second highest on record.

Kundzewicz et al. (2004) studied a set of 195 world-wide hydrological time series of maximum daily river flow, for every year, from holdings of the Global Runoff Data Centre (GRDC) in Koblenz, Germany. The analysis does not support the hypothesis of general growth of annual maximum river flows worldwide. Even if 27 cases of strong, statistically significant increase have been identified with the help of the Mann-Kendall test, there are 31 significant decreases as well, and most (137) time series do not show any significant changes.

However, highly skewed distributions render detecting changes in annual maxima of daily river flows difficult. As shown by Kundzewicz et al. (2004), it is not uncommon that the highest recorded annual maximum daily flow at a given station is considerably (even by the factor of 4.07 and 3.97) higher than the second highest value in the long time series of records.

As noted by Radziejewski & Kundzewicz (2004), tests are not able to detect a weak trend or a change, which has not lasted sufficiently long, but this cannot be interpreted as a demonstration of the absence of change. With the enhanced climate change, the changes of hydrological processes may be stronger and last longer, so that the likelihood of change detection may grow.

Kundzewicz et al. (2004) studied 70 long time series of annual maximum daily river flow from Europe, covering different periods. Since all analysed series start no later than in 1960, one can take the year 1961 as the starting point for a fourty-year common period for all data and then divide this common period into two twenty-year subperiods. It was found that the overall maxima (for the whole 1961-2000 period) occurred more frequently (46 times) in the later subperiod, 1981 – 2000 than in the earlier subperiod, 1961 – 1980 (24 times). This was despite the fact that not all time series last until the year 2000 (series at 15 stations end in 1999 and at 6 end in 1998). Hence, it may well be that even more maxima fall into the subperiod 1981-2000. ➤ *Figure 1* presents direction and significance of changes in annual maximum daily river flow at examined stations in Europe.

In a national-scale study (61 stations over a century) carried out for Sweden, Lindström & Bergström (2004) found a substantial recent increase in both annual discharge and flood magnitude, but it is not exceptional in the context of high flows experienced earlier.

Intensified extreme hydrological events have been associated with observed changes in climatic variability. Unprecedented increase of frequency, persistence, and intensity of El Niño (warm phase of ENSO) has been observed since the mid 1970s (IPCC 2001). This has been accompanied by higher probability of occurrence of wetter-than-usual conditions, and high river flow, in several regions of Southern Hemisphere. Also the NAO index has been high during the last two decades, with possible link to high river flows in Europe.

It is clear that the river flow process is controlled by several factors, climate being just one of them. Flood hazard and vulnerability tend to increase over many areas, due to a range of climatic and non-climatic impacts. The latter include impacts of changes in terrestrial systems (hydrological systems and ecosystems), and economic and social systems. Land-use changes, which induce land-cover changes, control the rainfall-runoff relation. Deforestation, urbanization, and reduction of wetlands empoverish the available water storage capacity in the catchment and increase the runoff coefficient, leading to growth in amplitude and reduction of the time-to-peak of a flood triggered by a "typical" intense precipitation (as indicated earlier, the nature of a "typical" intense precipitation event has also changed, becoming more intense). Urbanization has adversely influenced flood hazard in many watersheds by increase in the portion of impervious area. On average, 2 % of agricultural land has been lost to urbanization per decade in the EU. The timing of river conveyance may also have been altered by river regulation.

Van der Ploog et al. (2002) noted the increase in flood hazard in Germany. They attributed it to climate (wetter winters), engineering modifications, but also intensification of agriculture, large-scale farm consolidation, subsoil compaction, and urbanization. The urbanized area in the former West Germany has grown from 7.4 % in 1951 to 12.2 % in 1989.

Humans have been encroaching into unsafe areas thereby increasing the damage potential. This holds not only for informal settlements on flood plains around mega-cities in the developing world but also for human encroaching into flood-endangered areas in developed countries. An important factor influencing the flood hazard is an unjustified belief in absolute safety of structural defences. Even an over-dimensioned and well maintained dike does not guarantee complete protection. It can be overtopped when an extreme flood occurs. When a dike breaks, the damage may be higher than it would have been in a levee-free case.

Recently, a paper by Mudelsee et al. (2003) reported "no upward trends in the occurrence of extreme floods in Central Europe", hence demonstrating the lack of continuity between the observations of no increase (Mudelsee et al. 2003) and model projections for the future (Christensen & Christensen 2003),

Fig. 1

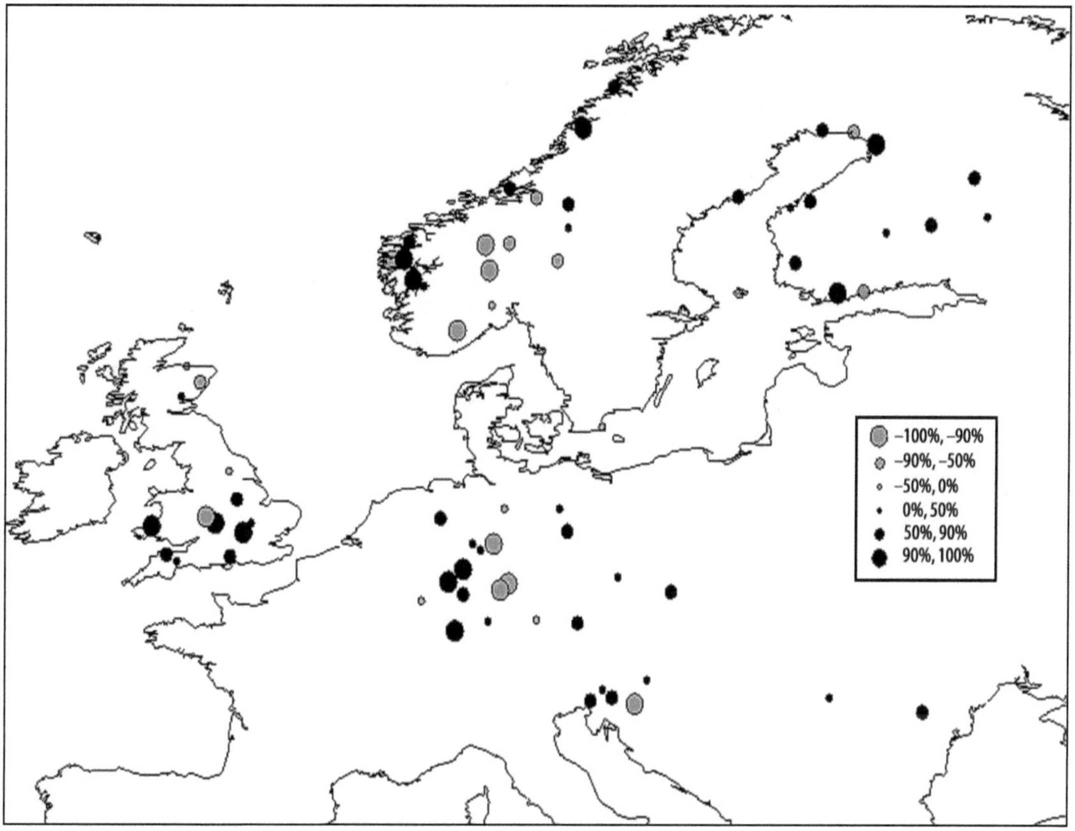

Direction and significance of changes in annual maximum daily river flow in Europe, based on a set of data provided by GRDC.

where increase in intense precipitation was demonstrated. The conclusions by Mudelsee et al. (2003) corroborate results of Bronstert (2003), who found that discharge rates of the Elbe river, corresponding to a range of recurrence intervals (from 2 to 200-year) in 20th century have been lower than in the 19th century.

Vulnerability to floods can be regarded as a function of exposure, sensitivity and adaptive capacity. Since, in many areas, exposure grows faster than adaptive capacity, the vulnerability increases too. Counter-intuitively, vulnerability of societies may grow even as they become wealthier, because technology helps populate and develop “difficult” areas, and societies become more exposed. High investment into maladaptation does not reduce the vulnerability!

It is estimated that 17 % of all the urban land in the USA, and 7 % of the total area of the conterminous US are located in the 100-year flood zone. About 10 % of the population of the USA and of the UK live in the 100-year flood zone. In Japan, half the total population and about 70 % of the total assets are located on floodplains, which cover only 10 % of the total country area. The percentage of flood-prone area is much higher in Bangladesh, and inundation of more than half of the country area is not rare (e.g. over two thirds of area of Bangladesh were inundated by the 1998 flood).

Projections of changes in extreme river flow for the 21st century

Climate change is likely to cause an increase of the risk of riverine flooding across much of Europe. However, changes in future flood frequency are complex, depending on the generating mechanism. Increasing flood magnitudes are projected where floods result of heavy rainfall and flood magnitudes generated by spring snowmelt may decrease (IPCC 2001a). Where snowmelt is the principal flood-generating mechanism, the time of greatest flood risk would shift from spring towards winter. Winter flood hazard is likely to rise for many catchments under many scenarios.

Menzel et al. (2005) examined flood frequency of the river Mosel at Cochem, comparing two thirty-year periods: 1961 – 1990 and 2061 – 2090. For the control period, they analysed observed discharge data (and fitted frequency curve) and modelled discharge series with three different input data sets: (1) observation records and (2 – 3) downscaled climate information from two GCMs: ECHAM4 and HadCM3. They also made calculations for the modelled future, using a hydrological model and downscaled climate information. For a specific discharge, the return interval is considerably lower in future climate, hence floods may become more frequent.

Milly et al. (2002) demonstrated changes in the risk of great floods (exceeding 100-year levels). For all (but one) large basins analysed, the control 100-year flood (100-year annual maximum monthly discharge) is exceeded more frequently as a result of CO_2 quadrupling. Probability of what has been a 100-year "flood" (quotation marks are used because maximum monthly discharge is not the most meaningful flood descriptor) is projected to increase drastically in the CO_2-quadrupling world. In some northern rivers, what was a 100-year "flood" in the control run, is projected to become much more frequent, even occurring every 2 to 5 years.

The risk from storm surge also appears to be changing. The global average sea level, which rose by 10 to 20 cm during the 20th century, is projected to continue to rise (IPCC 2001), with substantial adverse effects on low-lying lands and river deltas in flood prone areas: increased probability of storm surges and tidal flooding. Needless to say, that the sea-level rise itself is a dangerous occurrence, jeopardizing low-lying coastal areas, which may cause permanent inundations, resulting in massive relocation of people. Global warming has the potential to trigger large-scale singular events, such as the disintegration of the West Antarctic and Greenland ice sheets, in a time scale of multiple centuries, which would cause a significant sea-level rise (of the order of meters), and permanent inundation of large, now densely populated, areas. The probability of such developments in the near future is low, but should not be ignored given the severity of consequences. However, coastal flooding is beyond the scope of this article.

Conclusions

Higher and more intense precipitation has been already observed and this trend is expected to strengthen in the future, warmer world, directly impacting on flood risk. However, snowmelt and ice-jam related floods have been decreasing over much of Europe. These statements respond to the question posed in the title of the present paper. The impact of climate forcing on flood risk is complex, depending on the generating mechanism (rainfall vs snowmelt). In many places flood risk is likely to have grown and further growth is projected.

Floods have been identified by IPCC TAR (2001a) among regional reasons of concern. Yet, there is still a great deal of uncertainty in findings about climate change impacts on water-related extreme events. Only in some areas, the projected direction of change is consistent across different models and scenarios. As stated in IPCC (2001), "[t]he analysis of extreme events in both observations and coupled models is underdeveloped" and "the changes in frequency of extreme events cannot be generally attributed to the human influence on global climate." It is difficult to disentagle the climatic component in hydrological extremes from strong natural variability and man-made environmental changes. This remains an exciting scientific challenge.

Acknowledgements

The reported work has been a background activity of the author within the MICE (Modelling Impacts of Climate Extremes) project, financed within the Fifth Framework Programme of the European Union. It is also a contribution to the WatREx (Water-Related Extremes) project of the Potsdam Institute for Climate Impact Research (PIK).

References

Berz G (2001) Climatic change: Effects on and possible responses by the insurance industry. In: Lozán JL, Graßl H, Hupfer P (eds) Climate of the 21st Century: Changes and Risks. Office: Wissenschaftliche Auswertungen, Hamburg, pp 392 – 399

Bronstert A (2003) Floods and climate change: interactions and impacts. Risk Analysis 23:545 – 557

Christensen JH, Christensen OB (2003) Severe summertime flooding in Europe. Nature 421, 805

Easterling DR, Evans JL, Groisman PYa, Karl TR, Kunkel KE, Ambenje P (2000) Observed variability and trends in extreme climate events. Bull Am Met Soc 81:417 – 425

IPCC (Intergovernmental Panel on Climate Change) (2001) Climate Change 2001: The Scientific Basis. Contribution of the Working Group I to the Third Assessment Report of the Intergovernmental Panel on Climate Change. Cambridge University Press, Cambridge

IPCC (Intergovernmental Panel on Climate Change) (2001a) Climate Change 2001: Impacts, Adaptation and Vulnerability. Contribution of the Working Group II to the Third Assessment Report of the Intergovernmental Panel on Climate Change. Cambridge University Press, Cambridge

Kundzewicz ZW, Szamałek K, Kowalczak P (1999) The Great Flood of 1997 in Poland. Hydrol Sci J 44:855 – 870

Kundzewicz ZW, Takeuchi K (1999) Flood protection and management: quo vadimus? Hydrol Sci J 44:417 – 432

Kundzewicz ZW, Ulbrich U, Brücher T, Graczyk D, Leckebusch G, Menzel L, Przymusińska I, Radziejewski M, Szwed M (2005) Summer floods in Central Europe – climate change track? Natural Hazards (in press)

Kundzewicz ZW, Graczyk D, Maurer T, Przymusińska I, Radziejewski M, Svensson C, Szwed M (2004a) Change detection in annual maximum flow. Report WCASP-64, WMO/TD-No. 1239, WMO, Geneva, Switzerland

Lindström G, Bergström S (2004) Runoff trends in Sweden 1807 – 2002. Hydrol Sci J 49:69 – 83

Lins HF, Slack JR (1999) Streamflow trends in the United States. Geoph Res Letters 26:227 – 230

Menzel L, Thieken AH, Schwandt D, Bürger G (2005) Impact of climate change scenarios on the regional hydrology – Modelling studies in the German Rhine catchment. Natural Hazards (in print)

Milly PCD, Wetherald RT, Dunne KA, Delworth TL (2002) Increasing risk of great floods in a changing climate. Nature 415:514 – 517

Mudelsee M, Börngen M, Tetzlaff G, Grünewald U (2003) No upward trends in the occurrence of extreme floods in central Europe. Nature 421:166 – 169

Munich Re (1997) Flooding and Insurance. Munich Re, Munich, Germany

Palmer TN, Räisänen J (2002) Quantifying the risk of extreme seasonal precipitation events in a changing climate. Nature 415:512 – 514

Pielke RA, Downton M (2000) Precipitation and damaging floods. Trends in the United States, 1932 – 1997. J of Climate 13:3625 – 3637

Radziejewski M, Kundzewicz ZW (2004) Detectability of changes in hydrological records. Hydrol Sci J (in press)

Rudolf B, Rapp J (2003) The century flood of the river Elbe in August 2002: Synoptic weather development and climatological aspects. Quarterly Report of the Operational NWP Models of the Deutscher Wetterdienst, Special Topic, July 2003, pp 7 – 22

Trenberth KE (1998) Atmospheric moisture residence times and cycling: Implications for rainfall rates with climate change. Climatic Change 39:667 – 694

Trenberth KE (1999) Conceptual framework for changes of extremes of the hydrological cycle with climate change. Climatic Change 42:327–339

Van der Plog RR, Machulla G, Hermsmeyer D, Ilsemann J, Gieska M, Bachmann J (2002) Changes in land use and the growing number of flash floods in Germany. In: Steenvorden J, Claessen F, Willems J (eds) Agricultural Effects on Ground and Surface Waters: Research at the Edge of Science and Society. IAHS Publ No 273, pp 317–322

Bio-climatological Aspects of Summer 2003 Over France

Jean-Claude Cohen · Jean-Michel Veysseire · Pierre Bessemoulin

Summary

The extreme heat wave of the first two weeks of August 2003 occurred during the hottest summer period (June to August) of the last fifty years and followed a six month period of drought. Moreover, this heat wave was outstanding in duration (lasting for two weeks) and in geographic extension (over all parts of the country, including mountains and coastal regions) with absolute temperature records in 70 out of 180 stations. Its tragic health impacts, with 15,000 excess deaths, were probably strongly intensified by the persistently high night temperatures on the one hand, by high levels of pollution on the other hand: in Paris, with serial data files since 1873, morning temperatures on the 11th and 12th August were the highest ever registered, with 25.5 °C (previous record: 24 °C in 1976). Ozone (O_3) peaks were strong and frequent, accompanied with some NO_2 unusual peaks, probably due to the absence of bracing wind.

Nevertheless, this unique heat wave is consistent with climate change projections and more heat waves can be expected in the next few years or decades.

This heat wave affected most parts of Europe, yet France was the most strongly affected with Andalusia and Portugal, due to unusually thick hot air masses coming from North Africa and settling over Western Europe.

Météo-France issued a press release on 1st August announcing a progressive climb in temperatures for the following days over the whole country. On 4th August Météo-France offered on its website simple health advice and a revue of historical deadly heat waves. A further press release on 7th August included a health warning, especially directed towards elderly and sick persons. The progressive ending of the heat wave was announced on the 13th August.

In response to the heavy toll of this summer heat an early Heat Health Warning System is being established with French public survey agencies along with a common scheme for Cold Spells Warnings (previously running). This response includes a first announcement forecast for Health and Social Services professionals 4 to 7 days before the event, a warning forecast 1 to 3 days before the event including the media and general public, and an enhanced red warning on a four colours warning scale (green, yellow, orange and red) in case of pollution, strong summer humidity or strong winter winds.

The inclusion of other bio-meteorological warnings, such as UV index, pollen and pollution concentrations is also being considered.

The most outstanding heat wave for over 50 years

An extreme heatwave during a hot and dry summer

High temperatures already settling over France at the end of May 2003 showed only two short periods of attenuation, with mild to fresh temperatures at the beginning, then at the end of July. Then heat and its consequences (such as big fires, drought or pollution) were already a main focus in the news and media before the historic extreme heat wave of August.

Nevertheless, heat was until then perceived as being "friendly to people" with its cultural high popularity until the first days of August, when the population seemed to discover heat as an aggression and a threat: such a feeling was never or hardly experienced before in our mild climate culture.

If we compare daily mean summer temperatures over France, we find that the summer of 2003 (June to August three months period) was the hottest ever recorded during the last fifty years. The average daily maximum temperatures reached 28.6 °C to be compared with 27.1 °C in 1976, 26.3 °C in 1994 and 26.1 °C in 1983. Even the average of daily minimum temperatures reached 16.5 °C to be compared to 15.1 °C in 1994.

Previous summer heatwave records, respectively 1976 (long and intense heatwave), 1983 (more intense in the South East of France) and 1994, were 1.7 to 2.7 °C less hot in mean temperatures than during the 2003 summer. Compared now to mean summer temperatures for the 1971 – 2000 period, 2003 mean temperatures over the country were 3 to 7 °C hotter, especially over central and eastern regions.

Fig. 1

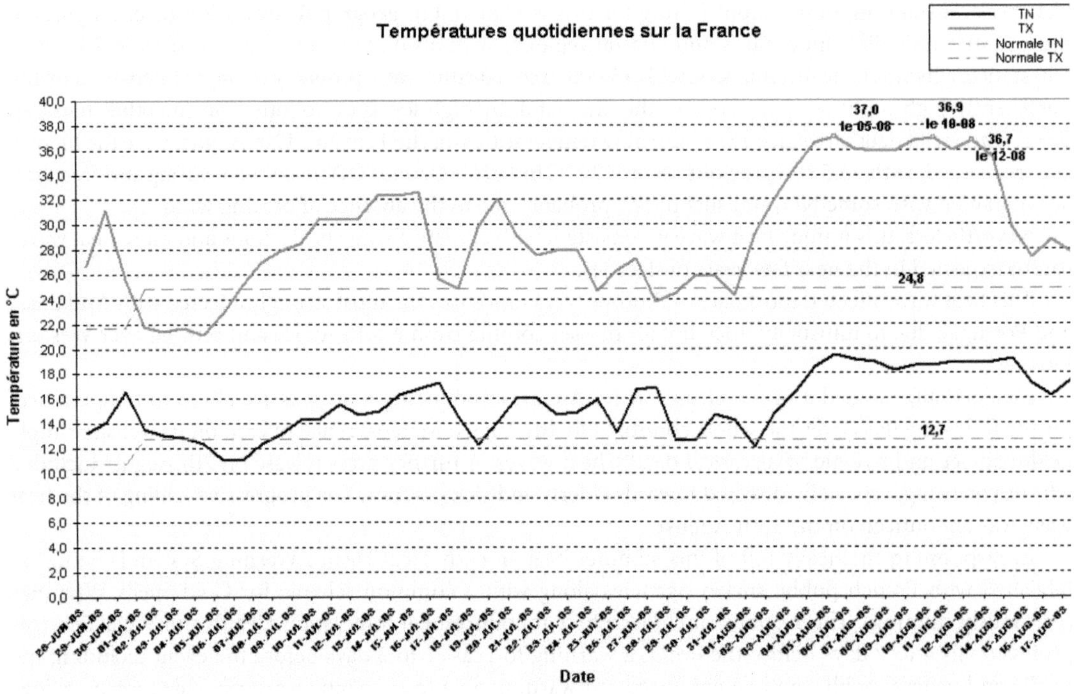

Chronology of daily minimum (T_N) and maximum (T_X) temperatures (average value over 22 stations across France) from 26th June to 17th August

Figure 2 a and b, Figure3

Nevertheless the first two weeks of August were the most outstandingly hot period that had ever occurred since 1950, if not for more than a century, and the most dramatic, as we know.

For two weeks, in all parts of the country, including mountains and coastal zones, the heatwave broke numerous temperature records and was associated with a tremendous toll of almost 15,000 excess deaths.

Fig. 2 a and b

A long, an intense and a dry heatwave. Daily mean temperatures over Paris since 1873 : among the 8 hottest records, 7 occured in 2003 (only the sixth occured on the 28th July 1947)

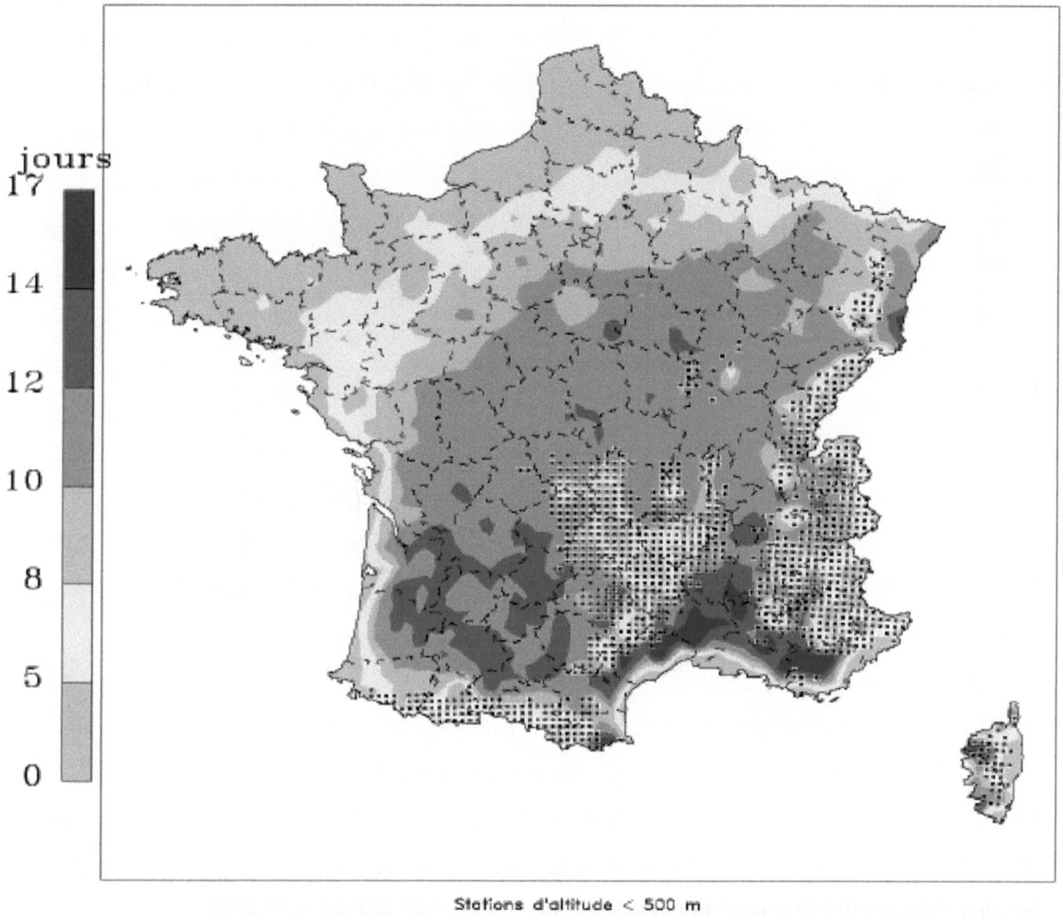

a **Number of days with a maximum temperature ≥ 35 °C from 01.08.2003 to 18.08.2003**

The meteorological situation

The general altitude circulation showed extremely hot air-masses (associated to the Azores High) moving north from the Sahara and settling over southern and central Europe affecting the whole country. This air-mass showed, in particular, exceptionally strong and steady ground to 700 hPa thickness (up to 3000 m above sea level) from August 1st to 13th: this meteorological index is usually considered to be one of the best indicators for estimating and forecasting dry heatwaves and has been used by Météo France forecasters.

Heat started to increase strongly from the 1st to the 5th of August, with maximum daily temperatures increasing from about 30 °C up to 40 °C within those five days; then it remained steady at this extremely high level, between 35 °C to 40 °C (maximum daily temperatures), until the 13th. Such hot sequences over 35 °C spanned up to 9 consecutive days in many regions, more especially over the central part of the country which was concerned by the strongest values of the ground – 700 hPa thickness index.

Fig. 2b

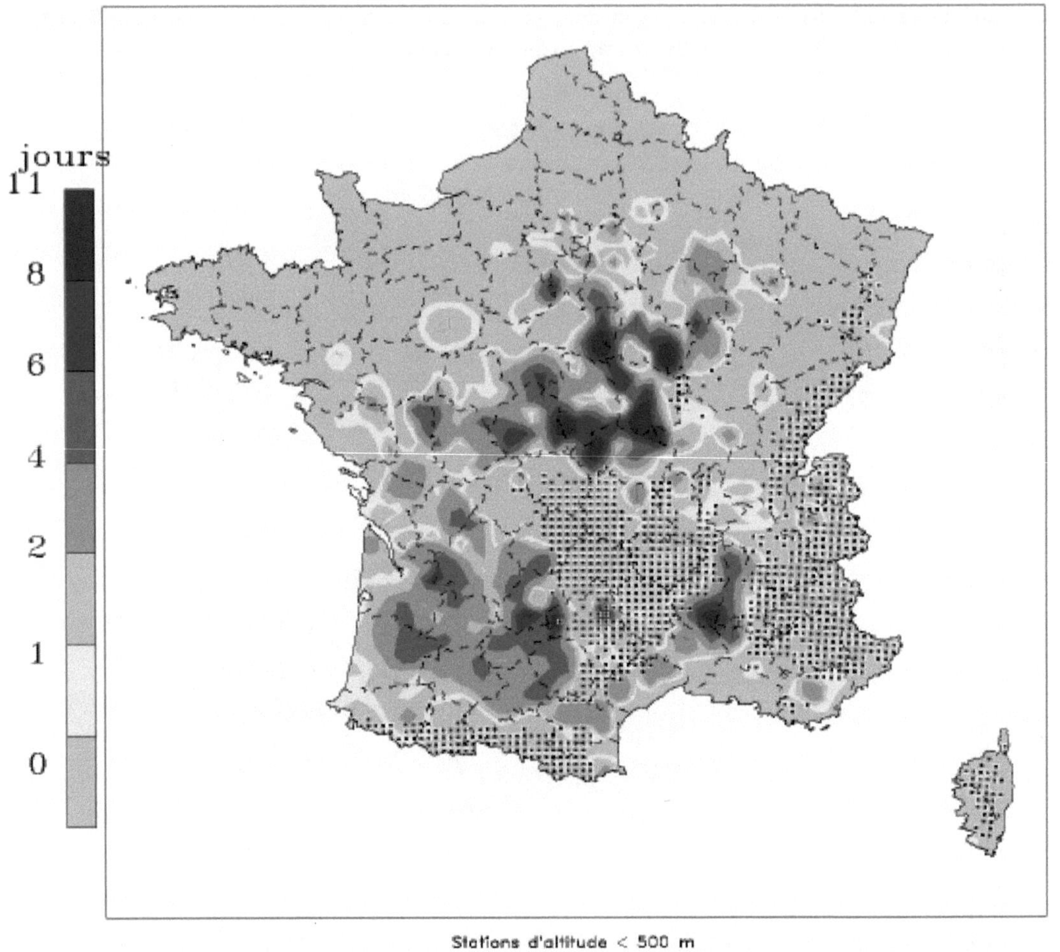

b **Number of days with a maximum temperature ≥ 40 °C from 01.08.2003 to 18.08.2003**

In Europe, if we compare minimum and maximum temperatures, Spain, Italy and Portugal were all strongly affected by the hot air masses. Yet it seems that only Andalusia and Portugal experienced such severe heat conditions as France. This is consistent with the position of the steady centre of maximum ground – 700 hPa thickness over central and Eastern parts of France.

Figure 4 a and b

Meteorological data over France

Temperatures over 35 °C were observed in two stations out of three, while 40 °C were reached in 15 % of the stations. Besides which, absolute heat records were broken in 70 stations out of 180.

In Paris, where observations have been recorded since 1873, the highest minimum (morning) temperature ever recorded was observed on the 11th and on the 13th (25.5 °C compared to the previous record of

Fig. 3

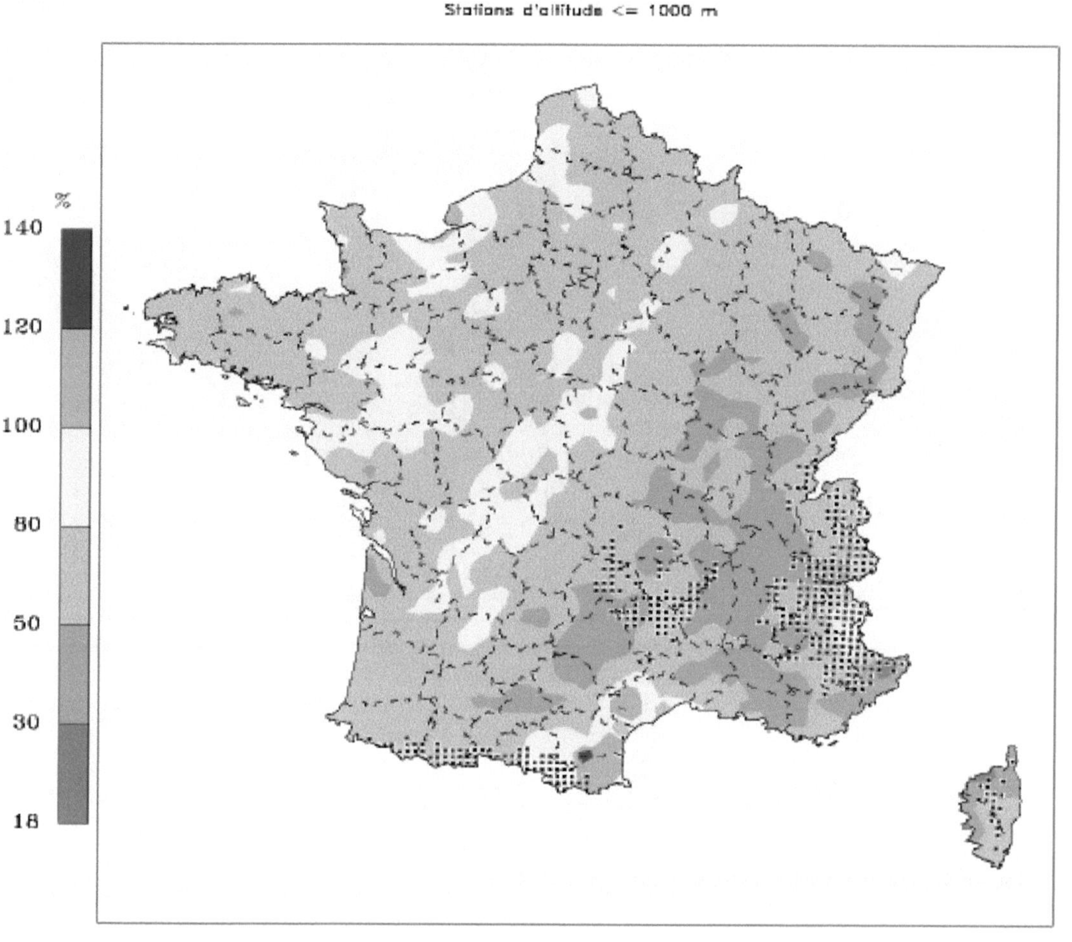

This dry heatwave occured after a six-month long drought. Ratio to average precipitation amount from 01/02/03 to 18/08/03.

24 °C in July1947); nine consecutive days showed temperatures over 35 °C compared with five days only in 1911 and four days in 1998 (the two previous duration records). This indicates that this extreme heatwave was the most intense for more than fifty years and probably for more than a century.

Figure 5

This meteorological context has been illustrated in particular by a bio-meteorological study that the Institut de Veille Sanitaire (the French Institute for Sanitary Survey) has been developing from observed meteorological data acquired by 13 representative selected stations of the Météo-France observations network.

■ **Fig. 4a**

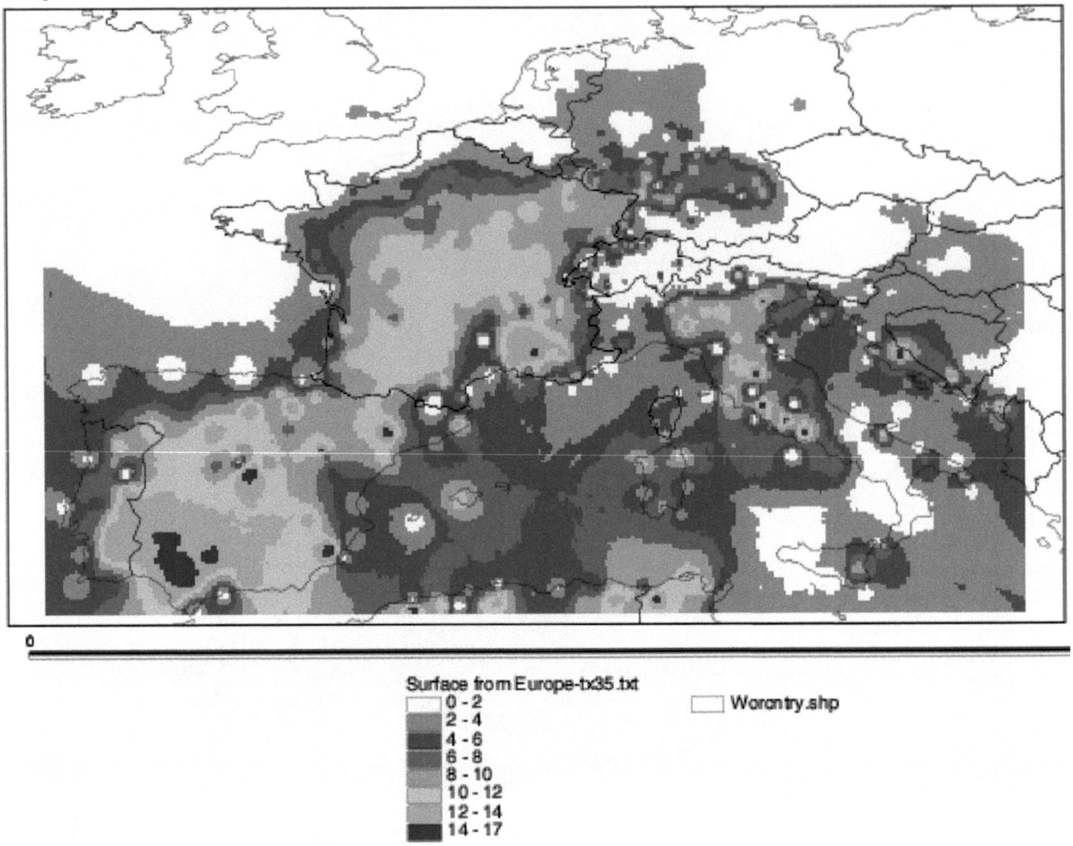

Number of days with a maximum temperature of ≥ 35 °C

This extreme heatwave is consistent with climate change projections

As is widely accepted now, the global mean temperature has been rising 0.6 to 0.8 °C since the beginning of the industrial era, even +1 °C over France and still a bit more over the south of the country. Projections estimate a continuous trend between 1.5 and 5.6 °C over this century.

Moreover, we have observed over the last ten years a linear growth of minimum temperatures, which is consistent with climate change projections.

Considering the significant increase in the frequency of extreme weather events all around the globe and in particular the serial absolute global temperature records that occurred one after the other during the1990s, we might consider that this August 2003 extreme heatwave is already a signal of actual climate change which seems to have become more sensitive since the early eighties.

Fig. 4b

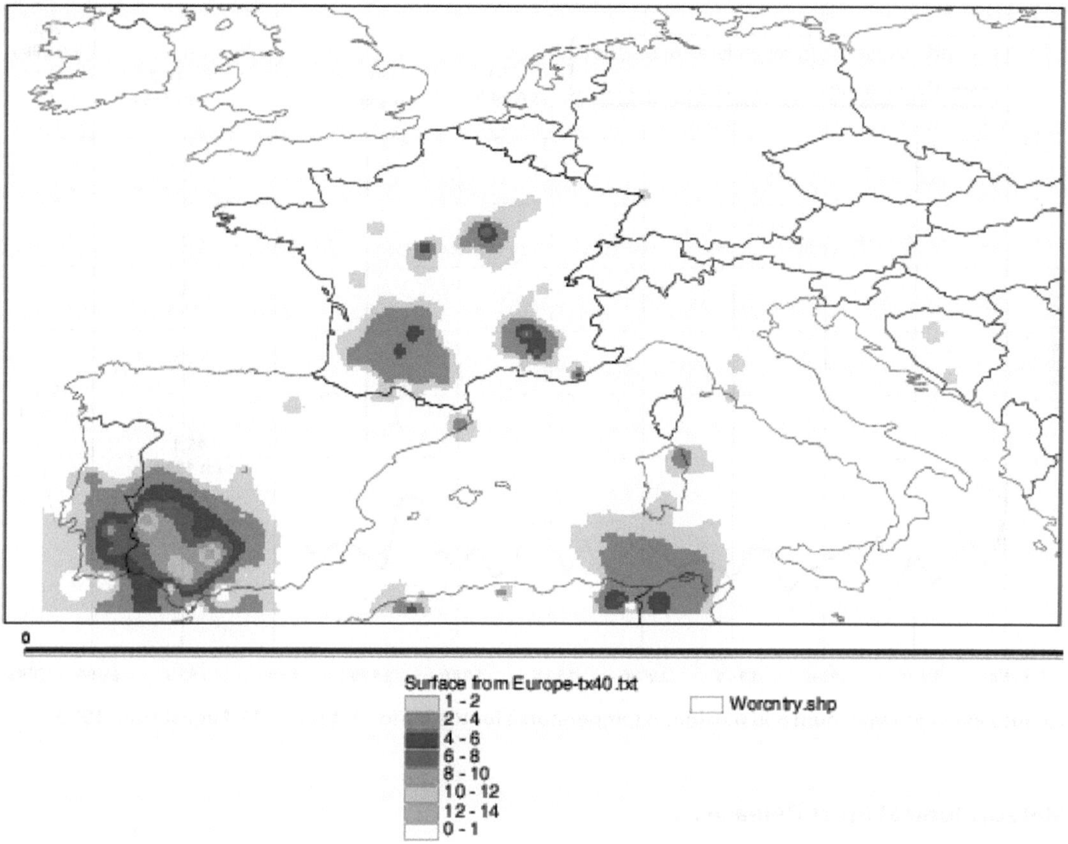

Number of days with a maximum temperature ≥ 40 °C

How did Météo France give the heat warning?

A previous twenty year long experience of Heat Health Watch Warning System

After the July 1983 heatwave over South-East France a first Heat Health Watch Warning System (HH-WWS) has been in operation since the mid-1980s in this region, on Prof. San Marco's initiative. Météo France announced sequences of more than two consecutives days with both minimum temperatures over 23 °C and maximum temperatures over 36 °C in the Marseilles area. When a warning was circulated, including basic health recommendations, medical agencies, emergency services, maternities, retirement homes, firemen and regional media could be alerted to take more care of people at risk, more especially elderly people and young babies.

Considering the previous experience of deadly heatwaves all around the world during the 1980s and 90s and taking into account our experience with Prof. San Marco, the Superior Council of Meteorology recommended in the year 2000 to increase the attention of the media and general public strongly in case of a heatwave with specific press release.

Fig. 5

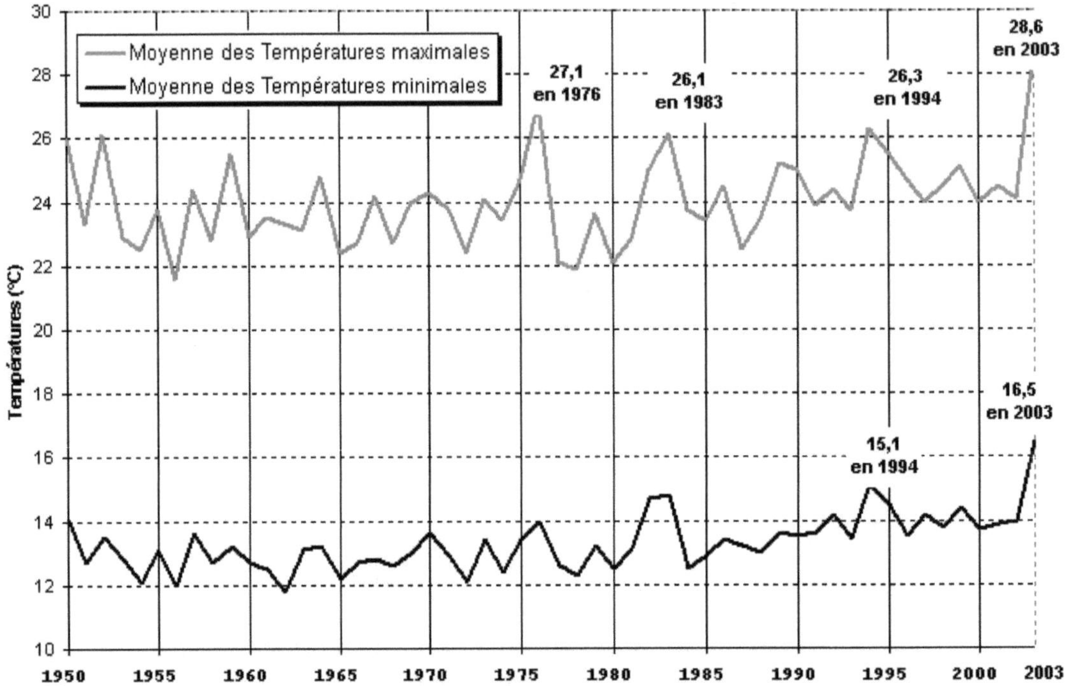

Country averaged minimum and maximum temperatures for the period 1st June – 11 August from 1950.

Meteorological Press Release …

Following this recommendation, the meteorological risk was announced on the 1st of August by a specific meteorological press release announcing the heatwave that was settling over the whole country for the entire week to come.

A further press release on the 7th August included a health warning, especially for the elderly and the sick. Its text, which was quite correctly reported in the media, announced: "... minimum temperatures over most parts of the country will reach 20 °C or more, locally up to 24 and 25 °C. Maximum temperatures will reach 36 to 40 °C. The persistence of high values of both minimum and maximum temperatures is exceptional and constitutes a significant health threat to sensitive persons (the elderly, the sick and young babies)."

The progressive ending of the heat wave was announced on the 13th August.

… and health recommendations

Meanwhile, on 4th August Météo-France offered on its website simple health advice and a review of historical deadly heat waves.

The basic set of health recommendations that was presented (on about five pages) has been widely reported and commented by the written press in particular.

This file on "heatwave and health" can still be consulted at the following web address:
http://www.meteo.fr/meteonet/actu/archives/dossiers/canicule/dos.htm

New perspectives for more efficient watch warning systems

Nevertheless, in August 2003, no procedure had yet been settled among national public health and emergency medical services to give any useful response to bio-meteorological warning. Moreover, it seems evident that, at least before the end of the first hot week, no one in France seemed to realise the full extent of this extreme sanitary disaster, even while it was already going on.

After the end of August, there were many contacts between Météo-France and the main French health and social authorities, namely the Ministry for Health (DGS), the Institut de Veille Sanitaire, the Ministry for Social Affairs, and various medical and emergency units.

The main outcome of these discussions is to include, from now on, heatwaves and cold spells in the current "Four Colour Meteorological Advisory Service" which have been running since the year 2001, as a starter for the departmental and national "alarm plan" with the Civil Security (French Ministry of the Interior).

The French "Four Colour Meteorological Advisory Service"

After the tragic flood of the Aude (south-east of France) in the middle of November 1999 and the two outstanding hurricanes the following month, Météo-France developed a new procedure of communication in case of extreme meteorological events, in collaboration with the Civil Security services of the French Ministry of the Interior. This new procedure aims to focus on the next 24 h meteorological risk on each of the French departments. Four levels of advisory warning (green, yellow, orange and red) indicate the intensity of the dangerous phenomena. Five of them were concerned in 2001 (torrential rains and floods, strong winds, thunder-storms, snow and ice, and avalanches) while heatwaves and cold spells, initially proposed by our services, were not recognized as relevant to Civil Security.

This is to change now, and both extreme temperatures sequences should be announced in a similar form to professionals and to the general public.

The new bio-meteorological advisory procedure

A HHWWS has been finalised in a common scheme procedure including also cold spells.

- A first forecast announcement is to be addressed to professionals (Health, Social and Civil Security correspondents) 4 to 7 days before the event;
- a warning forecast 1 to 3 days before the event is issued to professionals and to the media and general public;
- then a "four colours bio-meteorological advisory" for the general public the day before to focus public attention on an imminent health danger due to extreme persistent temperatures.

The general scheme of the procedure includes enhanced warnings in case of summer pollution or high humidity, in case of strong winter winds, and in case of persistent or enhanced extreme temperatures. This might be summarised as follows:

Tab. 1

Code (green; yellow; orange or red)	Valid	Heatwave	Cold spell
First announcement	D+4 to D+7	heat forecast	cold forecast
Warning 1	D+1 to D+3	confirmed forecast	confirmed forecast
Warning 2	D+1 to D+3	enhanced warning humidity, pollution (O_3, …)	enhanced warning strong wind, persistence …
Advisory	D to D+1	imminent risk	imminent risk

Several other bio-meteorological indicators have been proposed and our studies with the InVS still ongoing.

The "heat advisory procedure" will be presented as a departmental map of France including a basic set of health recommendations in case of orange or red colour.

Fig. 6

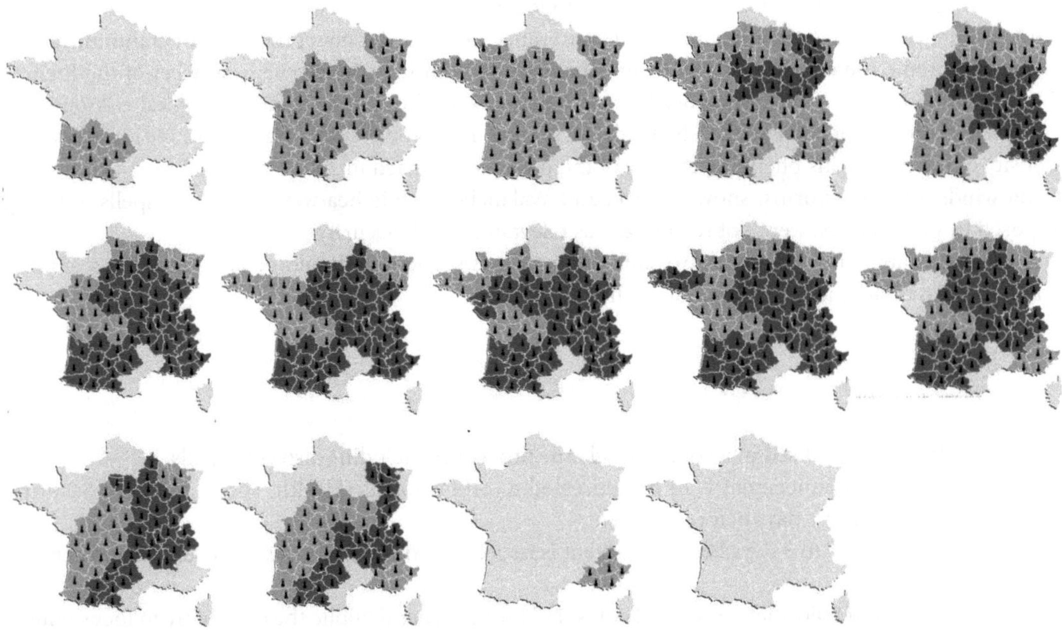

Set of 15 daily forecast advisory simulations for the August 2003 period, based on (minimum / maximum / persistence) criteria such as those used in the South-East of France after the 1983 heatwave

Météo France: A panel of biometeorological products and recommendations

New partners for new bio-meteorological services

The Ministry for Social Affairs

A strong partnership had previously been developed since the year 2002 with the Ministry for Social Affairs in the case of cold spells, for assistance procedures to homeless people in towns. A written agreement on "climatic urgencies" adopted in October 2003 includes a cold spells watch warning system (WWS) based on a daily forecast for the next three days of wind-chill index with selected thresholds corresponding to four levels of actions (no-action, alert, dangerous cold, extreme or exceptional cold).

The agreement includes also a Heat WWS for the summer period with specific announcements when unusual heat is likely to persist more than two days in a given department.

Fig. 7

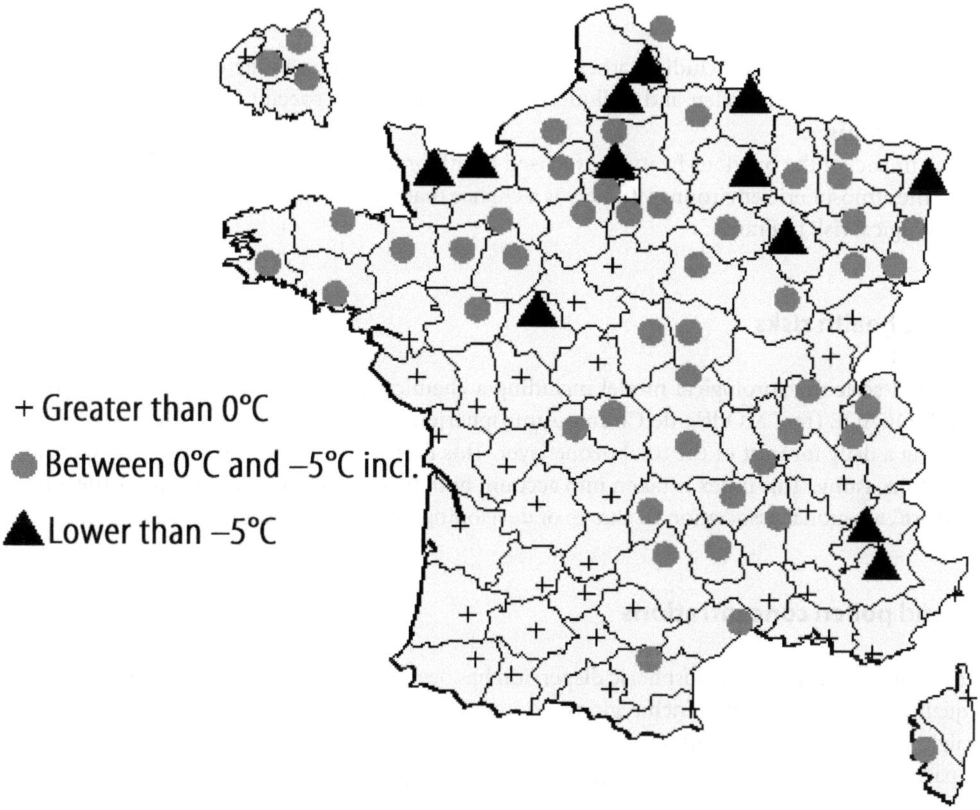

A specific web-page on " meteo.fr " is renewed daily for every departmental service of the Ministry, including a list of city wind-chill forecasts and a "four symbols" departmental map of France

The Institut de Veille Sanitaire (InVS) and the Ministry for Health (DGS)

A written agreement was adopted in January 2004 between Météo-France and the InVS, including technical cooperation in case of periods of extreme temperatures, as well as a scientific collaboration (both new research and programmes such as the PHEWE European programme in Paris).

A first scheme for cold spells and health warning concerns periods with unusually low wind-chill indexes persisting for more than two days over one of the French departments. This is still to be improved and a working group has been set up.

The DGS has been preparing a plan for a response to extreme temperatures.

More medical contacts with Météo France

More agreements and scientific collaborations are currently being developed with the National Institute for Medical Research (INSERM) and several emergency medical units (Samu).

More bio-meteorological products

Consideration is being given to including other warnings or forecasts in a range of bio-meteorological products including UV index, pollen and pollution concentrations, and specific medical risks for some weather-sensitive diseases.

The strong impact of the previous heatwave focused health professionals‘ attention on health risks associated with the atmospheric environment: there is a chance that tragedy might help to develop a culture of the meteorological risk in France.

UV index and health risks

Based on a small-scale meteorological model including a chemical representation of the global atmosphere called MOCAGE (for "MOdèle de Chimie Atmosphérique à Grande Echelle"), Météo-France has been developing a daily forecast of the total ozone layer. This forecast has been used since the year 2001 for UV index forecasting. This index is taken into account both by our services and by our partners from "Sécurité Solaire", a national association in charge of developing information on solar risks.

Pollution and pollen concentrations

Daily forecasts of an indicator of atmospheric dispersion has been used for years by our partners in charge of the air quality survey in several French regions.

This might lead to health risk announcements based on small-scale models forecasts (such as MOCAGE) in partnership with the Air Quality Agency and medical partners.

Weather-sensitive diseases

Cardio-vascular diseases

Previous statistical studies have led to two experiments concerned with bio-meteorological forecasts of myocardial infarction risk in collaboration with the Paris Emergency Unit (Samu de Paris). A daily forecast (including weather types and pollution level) allowed to announce correctly days with low clinical activity and were able to point out one day in two with high clinical activity.

Five daily weather change types are associated with higher clinical activity with cardiac diagnosis, such as the days when continental dry and cold air masses settle over Paris. It was shown that levels of pollution bring complementary information, explaining daily peaks that are not detected by the weather types.

After the strong impact of the August 2003 heatwave, new partnership programmes have been initiated and more experiments are being developed.

Ophthalmologic emergency and weather conditions

A recent statistical study on 100,000 cases of daily emergencies at the Quinze Vingt Ophthalmologic Hospital (Paris) showed several environmental conditions associated with higher clinical rates. Namely, hot weather, very dry as well as very wet conditions, grass and oak pollens and some pollution peaks could now be used, when automatically forecast, to predict ocular risks to ophthalmologists.

Conclusions

The heavy toll of the August 2003 outstanding heatwave and its political impacts focused both public and media attention on the necessity to give political consideration to extreme meteorological events such as heatwaves, cold spells and floods. The weak and inefficient impact on health and safety professionals of the bio-meteorological warnings that were announced by some weather offices such as those in France, Portugal or Germany pointed out:

- on the one hand the importance of preparing and testing operational responses to weather extremes including national, regional and local levels in partnership with Health professionals and Civil Security;
- on the other hand, it showed the natural responsibility for watch warning of every national weather service, capable of watching for weather risks 24 hours per day, 365 days per year and professionally used to emergency forecasting.

Moreover, political consideration, both at national and international or European levels, should be given to the increasing frequency of extreme weather events that has been affecting all parts of the world in recent years, including even countries with mild climates. These observed phenomena being consistent with climate change projections, more interest should be given to their impacts on public health.

A successful response to extreme weather events should include health recommendations corresponding to each kind of atmospheric danger. A general incitation to such an approach would help to develop and increase (even in countries like France that have not been used to giving consideration to such phenomena) a new culture for meteorological risks.

References

Bessemoulin P, Bourdette N, Courtier P, Manach J (2004): La Canicule d'août 2003 en France et en Europe. La Météorologie, 8 ème série, n° 46, 25 – 33

http://www.meteofrance.com/FR/index.jsp

Improving Public Health Responses to Extreme Weather Events

Kristie L. Ebi

Keywords

Extreme weather events, floods, heatwaves, climate change, Europe

Abstract

Recent advances in knowledge about the climate system have increased the ability of meteorologists to forecast extreme weather and climate events, such as floods and heatwaves. Public health agencies and authorities have had limited involvement in the development of early warning systems to take advantage of these forecasts to reduce the burden of disease associated with extreme events. Instead, public health has focused on surveillance and response activities to identify disease outbreaks following an extreme event. Although these systems are critical for detecting and investigating disease outbreaks, they are not designed for identifying and preventing many of the adverse health outcomes associated with extreme events. Designing and implementing effective disease prediction and preventions programs that incorporate advances in weather and climate forecasting have the potential to reduce illness, injury, and death. Critical components of an early warning system include the weather forecast, disease prediction models, and a response plan designed to pro-actively undertake activities to reduce projected adverse health outcomes. Because climate change may increase climate variability, early warning systems can both reduce current vulnerability to extreme events and increase the capacity to cope with a future that may be characterized by more frequent and more intense events.

Introduction

Public health is the science and art of preventing disease, prolonging life, and promoting health through the organized efforts of society (Committee of Inquiry into the Future Development of the Public Health Function 1988). "It is the combination of sciences, skills, and beliefs that is directed to the maintenance and improvement of the health of all people through collective or social actions. The programs, services, and institutions involved emphasize the prevention of disease and the health needs of the population as a whole. Public health activities change with changing technology and social values, but the goals remain the same: to reduce the amount of disease, premature death, and disease-produced discomfort and disability in the population." (Last 2001). Public health has been also defined by the WHO as "the art of applying science in the context of politics so as to reduce inequalities in health while ensuring the best health for the greatest number". (Yach 1996).

Measures to reduce disease and save lives are categorized into primary, secondary, and tertiary prevention. Primary prevention is the "protection of health by personal and community wide efforts" (Last 2001).

It aims to prevent the onset of disease in an otherwise unaffected population by preventing individual exposures to sufficient doses of an agent to initiate disease. Regulation of potentially hazardous environmental exposures, such as by setting limits on criteria air pollutants to prevent the onset of disease, is an example of primary prevention.

When primary prevention is not feasible, which is often the case with many naturally occurring environmental exposures, secondary prevention includes "measures available to individuals and populations for the early detection and prompt and effective intervention to correct departures from good health" (Last 2001). It focuses on preventive actions taken in response to early evidence of health impacts (e.g., strengthening disease surveillance and responding adequately to disease outbreaks such as the West Nile virus outbreak in the United States). Surveillance is necessary to gather early evidence of adverse health impacts.

Finally, tertiary prevention "consists of the measures available to reduce or eliminate long-term impairments and disabilities, minimize suffering caused by existing departures from good health, and to promote the patient's adjustment to irremediable conditions" (Last 2001). These measures include health-care actions taken to lessen the morbidity or mortality caused by the disease (e.g., improved diagnosis and treatment of cases of malaria).

Primary prevention is generally more effective and less expensive than secondary and tertiary prevention. Although disasters due to adverse weather and climate events cannot be entirely prevented, primary prevention, particularly development of early warning systems, can reduce the number of adverse health outcomes that occur during and following an event. Current primary prevention activities, where they exist, are generally limited to educational programs to inform the public of what to do (and not do) during and immediately following an event. These educational programs often were implemented in a city or region after an event caused injuries and deaths; few programs have been established proactively. Instead, public health activities have focused on surveillance and response systems (secondary prevention) to identify disease outbreaks following an event, such as an outbreak of waterborne disease following a flood.

Surveillance and response systems are a cornerstone of infectious disease prevention and health promotion activities. However, they are ineffective for identifying and preventing many of the adverse health outcomes associated with extreme weather and climate events. Even with improvements, surveillance systems will not provide a basis for effective public health response to an extreme weather event. In order to reduce the number of adverse health events due to extreme weather and climate events, effective prediction and prevention programs that incorporate recent advances in weather and climate forecasting need to be designed and implemented. The increasing ability to predict extreme events, particularly heatwaves, provides public health authorities and agencies with the opportunity to develop early warning systems to reduce the burden of disease associated with these events. Because climate change is projected to increase climate variability, early warning systems can both reduce current vulnerability to extreme events and increase the capacity to cope with a future that may be characterized by more frequent and more intense events (McGregor, this volume).

Forecasting Extreme Weather and Climate Events

The skill with which extreme weather events can be forecast has increased significantly over the past thirty years as more has been learned about the climate system. During this period, weather forecasting gradually increased from same-day forecasting to the three day advance forecast, and our understanding of the mechanics and teleconnections of El Nino Southern Oscillation (ENSO) now provides the capacity for seasonal and annual forecasting – assumed as recently as the 1970s to be the stuff of fancy not science (Nicholls 2002). In fact, Chen et al. (2004) recently suggested that El Nino events can be predicted two years in advance.

Public health has taken only limited advantage of the possibilities thus offered (Kovats et al. 1999; WHO 2004). For example, the timing and intensity of a number of diseases change during and after El Nino events. Coupling understanding of these disease relationships with forecasting of the timing of an El Nino event can be used to develop early warning systems that improve current public health responses. For example, Bouma and van der Kaay (1996) analysed historical malaria epidemics using an El Nino Southern Oscillation index, then predicted high-risk years for malaria on the Indian subcontinent. There was an increased relative risk of 3.6 for an epidemic in El Nino years in Sri Lanka and a relative risk of 4.5 for post El Nino years in the former Punjab province of India. Subsequently, associations were reported between El Nino events and malaria in other parts of the world, including South America (Bouma and Dye 1997; Bouma et al. 1997). This ability to predict malaria epidemics allows public health authorities to implement interventions, such as the distribution of bednets and anti-malarials, before surveillance would have detected an outbreak, thereby reducing disease and death. This more efficient allocation of resources allows for more effective use of scarce public health funds.

The value of early warning systems to prevent adverse health outcomes will increase in the future if projections of increased climate variability due to climate change are realized (McGregor, this volume). One likely consequence of increased climate variability will be surprises with regard to the timing, intensity, location, and duration of extreme events (Glantz 2004). Climate variability may already be increasing, resulting in extreme weather and climate events for which decision makers and the public are ill prepared. Increasing climate variability could affect the extreme event frequency-magnitude distribution in a number of ways (McGregor, this volume). One possibility is that the more rare or extreme events will be affected disproportionately such that only these events become more frequent. In this situation, the probable maximum flood or heatwave would increase, perhaps substantially. Another possibility is that the whole frequency-magnitude distribution might shift such that all events become more frequent. A third possibility is that only smaller events become more frequent. Thus, there is high uncertainty about the rate and intensity of any changes in climate variability in a particular location over a specified time period, but high certainty that without adequate preparation, extreme events will lead to increased morbidity and mortality. Extreme events cannot be prevented, but the vulnerability to these events can be reduced through a variety of measures, if appropriate warning mechanisms are in place.

Development of early warning systems in anticipation of increased climate variability due to climate change can be viewed as an application of the precautionary principle. The precautionary principle is an approach to public policy action that can be used in situations of potentially serious or irreversible threats to health or the environment, where there is a need to act to reduce potential hazards before there is strong proof of harm. The precautionary principle and its application to environmental hazards and their uncertainties began to emerge as a foundation for decision-making in the 1970s; it has been increasingly included in national legislation and international treaties. To this end, on 2 February 2000, the European Commission approved a communication on the precautionary principle that provided guidelines for its application (CEC 2000). Elements of a precautionary approach include research and monitoring for the early detection of hazards, cost-benefit analysis of action and of no action, and the taking of action before full proof of harm is available if impacts could be serious or irreversible (Boehmer-Christiansen 1994; EEA 2001). Application of the precautionary approach should trigger research to provide more certainty about the exposure-response relationship.

Because the character and intensity of extreme weather and climate events are changing, it would be useful to develop scenarios identifying potential impacts that may not be anticipated by decision makers and the public. For example, if a major heatwave were to result in the shutdown of power plants that supply electricity to another country, and if the recipient country did not have backup sources of power, then a range of impacts can be imagined for which responses would need to be developed. These scenarios can then be used to inform development of early warning systems, including specific interventions to be implemented.

Limitations of Surveillance and Response for Extreme Weather Events

Surveillance aims to continually scrutinize all aspects of the occurrence and spread of disease that are pertinent to effective control (Last 2001). Surveillance activities include the systematic collection of health, disease, and exposure information; the analysis of disease and exposure patterns; interpretation of trends; and distribution of results to responsible agencies, whether local, regional, national or global, to identify and implement appropriate responses to disease events (Wilson and Anker 2005). Surveillance is the means by which public health agencies keep themselves informed about the health status of the populations that they serve. Surveillance data are used to monitor levels and trends in disease occurrence, characterize geographical spread of disease over time, detect and investigate outbreaks as they occur, recognize new strains of pathogens, etc. Surveillance data are used to detect and respond to outbreaks of infectious diseases, such as West Nile virus. Typically, many cases need to occur before the surveillance systems can detect that an outbreak is in progress. Unfortunately, it is not uncommon for actions to be initiated after the peak of the epidemic has passed.

In the case of extreme weather and climate events, surveillance is needed in the time period immediately surrounding an event to determine if the event is associated with an increase in disease, such as diarrheal disease following a flood, in order to institute appropriate measures (such as a boil water alert). In addition, surveillance of age and cause-specific deaths over time are needed to calculate baseline or normal mortality rates in order to recognize unusual increases in mortality over what would have been expected for that time and place (Wilson and Anker 2005). Knowledge of the usual pattern of deaths allows calculation of the number and type of excess deaths resulting from a flood, heatwave, or cold spell. However, because there is a considerable lag between when deaths occur and when the data are available for analysis by public health authorities, surveillance is not effective for determining during an extreme weather event if it is causing excess deaths. For example, mortality begins to rise within one day of a heatwave, but these data are often not available for days to months.

The limitations of surveillance were demonstrated during the 2003 European heatwave. This heatwave took French public authorities by surprise (Abenheim, this volume). There are a number of reasons why this heatwave caused nearly 15,000 excess deaths in about a two-week period, from characteristics of the event itself to surveillance systems not designed to detect and respond to a heatwave. Health surveillance systems did not provide authorities with information quickly enough to detect the increased number of deaths in a timely manner. For example, a retrospective assessment determined there had been about 3900 deaths at the time when 10 deaths had been reported via normal surveillance. In fact, it would have been difficult for the current surveillance system to determine the size of the problem: the number of emergency ward visits in August (1900 visits) was not higher than usual (2100 visits in 2000); and the number of deaths attributed to the heat was less than one per nursing home. Surveillance systems are designed to detect outbreaks of a variety of usual or new infectious diseases, such as measles, but are not designed to determine if more deaths from cardiovascular disease are occurring today or this week. In addition, health surveillance systems are not designed to cope with the large numbers of people who were at risk for mortality from chronic diseases, such as cardiovascular and respiratory diseases; it is estimated in France that there were 6 million people at risk, of which 1 million were at very high risk. There were 500,000 at very high risk and isolated. Finally, emergency public health interventions have not been designed to address sudden increases in endemic and common diseases, such as cardiovascular disease. There were no widely available and efficient measures for reducing heat-related mortality. Air conditioning may have saved some lives, but is generally not available, particularly for the populations at highest risk, such as the elderly in nursing homes.

The French public health surveillance system is being re-evaluated to identify limitations and areas for improvement (Abenheim, this volume). But improvements to surveillance systems can only partly address the adverse health impacts of extreme weather and climate events. There needs to be increased focus on prediction and prevention if further disasters are to be avoided. Surveillance and early warning systems, coupled with effective response capabilities, can reduce current and future vulnerability.

Early Warning Systems Can Reduce Current and Future Vulnerability to Extreme Weather Events

Whereas surveillance systems are intended to detect, measure, and summarize disease outbreaks as they occur, early warning systems for extreme weather and climate events are designed to alert the population and relevant authorities that meteorological conditions are such that adverse health events could result. Early warning systems can be very effective in preventing deaths, diseases, and injuries as long as effective prediction is coupled with adequate communication strategies and timely response capabilities . For extreme weather events, forecasting is needed of both the event itself, which is the domain of meteorology, and prediction of the health impacts that could occur. In theory, an effective early warning system should both reduce vulnerability and increase preparedness (Committee on Climate Ecosystems, Infectious Disease, and Human Health 2001).

The development of an early warning system assumes that the responsible agencies have agreed upon what constitutes a risk. Risk can be viewed as a combination of the probability that an adverse event will occur and the consequences of that event (USPCC RARM 1997). Both factors need to be considered when evaluating risk because the public health responses developed for high probability events with low consequences will differ from the responses for low probability events with high consequences. An example of a high probability event with low consequences is a heavy rain event not associated with flooding. Responses, including early warning systems, for the low probability events with high consequences can be considered within the context of the precautionary principle.

An event that has the potential to cause harm is considered a hazard. The consequences of a hazardous event in the public health sector are measured in terms of the burden of associated morbidity and mortality. As discussed elsewhere in this volume, extreme weather and climate events have caused severe consequences over the past few years. However, as discussed in Hajat et al. (2003) the health risks associated with flooding are surprisingly poorly characterized. The dearth of good quantitative data results in uncertainty of the full range of potential health impacts associated with flooding events. There is better understanding of the health risks of heatwaves, but more information is needed on effective interventions to reduce morbidity and mortality. In order to more effectively target intervention programs for extreme weather and climate events, better understanding is needed of vulnerability risk factors. Projections of more, and more intense, extreme weather and climate events as a consequence of climate change increase the importance of improving our understanding of population vulnerability and interventions that can effectively reduce that vulnerability.

The principal components of public health early warning systems are disease prediction and response (Woodruff 2005).

Disease Prediction

A disease prediction model needs to be accurate, specific, and timely. To do so, there must be knowledge of disease dynamics and the ability to obtain reliable and up-to-date information on both the health outcome and the factors critical to disease incidence. This knowledge of disease dynamics applies primarily to communicable diseases, such as those that can follow an extreme weather event. For example, the disease outbreaks reported after flooding events have well-described etiologies. Such disease outbreaks are rare in Europe due to the public health infrastructure, including water treatment and sanitation. However, fever and waterborne disease have been reported following flood events (Hajat et al. 2003). One example of an outbreak occurred when cases of leptospirosis were reported in the Czech Republic following flooding in 1997; however, the quality of data appears to be poor (Kriz et al. 1998). Analysis of waterborne disease outbreaks in Finland over the period 1998–1999 found that thirteen of fourteen outbreaks were associated with groundwater that was not disinfected, mostly related to flooding (Miettinen et al. 2001).

The association between elevated ambient temperatures and morbidity is well documented. Historically, cardiovascular diseases have accounted for 13 % to 90 % of the increase in overall mortality during and following a heatwave, cerebrovascular disease 6 % to 52 %, and respiratory diseases 0 % to 14 % (Kilbourne 1997). Heatwaves also increase the rate of nonfatal illnesses. Disease prediction models have been constructed within the context of heatwave early warning systems (Kalkstein 2003; Sheridan and Kalkstein 1998).

However important, disease prediction models are not sufficient for developing an early warning system (Woodruff 2005). Consideration needs to be taken of confounding or modifying factors that affect the potential for an outbreak. An effective response plan also is required to reduce the predicted burden of disease. The second component of disease prediction is the availability of reliable and timely information on critical factors in the exposure-disease relationship. As discussed above, surveillance systems are relatively effective for communicable diseases. Surveillance systems are relied upon after flooding to detect increases in diarrheal and other diseases. However, also discussed above, surveillance systems are not designed to detect increases in common chronic diseases. Different approaches need to be developed to determine if deaths increase during a heatwave, such as obtaining reports from a random sample of funeral directors to determine if more deaths than usual are occurring. This implies that the funeral directors have analyzed historical data to determine, for a particular time period, the expected or baseline number of deaths. This also means that thresholds need to be established, beyond which specific responses are taken. For example, one excess death per week is unlikely to be sufficient to suggest an increase in mortality. But should actions be taken at two, three, four, etc.? Thresholds need to balance between recommending actions when they are not needed, and not recommending actions when they would reduce morbidity and mortality.

It is important to consider the sensitivity and specificity of a predictive model in the design of public health responses. All models have multiple sources of uncertainty, from data uncertainties to incomplete understanding of disease etiology. Uncertainties need to be characterized because they have implications for the design of response activities. For example, no early warning system can ever be completely accurate. False positives (issuing a warning when none was required) and false negatives (not issuing a warning when one was needed) have consequences, not only in terms of increased morbidity and mortality, but also in terms of public willingness to rely on subsequent warnings. Incorporating understanding of these uncertainties, and their associated costs, into the design of an early warning system can improve its effectiveness.

Even when models are available that are reasonably predictive and surveillance data provides timely information, models are still of limited value until they are actually used to direct disease prevention efforts (Woodruff 2005).

Response

Response is the second principal component of a public health early warning system. The design and implementation of response activities needs to be within the context of the cultural, social, economic, and political constraints of a particular region. The components of response include:

- a response plan, detailing thresholds for action;
- available and effective interventions;
- economic assessment of the cost-effectiveness and affordability of the system;

- communication strategy; and
- involvement of all relevant stakeholders in the process (Woodruff 2005).

A variety of response plans exist or are being developed for heatwaves and floods. Emergency management agencies in most countries have flood plans in place, but these focus on post-disaster response and often do not include adequate pre-disaster planning. Early warnings of flooding risk have been shown to be effective in reducing flood-related deaths when the warning is coupled with appropriate responses by citizens and emergency responders (Malilay 2001). An example of the effectiveness of response plans is the differences between the 1993 – 94 flooding along the Rhine and Meuse in Germany and the 1995 flooding along the same rivers (Estrela et al. 2001). The two floods had similar characteristics, although the 1993 – 94 flood had a second peak discharge. Persistent high precipitation caused both events: in December 1993, the accumulated precipitation was more than double the long-term average for that month. Ten people lost their lives in the 1993 – 94 flood and the total damage was estimated at USD 900 million for Belgium, Germany, France, and the Netherlands (Bayrische Ruckversicherung 1996a,b; Estrela et al. 2001). The economic cost of the flood damage in Germany in 1995 was half this amount because people were aware of the risks and were better prepared.

There are several areas in which flood plans can be improved. As discussed in Hajat et al. in this volume, current plans focus on the larger events even though more frequent smaller events also have important health impacts, and have not incorporated consideration of the mental health impacts that can follow a flooding event. The mental health impacts of a flood may be considerable, and may last for months to possibly even years after the flood event. Currently it is unclear whether reduction of mental health impacts will respond best to psychological and/or pharmacological interventions delivered through health services, or whether the interventions would best be targeted at providing financial or other assistance with recovery activities.

Although periodic heatwaves have led to excess morbidity and mortality, few Ministries of Health have made heatwave prediction and response a high priority. Prior to the 2003 heatwave, high ambient temperatures were considered a rare problem about which public health authorities could do little. Few heat/health warning systems were in place in Europe and most of those that were had only recently been implemented. Initial funding for several of these systems came from the World Meteorological Organization. Systems recently established in Lisbon and Rome were able to notify public health authorities and/or the population of the hazardous conditions during the 2003 heatwave, which presumably saved lives – as has been found in other cities. For example, the combination of a heat watch warning system and a response plan to reduce the exposure of vulnerable population groups to extreme heat likely led to the substantial reduction in deaths from extreme heat in the midwestern United States in 1999 as compared with a similar heatwave in 1995 (Palecki et al. 2001). It took an extreme event such as that in 2003 to demonstrate to Ministries of Health and others that heatwaves are dangerous, with the potential to cause large numbers of excess deaths, and that effective early warning systems need to be designed and implemented in advance of the next heatwave if additional lives are not to be lost.

The characteristics of an early warning response plan are informed by the political, social, and cultural setting in which the system operates (Glantz 2004). Therefore, an effective plan developed for one community, region, or country, may not have the same degree of effectiveness if implemented elsewhere. The appropriateness of the plan components, including thresholds of action and the interventions that these trigger, need to be evaluated and modified to maximize response effectiveness for the region for which it is designed. For example, some heatwave early warning systems in the United States have two levels of health alerts: a watch, which indicates that meteorological conditions are such that a heatwave could arise; and a warning, which indicates that a heatwave has started (Kalkstein et al. 1996). In essence, a watch means to be prepared, and a warning means that an extreme event is in progress so response plans should be activated. 'Watch' and 'warning' have been used by the United States National Weather Service in early warning systems for other climatic events, such as tornados, so the public understands their meaning. The same may not be true in other regions and countries.

Response plans require the explicit definition of thresholds for action. For example, a heat/health warning system needs to define what constitutes a heatwave. Such a definition is contextual within a population; what is considered a hot day in Stockholm differs from what is considered a hot day in Athens. Thresholds need to consider not only at what point interventions should be implemented to protect the health of the entire population, but also when interventions should be implemented to protect various vulnerable sub-groups, such as the elderly or tourists. As noted above, there may be more than one threshold for action. In addition, there needs to be sufficient lead-time for response plans to be activated once the threshold is reached.

Development of an early warning system is predicated on the availability of effective interventions to reduce the burden of disease. Further, interventions should be specific to a vulnerable group. The elderly, disabled, children, ethnic minorities, and those on low incomes may be at increased risk during and after an extreme event. Although there will be common interventions, such as lowering body temperature to prevent the onset of heat stress, there should be messages specific to particularly vulnerable groups. For example, the messages designed to alert parents to the risk of leaving infants unattended in automobiles during a heatwave should be different than messages designed to motivate seniors to spend time at a cooling center. Tourists are often not considered in the design of response plans, yet can be particularly vulnerable because they are unlikely to know what specific actions to take and may be difficult to inform because of language barriers. A variety of interventions are in use in flood and heatwave early warning systems. Unfortunately, their effectiveness has generally not been evaluated.

Response plans need to assess the cost-effectiveness of the system. This includes knowledge of the effectiveness of specific interventions; unfortunately little research has been conducted in this area. One study looked at the cost-effectiveness of the Philadelphia Hot Weather-Health Watch/Warning System (Ebi et al. 2004). This system was initiated in 1995 to alert the city's population to take a variety of precautionary actions during a heatwave. It was estimated that issuing a warning saved, on average, 2.6 lives for each warning day and for three days after the warning ended; the system saved an estimated 117 lives over a three-year period. Estimated dollar costs for running the system were small compared with estimates of the value of a life. Unfortunately, data were not available on the specific interventions included in the system, so their individual effectiveness could not be evaluated. This is an area in which further research is much needed.

The existence of the response plan, along with the interventions included in the plan, need to be communicated to all potentially affected groups. A communication strategy is a critical element of the response plan. This strategy needs to define who is responsible for the communication program; when the program should be initiated; who are the key audiences; what are the key messages to be delivered and how they should be delivered; and how the effectiveness of the communications will be monitored and evaluated (Kovats et al. 2003).

Finally, a response plan needs to be developed and operated in a transparent manner that includes all stakeholders in the process (Glantz 2004). Stakeholders include the agencies and/or organizations that will fund the development and operation of the system; the groups who will be expected to take actions (such as emergency responders and others); and those likely to be affected by the extreme event. Such involvement increases the likelihood of success of the system. In addition, including those who have been affected by extreme events in the past may reveal local knowledge about responses and their effectiveness.

Conclusion

The skill with which extreme weather and climate events, such as floods and heatwaves, can be forecast has increased significantly over the past thirty years as more has been learned about the climate system. However, public health agencies and authorities have taken only limited advantage of the possibilities thus offered. Instead, the focus of public health activities has been on surveillance and response systems to identify disease outbreaks following an event. Although surveillance and response systems are a cornerstone of disease prevention and health promotion activities, they are ineffective for identifying and preventing many of the adverse health outcomes associated with extreme events. In order to reduce the number of adverse health events due to extreme weather, effective prediction and prevention programs need to be designed and implemented that incorporate advances in weather and climate forecasting. Because climate change may increase climate variability, early warning systems can both reduce current vulnerability to extreme events and increase the capacity to cope with a future that may be characterized by more frequent and more intense events.

References

Bayrische Ruckversicherung (1996a) 13 months later. The January 1995 floods. Bayrische Ruckversicherung. Special Issue 17

Bayrische Ruckversicherung (1996b) The 'Christmas floods' in Germany 1993 – 1994. Bayrische Ruckversicherung. Special Issue 16

Boehmer-Christiansen S (1994) The precautionary principle in Germany: enabling government. In: O'Riordan T, Cameron J (eds) Interpreting the precautionary principle. Cameron and May, London

Bouma MJ, Dye C (1997) Cycles of malaria associated with El Niño in Venezuela. JAMA 278:1772 – 4

Bouma MJ, Poveda G, Rojas W, Chavasse D, Quinones M, Cox J, Patz J (1997) Predicting high-risk years for malaria in Colombia using parameters of El Niño Southern Oscillation. Tropical Medicine and International Health 2:1122 – 7

Bouma MJ, van der Kaay HJ (1996) El Niño Southern Oscillation and the historic malaria epidemics on the Indian subcontinent and Sri Lanka: an early warning system for future epidemics? Tropical Medicine and International Health 1:86 – 96

Chen D, Cane MA, Kaplan A, Zebiak SE, Huang D (2004) Predictability of El Nino over the past 148 years. Nature 428:733 – 736

Commission of the European Community (CEC) (2000) Communication from the Commission on the precautionary principle, Brussels

Committee of Inquiry into the Future Development of the Public Health Function (1988) Public Health in England. Cmnd 289. HMSO, London

Committee on Climate, Ecosystems, Infectious Disease, and Human Health, Board on Atmospheric Sciences and Climate, and National Research Council (2001). Under the Weather: Climate, Ecosystems, and Infectious disease. National Academies Press, Washington

Ebi KL, Teisberg TJ, Kalkstein LS, Robinson L, Weiher RF. Heat watch/warning systems save lives: estimated costs and benefits for Philadelphia 1995 – 1998. Bulletin of American Meteorological Society (BAMS) 2004; 85(8):1067 – 1073

EEA (2001) Late lessons for early warnings: the precautionary principle, 1896 – 2000. European Environment Agency, Copenhagen

Estrela T, Menendez M, Dimas M, Marcuello C et al. (2001) Sustainable water use in Europe. Part 3. Extreme hydrological events: floods and droughts. Environment issue report No. 21. European Environment Agency, Copenhagen

Glantz MH (2004). Usable Science 8: Early Warning Systems: Do's and Don'ts. Report of a workshop 20 – 23 October 2003 in Shanghai, China. National Center for Atmospheric Research. www.esig.ucar.edu/warning

Hajat S, Ebi KL, Kovats S, Menne B, Edwards S, Haines A (2003) The human health consequences of flooding in Europe: a review. Applied Environmental Science and Public Health 1:13 – 21

Kalkstein LS (2003). Description of our heat/health watch-warning systems: their nature and extent, and required resources. Final report to Stratus Consulting Company, Boulder, Colorado, 31 pp

Kalkstein LS, Jamason PF, Greene JS, Libby J, Robinson L (1996)

The Philadelphia Hot Weather-Health Watch/Warning System: development and application, Summer 1995. BAMS 177, 1519–1528

Kilbourne EM (1997) Heat waves and hot environments. The Public Health Consequences of Disasters, E.K. Noji, Ed., Oxford University Press, New York, 245–269

Kovats RS, Bouma M, Haines A (1999) El Niño and Health. (WHO/SDE/PHE/99.4), Geneva: WHO

Kovats RS, Ebi KL, Menne B (2003) Methods for Assessing Human Health Vulnerability and Public Health Adaptation to Climate Change. WHO/Health Canada/UNEP

Kriz B, Benes C, Castkova J, Helcl J (1998) Monitoring the epidemiological situation in flooded areas of the Czech Republic in 1997. In: Konference DDD '98; Kongresove Centrum Lazeoska Kolonada Podibrady, 11.–13. Kvitna 1998. Prodebrady, Czech Republic

Last JM (2001) A Dictionary of Epidemiology, 4th edn. Oxford University Press, Oxford

Last JM (1998). Public Health and Human Ecology, 2nd edn. Prentice Hall International, Inc., London, p 9

Malilay J (1997) Floods. In: Noji E (ed) Public Health Consequences of Disasters. OUP, New York, p 287–301

Miettinen IT, Zacheus O, von Bonsdorff CH, Vartiainen T (2001) Waterborne epidemics in Finland in 1998–1999. Water Sci Technol 43:67–71

Nicholls N (2002) Climatic outlooks: from revolutionary science to orthodoxy. In: Climate and Culture, Canberra, 25–27 September. Australian Academy of Science

Palecki MA, Changnon SA, Kunkel KE (2001) The nature and impacts of the July 1999 heat wave in the Midwestern United States: learning from the lessons of 1995. BAMS 82:1353–1367

Sheridan SC, Kalkstein LS (1998) Heat watch-warning systems in urban areas. World Resource Rev 10:375–383

USPCC RARM (1997). Framework for Environmental Health Risk Management, Final Report Volume 1, US Presidential/Congressional Commission on Risk Assessment and Risk Management, Washington DC

Wilson ML, Anker M (2005) Disease surveillance in the context of climate stressors: needs and opportunities. In: Ebi KL, Smith JB, Burton I. Integration of Public Health with Adaptation to Climate Change: Lessons Learned and New Directions. In press

Woodruff R (2005) Epidemic early warning systems. In: Ebi KL, Smith JB, Burton I. Integration of Public Health with Adaptation to Climate Change: Lessons Learned and New Directions. In press

World Health Organization (2004) Using Climate to Predict Infectious Disease Outbreaks: A Review. WHO/SDE/OEH/04.01. WHO, Geneva

Yach D (1996) Redefining the scope of public health beyond the year 2000. Current Issues in Public Health 2:247–252

Temperature Extremes and Health Impact

Cold Extremes and Impacts on Health

J. Hassi

Abstract

A special feature of the future climatic change is the climate variability in particular the frequency and intensity of extreme conditions. Cold spells will remanin a problem within Europe even under the circumstances of climatic changes. With Britain's predicted increase in environmental temperature by 2 °C during the next 50 years, seasonal mortality during the cold months of the year still will present the majority of excess mortality. Epidemiological evidence has indicated a causal relationship between mortality and cold weather. The most important diseases associated with cold-related excess mortality are ischaemic heart disease, cerebro-vascular disease and respiratory disease, especially influenza. Body cooling may offer a better explanation for the cold-related excess mortality than environmental temperature.

The goals of public health activities related to the health impact of cold extremes are to reduce premature deaths, the amount of disease and injuries, disease-produced discomfort, sickness and disability in the population. In order to evaluate the prevention of cold exposure-related excess mortality, we need the collaboration between health care, weather services and other officials to produce usable preventive action models. The definition of public health programmes aimed at preventing cold-related mortality needs further research. The prevention of cold injuries and illnesses is more the responsibility of health care providers and it requires practical information, education and professional support.

Introduction

There is scientific consensus that the composition of the atmosphere is changing, thereby altering the radiation balance of the Earth's atmospheric system and producing global warming. Climate change during the 21st century is predicted to be rapid and large, although there are many associated uncertainties. A special feature of the predicted change is the variability of the future climate and especially the occurrence of extreme conditions. The latter include the occurrence of anomalously high and low temperatures, precipitation amounts and wind speeds. Even in the circumstances of climate warming, extreme cold spells remain possible which could be detrimental to health. Mortality is subject to seasonality. In many temperate countries 'all-cause mortality' as well as cardiovascular and respiratory mortality are higher during winter months. Some epidemiologist's use the notion of excess winter mortality to describe this seasonal phenomenon. Most European countries suffer from 5 – 30 % excess winter mortality [19].

The average annual excess mortality related to cold climate has decreased [27] and will decrease even further in the future, while excess mortality related to heat will increase. Despite this, with, for example, Britain's predicted increase in environmental temperature by 2 °C during the coming 50 years, mortality related to cold climate will still represent the majority of mortality excess related to extreme temperatures [26].

Seasonal changes in human health have been recognized for more than 2000 years [2] and excess winter mortality has been known for hundreds of years [45]. In more recent times scientific research has documented dramatic rises in mortality every winter, and smaller rises during unusually hot summers. Mortality figures based on monthly, rather than daily, mortality statistics under-estimate the problem [26]. Special attention has been given to excess winter mortality in Europe during the last years [19,43] and deaths from excessive cold have also been epidemiologically quantified [13].

Less recognition has been given to the public health actions to prevent the negative health impacts of cold temperature [35]. In evaluating the need for preventive and protective public health actions, it is important to recognize not only excess cold mortality, but also cold injuries, illnesses and physiological cold stress, all of which affect health and performance limitations.

Cold mortality at different daily temperatures

Environmental temperatures are associated with increased daily mortality, not only in countries having cold winters but also in warmer countries with milder winters, such as in the Mediterranean countries. The relationship is generally a non-linear J, or V shaped. The lowest mortality related to daily environmental temperature is seen at higher environmental temperatures in countries with the warmest summers (➔ *Fig. 1)* [9, 10, 44]. This suggests that different populations adapt to their environmental temperatures. Daily mortality in Britain is lowest when the outdoor temperature is around +18 °C. Reduction in temperature below this level cause about 40,000 excess deaths per year relative to the number observed at the temperature of minimum mortality. Apart from the special case of heat-waves, warm weather in Britain causes around 1000 extra deaths per year [26].

The Eurowinter group studied whether increases in mortality per 1 °C fall in temperature differ across various European regions and related any differences to average winter climate and to measures to protect against cold. The findings were that the percentage increases in all-cause mortality per 1 °C fall in temperature below 18 °C were greater in warmer regions than in colder regions, e.g. Athens 2.15 % [95 % CI 1.20 – 3.10] versus south Finland 0.27 % [0.15 – 0.40]). At an outdoor temperature of 7 °C, the mean living-room temperature was 19.2 °C in Athens and 21.7 °C in South Finland; 13 % and 72 % of people in these regions, respectively, wore hats when outdoors at 7 °C.

Multiple regression analyses (with allowance for sex and age, in the six regions with full data) showed that high indices of cold-related mortality were associated with high mean winter temperatures, low living-room temperatures, limited bedroom heating, low proportions of people wearing hats, gloves, and anoraks, and inactivity and shivering when outdoors at 7 °C ($p < 0.01$ for all-cause mortality and respiratory mortality; $p > 0.05$ for mortality from ischaemic heart disease and cerebrovascular disease) [43]. The Eurowinter group noted that a lower proportion of people wear hats and gloves in slightly cool outdoor temperatures (+7 °C) [43]. It could be hypothesized that the higher cold-related mortality of populations in regions where extreme cold occurs less frequently is associated with a poorer adaptation to the cold.

Deaths outside of hospitals in the USA are strongly associated with cold temperatures, irrespective of gender. Stronger associations with cold are seen for those aged less than 65 years. The strong out-of-hospital effect modifier was more than five times greater than in-hospital deaths, supporting the biological plausibility of the association [33]. In Europe, exposure to cold increases the risk of coronary and cerebral thrombosis in elderly people with atheromatous arteries [26]. In recent years, epidemiological evidence has accumulated indicating a causal relationship between mortality and cold weather and today a critical discussion of weather as a confounder is not as common as it used to be.

Exposure to cold has an important and partially direct effect on daily mortality [1]. The most important diseases associated with cold related excess mortality are ischaemic heart disease, cerebro-vascular disease and respiratory disease [23], especially influenza [1]. Approximately half of the excess deaths are

Fig. 1

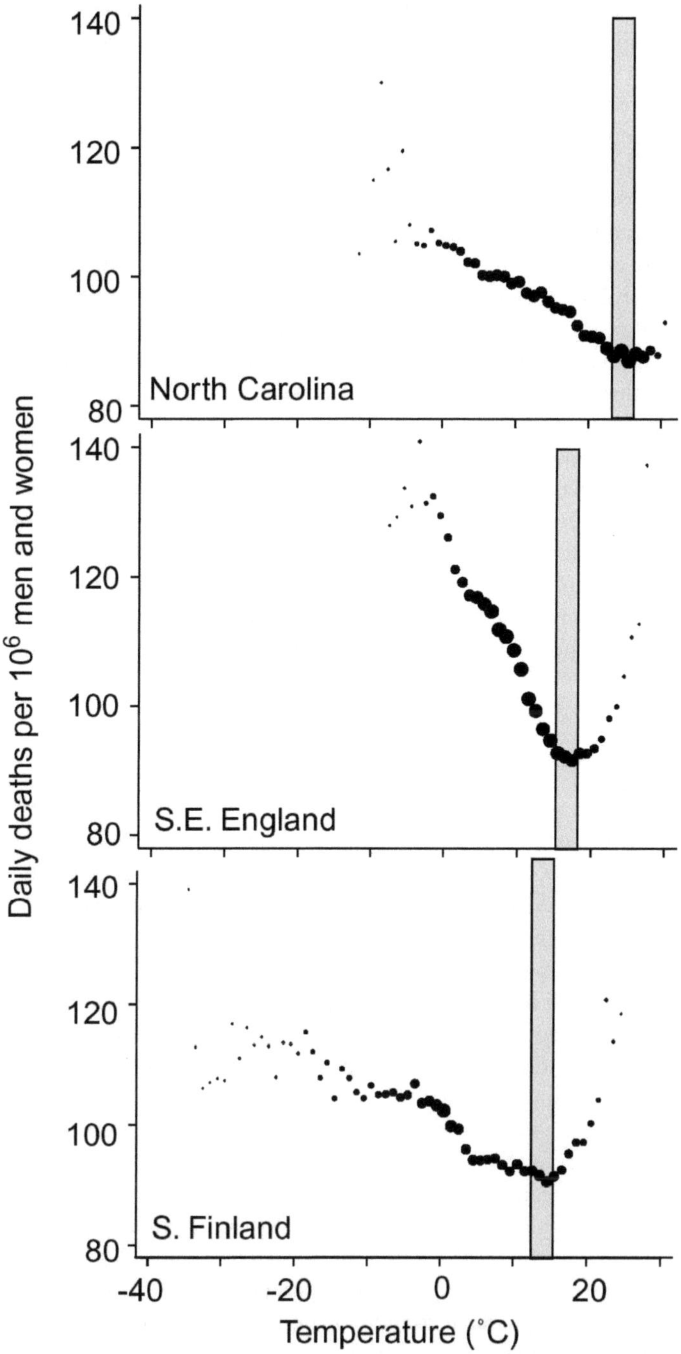

Mortality at different mean daily temperatures. Pooled data for each region at age 55+, 1971 – 1997. The areas of circles are proportional to the number of days at each temperature. (Donaldson GC, Keatinge WR, Nayha S. Changes in summer temperature and heat-related mortality since 1971 in North Carolina, South Finland, and Southeast England. Environmental Research 200; 91(1):1-7 as figure 1)

due to coronary thrombosis, with a peak two days following a cold spell. Some later coronary deaths occur secondarily to respiratory disease. One-forth of excess mortality is caused by respiratory disease, peaking about 12 days after the peak of a cold spell [26]. The variance of winter-time temperatures modifies the effects of very cold days on respiratory mortality [5]. Associations between ischaemic heart disease deaths and respiratory deaths that are seen in the temperature range of 0 - 15 °C, were reversed for temperatures a few degrees below 0 °C, probably due to multifactorial causes [8].

Theories about the mechanisms of cold-related cardiovascular mortality have tended to focus on changes in the circulation and haemostasis that may pre-dispose to thrombotic events [3, 32, 42, 46, 47] as a result of body cooling [14, 41]. Coronary and cerebral mortality are related to a strong increase of blood pressure as a consequence of cold exposure and associated peripheral blood vessels constriction, decreasing blood volume and the activation of thrombogenic factors. This may lead to thrombosis in cerebral and coronary vessels causing deaths. These and respiratory infections explain the majority of winter deaths.

Keatinge and colleagues place more emphasis on personal behaviours, and have argued that much of the excess winter mortality from arterial thrombosis is related to cold exposure from "brief excursions outdoors rather than to low indoor temperature"[43, 25].

The discrepancy of the findings of excess cold mortality associated with the level of environmental cold exposure indicates some other, more direct, causal association. The skin temperatures associated with environmental cold exposure commonly decrease, especially in the extremities. During cold winter months even in well-insulated Nordic homes, foot skin temperatures of the elderly are lower than in summer months. These results indicate that man is commonly cold stressed during cold exposure [28]. Body cooling may act as one mediator of cold environmental temperature offering an explanation for cold related excess mortality better than the different possible causal pathways for cold related excess mortality are widely discussed [4].

Winter mortality

The rate of mortality increases seasonally every winter. On average this excess winter mortality in most European countries varies from 5 % to 30 % [6, 19, 31]. However, within a single country excess winter mortality may vary by as much as 40 % from year to year [12]. Cardiovascular and respiratory causes of death are most strongly linked to seasonal changes in temperature. Elderly people and those with impaired health or suffering from social conditions are the most susceptible to the impact of weather changes [2]. During the last decades there has been a decline in influenza epidemics, which account for less than 5 % of the excess winter mortality in Britain [26]. Countries with mild climates show higher levels of excess winter deaths associated with socio-economic conditions and housing standards as explaining variables [43]. Countries with high levels of income poverty and inequality have the highest coefficient of seasonal variation in mortality [12]. During cold periods the availability of continuous electricity and heating is essential for human well-being. There are numerous examples of accidents in electricity supply and cut-offs in the Russian Federation and many eastern European countries. In the case of heat supply cut-offs, when outdoor temperatures are below 10 °C (this period lasts 3 – 5 months a year in the Northern countries), indoor temperatures fall below 15 °C. For example, during the winter of 2002 – 2003, 20 Russian cities had problems with residential heating systems. More than 1 million people experienced temperature discomfort, which lasted for longer than a month. (Revich, personal communication). Freezing to death and accidental hypothermia resulting in frostbites is extremely rare as a consequence of lack of heating in homes; the rare extreme cases have been observed during the very cold winters in places like Northern Russian Federation and the Baltic States.

Winter rainfall and excess deaths are significantly associated (a regression coefficient of 0.54, $p < 0.001$) [19]. Cold spells increase mortality [8, 39], but their relative importance is unknown. A recent study conducted by Wilkinson et al, (forthcoming) in the United Kingdom, assessed factors for winter excess mortality in the over 75 years old. A relationship between winter mortality and cold was found, however the findings suggest that the risk of excess winter death is more widely distributed within the elderly population than hitherto has been appreciated, and is less related to factors such as access to heating than had been thought. This may limit the potential health benefits of government energy efficiency initiatives that are specifically targeted at low-income households.

Hypothermia and cold-induced injuries

Accidental hypothermia is known to be a hazard to elderly people in temperate and cold climates [37]. Hypothermic deaths are seen as a special consequence of unsatisfactory dwellings. The appearance of outdoor hypothermic deaths is strongly related to the awareness of hypothermia and protection against cold. The typical victim of outdoor hypothermic death in Finland is a resident of an elderly persons' home forgetting his/her way back home from their walking trips during cold spells.

The total injury rate may change as a consequence of both direct or indirect effects of cold. Causal relationships between different injury sources and accident types, the nature of the injury and the degree of the disability sustained from injury, may also have different pathways (➔ *Fig. 2)*.

In the U.S.A. the majority of occupational outdoor cold exposure injuries occur during the few coldest winter days. Wind speed strongly increases the injury rate. Freezing, strains and sprains are commonly represented among the cold exposure injuries. Cold exposure injuries display a strong negative relationship with temperature [40]. Occupational slips and falls exhibit a similar negative correlation with temperature. In the mining industry in the U.S.A. the higher rates are linked to about 0 °C [16]. Unintentional injury occurs least frequently at a temperature of about +20 °C and increases at lower and higher ambient temperatures [36]. Cold environmental temperature is usually a secondary source of injury, rather than a primary one. Cold exposure injuries are reported with much higher frequencies in questionnaire studies than in records, which tend to underestimate this type of injury [34]. Accidental cold exposure occurs mainly outdoors, in socially deprived people, workers, alcoholics, the homeless, the elderly in temperate cold climates [37]. The risk of suffering frostbite increases with age [21].

The onset of frostbite resulting from exposure to cold air appears at an environmental temperature of −11 °C. Wind, high altitude and wet clothing lead to onset of freezing injury at higher environmental temperatures [7]. The incidence of more serious frostbite requiring hospital treatment increases at temperatures of −15 °C and below. Such injuries have been reported to be more common in metropolitan areas than in other types of living resorts in Finland. The risk of suffering frostbite increases with age. [21].

Injury from frostbite is comparable to burns with respect to the immediate consequences and may range in severity from mild to more severe functional limitation of the injured area, to sick leave, or, in some cases, to hospitalization [21]. The latent symptoms of frostbite which are most common include a local hypersensitivity to cold and pain in the injured area, cold-induced sensations and disturbances of muscular function, and excessive sweating. These latent symptoms have been shown to have a negative impact on occupational activities [13]. Cold injuries need to be recognized as one of the health impacts of cold weather.

Fig. 2

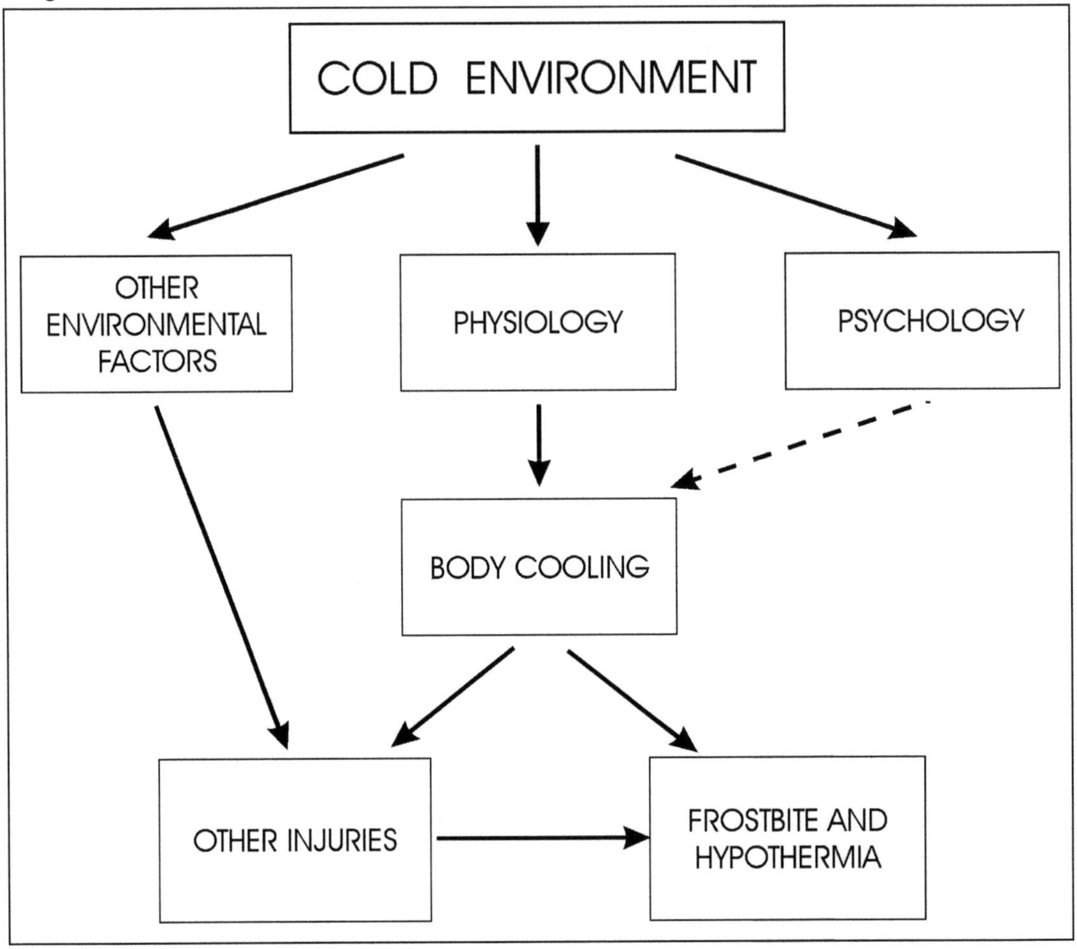

Associations between environmental cold and injuries.

Actions of Public Health for Preventing Cold-Related Health Impacts

Predicting and alerting a city's residents of imminent dangerous weather conditions, especially during the summer, has been practiced in large metropolitan areas of the USA: health-watch warning systems are based on meteorological forecast variables [22], but have been only slightly modified for cold spells.

Warm housing is a key element in the prevention of excess cold mortality [26]. Physical changes in housing design cannot be expected to happen very rapidly and probably do not represent the most effective preventive action to be taken against cold-related excess mortality [24]. Modelling of seasonal variations in mortality and domestic thermal efficiency indicates that low energy efficiency could explain 5 % of excess winter deaths [19].

The responses of people to their thermal environments during the winter months in Norway indicate that they are more cold-stressed at this time of the year [28]. Environmental cold may provoke cardiopulmonary symptoms independent of sex and age [17] and also provoke other types of symptoms [15].

Protection against outdoor body-cooling is one method of preventing excess cold mortality. In northern countries, people are more experienced and successful in using effective cold-protective clothing [43]. In the Nordic countries, some intervention programmes have been established aiming to develop behavioural changes in cold-exposed workplaces, and some methodological tools have been produced for the prevention of cold stress through the actions of health care personnel [18, 20, 29, 30, 38]. They have been used as a base to develop an ISO standard for working practices in cold environments: ISO CD 15743 (2002).

There is a need to define preparative actions for preventing cold-related impacts on health. The definition of public health programmes to prevent cold-related mortality needs further research in order to evaluate: 1) the populations at risk, 2) the lag time of the effect, 3) the effect of cardio-vascular and respiratory morbidity, 4) the role of respiratory mortality, and 5) the significance of other meteorological variables [2]. To prevent cold exposure-related excess mortality, we need collaboration between health care, meteorologists and other officials to produce usable preventive action models. The prevention of cold injuries and illnesses is very much the responsibility of health care providers, requiring practical information, education and professional support.

Conclusions

In the research published in recent years epidemiological evidence has accumulated indicating a causal relationship between mortality and cold weather. Evidence of the importance of cold-related sicknesses and injuries is growing. The importance of cold- related health impacts in the whole of Europe today and in the future is not well recognized in the public health sector. There is a need to evaluate more by research cold induced morbidity, mortality and meteorological variables. Despite some models and standards for prevention of cold stress, we need more collaboration between health care, weather broadcasting and other officials to produce usable preventive action models and their national applications in Europe.

References

1. Aylin P, Morris S, Wakefield J, Grossinho A, Jarup L, Elliott P (2001) Temperature, housing, deprivation and their relationship to excess winter mortality in Great Britain, 1986 – 1996. Int J Epidemiol 30(5):1100 – 1108
2. Ballester F, Michelozzi P, Iñiguez C (2003) Weather, climate, and public health. J Epidemiol Community Health 57(10):759 – 760
3. Bøkenes L, Alexandersen TE, Østerud B, Tveita T, Mercer JB (2000) Physiological and haematological responses to cold exposure in the elderly. Int J Circumpolar Health 59:216 – 21
4. Bøkenes L, Alexandersen TE, Tveita T, Østerud B, Mercer JB (2004) Physiological and hematological responses to cold exposure in young subjects. Int J Circumpolar Health 63(2):115 – 128
5. Braga AL, Zanobetti A, Schwartz J (2002) The effect of weather on respiratory and cardiovascular deaths in 12 U.S. cities. Environ Health Perspect 110(9):859 – 863
6. Curwen M (1990) Excess winter mortality: a British phenomenon? Health Trends 4:169 – 175
7. Danielsson U (1996) Windchill and the risk of tissue freezing. J Appl Physiol 81(6):2666 – 2673
8. Donaldson GC, Keatinge WR (1997) Early increases in ischaemic heart disease mortality dissociated from and later changes associated with respiratory mortality after cold weather in south east England. J Epidemiol Community Health 51(6):643 – 648
9. Donaldson GC, Tchernjavskii VE, Ermakov SP, Bucher K, Keatinge WR (1998) Winter mortality and cold stress in Yekaterinburg, the Russian Federation: interview survey. BMJ 316(7130):514 – 518
10. Donaldson GC, Keatinge WR, Näyhä S (2003) Changes in summer temperature and heat-related mortality since 1971 in North Carolina, South Finland, and Southeast England. Environ Res 91(1):1 – 7

11. Doyle R (1998) Deaths from excessive cold and excessive heat. Sci Am 278(2):16–17
12. Eng H, Mercer JB (1998) Seasonal variations in mortality caused by cardiovascular diseases in Norway and Ireland. J Cardiovasc Risk 5(2):89–95
13. Ervasti O, Hassi J, Rintamäki H, Virokannas H, Kettunen P, Pramila S, Linna T, Tolonen U, Manelius J (2000) Sequelae of moderate finger frostbite as assessed by subjective sensations, clinical signs, and thermophysiological responses. Int J Circumpolar Health 59(2):137–145
14. Goodwin J, Taylor RS, Pearce VR, Read KL (2000) Seasonal cold, excursional behaviour, clothing protection and physical activity in young and old subjects. Int J Circumpolar Health 59:195–203
15. Hassi J, Juopperi K, Remes J, Näyhä S, Rintamäki H (1999) Cold exposure and cold-related symptoms among Finns aged 25–64 years. In: Second International Conference on Human-Environment System; Yokohama, Japan, pp 271–274
16. Hassi J, Gardner L, Hendricks S, Bell J (2000) Occupational injuries in the mining industry and their association with statewide cold ambient temperatures in the USA. Am J Ind Med 38(1):49–58
17. Hassi J, Remes J, Kotaniemi JT, Kettunen P, Näyhä S (2000) Dependence of cold-related coronary and respiratory symptoms on age and exposure to cold. Int J Circumpolar Health 59(3–4):210–215
18. Hassi J, Raatikka V-P, Huurre M(2003) Health-check questionnaire for subjects exposed to cold. Int J Circumpolar Health 62(4):436–443
19. Healy JD (2003) Excess winter mortality in Europe: a cross country analysis identifying key risk factors. J Epidemiol Community Health 57(10):784–789
20. Holmér I (2000) Strategies for prevention of cold stress in the elderly. Int J Circumpolar Health 59(3–4):267–272
21. Juopperi K, Hassi J, Ervasti O, Drebs A, Näyhä S (2002) Incidence of frostbite and ambient temperature in Finland, 1986–1995. A national study based on hospital admissions. Int J Circumpolar Health 61(4):352–362
22. Kalkstein LS, Jamason PF, Greene JSc, Libby J, Robinson L (1996). 1996: The Philadelphia Hot Weather-Health Watch/ Warning System: Development and Application, Summer 1995. Bulletin of the American Meteorological Society 77(7):1519–1528
23. Keatinge WR, Coleshaw SRK, Easton JC, Cotter F, Mattock MB, Chelliah R (1986) Increased platelet and red cell count, blood viscosity, and plasma cholesterol levels during heat stress, and mortality from coronary and cerebral thrombosis. Am J Med 81:795–800
24. Keatinge WR (1986) Seasonal mortality among elderly people with unrestricted home heating. Br Med J (Clin Res Ed) 293(6549):732–733
25. Keatinge WR, Coleshaw SR, Holmes J (1989) Changes in seasonal mortalities with improvement in home heating in England and Wales from 1964 to 1984. Int J Biometeorol 33:71–76
26. Keatinge WR (2002) Winter mortality and its causes. Int J Circumpolar Health 61(4):292–299
27. Kunst AE, Looman CW, Mackenbach JP (1988) Medical care and regional mortality differences within the countries of the European community. Eur J Popul 4(3):223–245
28. Mercer JB (2003) Cold – an underrated risk factor for health. Environ Res 92(1):8–13
29. Mäkinen T, Hassi J, Påsche A, Abeysekera J, Holmér I (2002) Project for developing a cold risk assessment and management strategy for workplaces in the Barents region. Int J Circumpolar Health 61(2):136–141
30. Mäkinen TM, Hassi J (2002) Usability of isothermal standards for cold risk assessment in the workplace. Int J Circumpolar Health 61(2):142–153
31. Näyhä S (2000) Seasonal variation of deaths in Finland – is it still diminishing? Int J Circumpolar Health 59(3–4):182–187
32. Neild PJ, Syndercombe-Court D, Keatinge WR, Donaldson GC, Mattock M, Caunce M (1994) Cold-induced increases in erythrocyte count, plasma cholesterol and plasma fibrinogen of elderly people without a comparable rise in protein C or factor X. Clin Sci (Lond) 86:43–8
33. O'Neill MS, Zanobetti A, Schwartz J (2003) Modifiers of the temperature and mortality association in seven US cities. Am J Epidemiol 15;157(12):1074–1082
34. Pekkarinen A. 1994. Occupational accidents occurring in different physical environments with particular reference to indoor and outdoor work. Doctoral dissertation, Oulu, Finland, p 148
35. Poikolainen K, Eskola J (1988) Health services resources and their relation to mortality from causes amenable to health care intervention: a cross-national study. Int J Epidemiol 17:86–89
36. Ramsey JD, Burford CL, Beshir MY, Jensen RC (1983). Effects of workplace thermal conditions on safe work behavior. J Safety Res 14:105–114
37. Ranhoff AH (2000) Accidental hypothermia in the elderly. Int J Circumpolar Health 59(3–4):255–259

38. Risikko T, Mäkinen TM, Pasche A, Toivonen L, Hassi J (2003) A model for managing cold-related health and safety risks at workplaces. Int J Circumpolar Health 62(2):204 – 215
39. Saez M, Sunyer J, Castellsague J, Murillo C, Anto JM (1995) Relationship between weather temperature and mortality: a time series analysis approach in Barcelona. Int J Epidemiol 24(3):576 – 582
40. Sinks T, Mathias CG, Halperin W, Timbrook C, Newman S (1987) Surveillance of work-related cold injuries using workers' compensation claims. J Occup Med 29(6):504 – 509
41. Smolander J (2002) Effect of cold exposure on older humans. Int J Sports Med 23:86 – 92
42. Stout RW, Crawford VL, McDermott MJ, Rocks MJ, Morris TC (1996) Seasonal changes in haemostatic factors in young and elderly subjects. Age Ageing 25:256 – 8
43. The Eurowinter Group; Keatinge WR, Donaldson GC, Bucher K, Jendritzky G, Cordioli E, Martinelli M, Katsouyanni K, Kunst AE, McDonald C, Nayha S, Vuori I (1997) Mortality from ischaemic heart disease, cerebrovascular disease, respiratory disease, and all causes in warm and cold regions of Europe. Lancet 10;349(9062):1341 – 1346
44. The Eurowinter Group; Keatinge WR, Donaldson GC, Bucher K, Jendritzky G, Cordioli E, Martinelli M, Katsouyanni K, Kunst AE, McDonald C, Nayha S, Vuori I (2000) Winter mortality in relation to climate. Int J Circumpolar Health 59(3 – 4):154 – 159
45. Wargentin P (1767) Uti hvilka Månader flera Människor årligen födas och dö i Sverige. Kungl. Vetenskaps-Academiens Handlingar, pp 249 – 258
46. Woodhouse PR, Khaw KT, Plummer M (1993a) Seasonal variation of blood pressure and its relationship to ambient temperature in an elderly population. J Hypertens 11:1267 – 74
47. Woodhouse PR, Khaw KT, Plummer M (1993b) Seasonal variation of serum lipids in an elderly population. Age Ageing 22:273 – 8

Temperature Regulation, Heat Balance and Climatic Stress

George Havenith

Keywords

thermoregulation, heat, cold, comfort, vapour, age, stress, strain, mortality, morbidity

Abstract

This paper discusses human thermoregulation and how this relates to health problems during exposure to climatic stress. The heat exchange of the body with the environment is described in terms of the heat balance equation which determines whether the body heats up, remains at stable temperature, or cools. Inside the body the thermoregulatory control aims at creating the right conditions of heat loss to keep the body temperature stable. In the heat the main effector mechanism for this is sweating. The heat balance is affected by air temperature, radiant temperature, humidity and wind speed as climatic parameters and by activity rate, clothing insulation, and sweat capacity as personal parameters. Heat tolerance is discussed in the light of personal characteristics (age, gender, fitness, acclimatisation, morphology and fat) indicating age and fitness as most important predictors. Heat related mortality and morbidity are strongly linked to age.

1 Introduction

Recent extreme weather events have been linked to increased morbidity and mortality. Longer periods of hot weather, especially when little relief is given at night have hit mainly the older population. In order to understand the link between the climate, the stress it poses for the human, and the way the physiological strain experienced by the human is linked to age and other personal characteristics (e.g. a higher mortality was observed in females than in males) this chapter will provide some background knowledge on the parameters relevant to heat stress and strain.

In the evolutionary sense, man is considered a tropical animal. Our anatomy as well as our physiology is geared towards life in moderate and warm environments. There, we can maintain our bodily functions, especially thermoregulation, without artificial means. The goal is to keep the body temperature within acceptable limits and the success of the effector actions will very much depend on the climate conditions and the person's clothing and work load. The interaction of the person with the climate is represented by the heat balance equation, which will be discussed in detail. How the person reacts is governed by the human thermoregulatory system. Finally, individual characteristics of the person will affect his or her ability to thermoregulate and affect the risk of heat or cold stress. Problems, in terms of morbidity and mortality, occur when thermoregulation is impaired in conditions of high levels of heat or cold stress.

2 Temperature Regulation

In a neutral climate, at rest, the human body regulates its temperature around 37 °C. This is by no means an exactly fixed temperature for all humans. Over a population, when measured in the morning after bed-rest, the mean will be around 36.7 °C, with a standard deviation of 0.35 °C (calculated from data of Wenzel and Piekarski (1984)). During the day, the temperature will increase (typically by about 0.8 °C), peaking in the late evening, and declining again until early morning due to the circadian rhythm. Also, exercise will cause an increase in body temperature; with temperatures around 38 °C typical for moderate work and values up to 39 °C and occasionally above 40 °C for heavy exercise (e.g. marathon). Short term increases up to 39 °C are seldom a problem to the body and should be considered a normal phenomenon in thermoregulation of a healthy person.

In fever, an increase in body temperature is observed as well. This increase differs from that in exercise in that the increase in temperature due to fever is defended by the body, whereas the increase in temperature induced by exercise is not. Thus, when a fever of 38.5 °C is present, cooling the body will lead to activation of heat conservation mechanisms by the body (shivering, vasoconstriction) to keep the temperature at that level. In exercise, the body would continue sweating until the body temperature is back to neutral levels.

Fig. 1

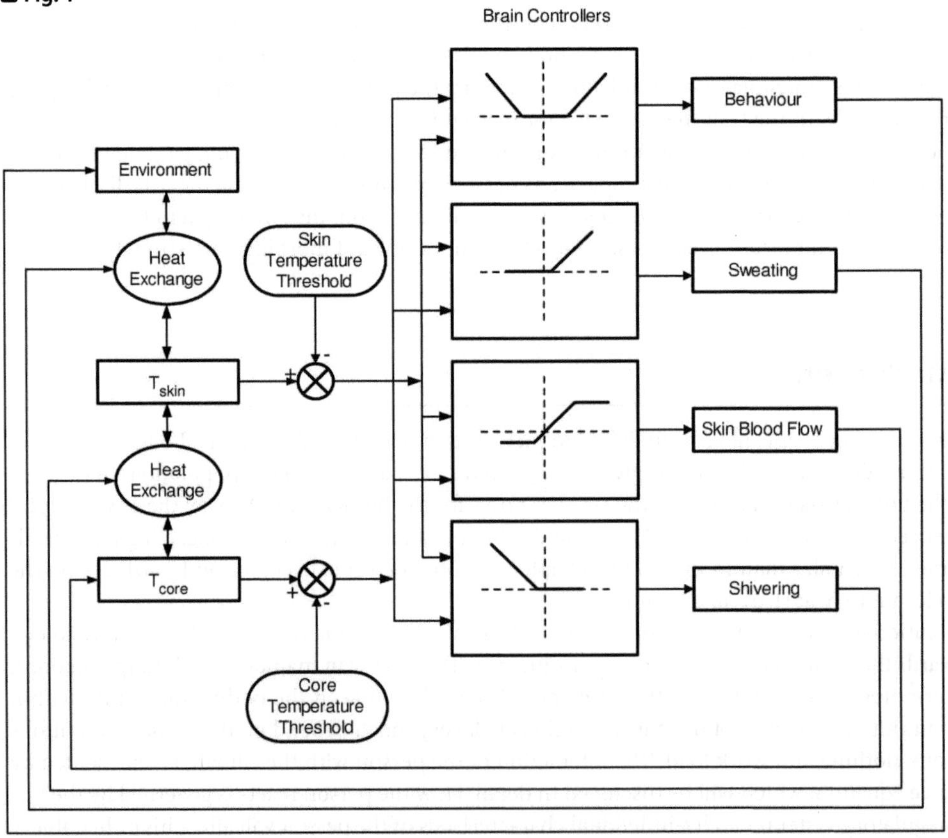

Schematic representation of the thermoregulatory control system. Tcore = body core temperature; Tskin = mean skin temperature; brain controller graphs show reaction of effector (Y-axes) to error signal (x-axes) (copyright G.Havenith, 2002)

An example of how the body's temperature regulation could be represented is given in *Figure 1*. Here we have a body, which is represented by a body core temperature and by a skin temperature. Afferent signals representing these body temperatures are relayed to the control centres in the brain. There they are compared to a reference signal, which could be seen as a single thermostat setpoint, or as a number of thresholds for initiating effector responses. Based on the difference between actual temperature and the reference value (the error signal), various effector responses can be initiated. The main ones are sweating and vasodilation of skin vessels (if body temperatures are higher than the reference, i.e. a positive error) and shivering and vasoconstriction (negative error). Sweat evaporation will cool the skin, shivering will increase heat production and heat the core, and vasodilation and constriction will regulate the heat transport between core and skin.

Of course this is a simplified model, as many different thermosensitive regions of the body have been identified, and many different and more complex models are possible.

When we think of clothing we can think of it as an additional, behavioural effector response. If we can freely choose our clothing, which often is not the case due to cultural or work restraints, we adjust our clothing levels to provide the right amount of insulation to allow the other effector responses to stay within their utility range. The main effect of clothing will be its influence on the heat exchange between the skin and the environment. To understand these effects we will need to analyse the heat flows that exist between the body and its environment, in other words we have to look at the body's heat balance.

3 Heat Balance

Normally the body temperature is quite stable. This is achieved by balancing the amounts of heat produced in the body with the amounts lost.

! Stability: Heat Production = Heat Loss

In *Figure 2*, a graphical representation of all the heat inputs to and outputs from the body is presented (Havenith, 1999).

Heat production is determined by metabolic activity. When at rest, this is the amount needed for the body's basic functions, e.g. respiration and heart function to provide body cells with oxygen and nutrients. When working however, the need of the active muscles for oxygen and nutrients increases as does the metabolic activity. When the muscles burn these nutrients for mechanical activity, part of the energy they contain may be liberated outside the body as *external work*, but most of it is released in the muscle as heat. The ratio between this external work and the energy consumed is called the *efficiency* with which the body performs the work. This process is similar to what happens in a car engine. The minor part of the fuel's energy is actually effective in the car's propulsion, and the major part is liberated as waste heat. The body, as the car engine, needs to get rid of this heat; otherwise it will warm up to lethal levels. As an example: if no cooling would be present, a person working at moderate levels (metabolic rate 450 Watt) would show an increase in body temperature around 1 °C every 10 minutes.

For most tasks, as e.g. walking on a level, the value for the external work (energy released outside the body) is close to zero. Only the heat released by friction of shoes etc. is released outside the body, whereas all other energy used by the muscles ends up as heat within the body. In the cold, additional heat is produced by shivering: muscle activity with zero efficiency. The basal metabolic rate and heat production can be increased up to fourfold in this way.

For *heat loss* from the body, between skin and environment, several pathways are available. For each pathway the amount of transferred heat is dependent on the driving force (e.g. temperature or vapour

Fig. 2

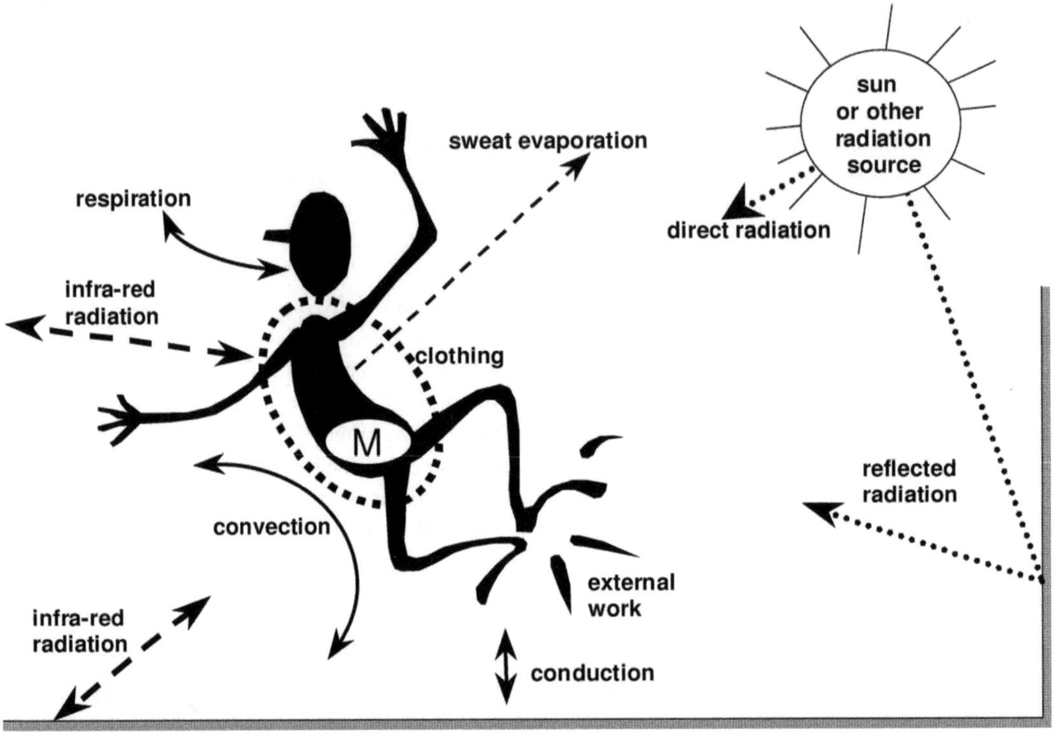

Schematic representation of the pathways for heat loss from the body. M = metabolic heat production (reproduced with permission, Havenith, 1999)

pressure gradient), the body surface area involved and the resistance to that heat flow (e.g. clothing insulation).

! Heat Loss = gradient x surface area / resistance

A minor role is taken by *conduction*. Only for people working in water, in special gas mixtures (prolonged deep-sea dives), handling cold products or in supine positions, does conductivity becomes a relevant factor.

More important for heat loss is *convection*. When air flows along the skin, it is usually cooler than the skin. Heat will therefore be transferred from the skin to the air around it. Heat transfer through electromagnetic (mainly infra-red) *radiation* can be substantial. When there is a difference between the body's surface temperature and the temperature of the surfaces in the environment, heat will be exchanged by radiation.

Finally, the body possesses another avenue for heat loss, which is heat loss by *evaporation*. Any moisture present on the skin (usually sweat) can evaporate, with which large amounts of heat can be dissipated from the body.

Apart from convective and evaporative heat loss from the skin, these types of heat loss also take place from the lungs by *respiration*, as inspired air is usually cooler and dryer than the lung's internal surface. By warming and moisturising the inspired air, the body loses an amount of heat with the expired air, which can be up to 10 % of the total heat production.

For body temperature to be stable, heat losses need to balance heat production. If they do not, the body heat content will change, causing body temperature to rise or fall. This balance can be written as:

Store = Heat Production – Heat Loss
= (Metabolic Rate – External Work) –
(Conduction + Radiation + Convection + Evaporation + Respiration)

Thus if heat production by metabolic rate is higher than the sum of all heat losses, Store will be positive, which means body heat content increases and body temperature rises. If Store is negative, more heat is lost than produced. The body cools.

It should be noted that several of the "heat loss" components might in special circumstances (e.g. ambient temperature higher than skin temperature) actually cause a heat gain, as discussed earlier.

4 Relevant parameters in heat exchange

The capacity of the body to retain heat or to lose heat to the environment is strongly dependent on a number of external parameters (Havenith, 2004):

4.1 Temperature

The higher the air temperature, the less heat the body can lose by convection, conduction and radiation (assuming all objects surrounding the person are at air temperature). If the temperature of the environment increases above skin temperature, the body will actually gain heat from the environment instead of losing heat to it. There are three relevant temperatures:

Air temperature. This determines the extent of convective heat loss (heating of environmental air flowing along the skin or entering the lungs) from the skin to the environment, or vice versa if the air temperature exceeds skin temperature.

Radiant temperature. This value, which one may interpret as the mean temperature of all walls and objects in the space where one resides, determines the extent to which radiant heat is exchanged between skin and environment. In areas with hot objects, as in steel mills, or in work in the sun, the radiant temperature can easily exceed skin temperature and results in radiant heat transfer from the environment to the skin.

Surface temperature. Apart from risks for skin burns or pain (surface temperature above 45 °C), or in the cold of frostbite and pain, the temperature of objects in contact with the body determines conductive heat exchange. Aside from its temperature, the object's properties, as e.g. conductivity, specific heat and heat capacity, are also relevant for conductive heat exchange.

4.2 Air humidity

The amount of moisture present in the environment's air (the moisture concentration) determines whether moisture (sweat) in vapour form flows from the skin to the environment or vice versa. In general the moisture concentration at the skin will be higher than in the environment, making evaporative heat loss from the skin possible. As mentioned earlier, in the heat, evaporation of sweat is the most important avenue for the body to dissipate its surplus heat. Therefore situations where the gradient is reversed (higher moisture concentration in environment than on skin) are extremely stressful (condensation on skin) and allow only for short exposures. It should be noted that the moisture concentration, not the relative humidity, is the determining factor. The definition of relative humidity is the ratio between the actual amount of moisture in the air and the maximal amount of moisture air can contain at that temperature (i.e. before condensation will occur). Air that has a relative humidity of 100 % can thus contain different amounts of moisture, depending on its temperature. The higher the temperature, the higher the moisture content at equal relative humidities. Also: at the same relative humidity, warm air will contain more moisture (and allow less sweat evaporation) than cool air. The fact that relative humidity is not the determining factor in sweat loss can be illustrated with two examples:

- In a 100 % humid environment the body can still evaporate sweat, as long as the vapour pressure is lower than that at the skin. So, at any temperature below skin temperature the body can evaporate sweat, even if the environment is 100 % humid.
- Sweat evaporation (at equal production) will be higher at 21 °C, 100 % relative humidity (vapour pressure is 2.5 kPa) than at 30 °C, with 70 % relative humidity (vapour pressure = 3 kPa) as the vapour pressure gradient between wet skin (vapour pressure 5.6 kPa) and environment at 21 °C is 5.6 – 2.5 = 3.1 kPa, whereas at 30 °C and, 70 % relative humidity it is only 5.6 – 3 = 2.6kPa.

4.3 Wind speed

The magnitude of air movement effects both convective and evaporative heat losses. For both avenues, heat exchange increases with increasing wind speed. Thus in a cool environment the body cools faster in the presence of wind, in an extremely hot and humid environment, it will heat up faster. In very hot, but dry environments the effects on dry and evaporative heat loss may balance each other out.

4.4 Clothing insulation

Clothing functions as a resistance to heat and moisture transfer between skin and environment. In this way it can protect against extreme heat and cold, but at the same time it hampers the loss of superfluous heat during physical effort. For example, if one has to perform hard physical work in cold weather clothing, heat will accumulate quickly in the body due to the high resistance of the clothing for both heat and vapour transport.

In cases where no freedom of choice of clothing is present, clothing may increase the risk of cold or heat stress. Examples are cultural restrictions (long, thick black clothes for older females in southern European countries) or work requirements (protective clothing; dress codes).

The environmental range for comfort, assuming no clothing or activity changes are allowed, is quite narrow. For light clothing with low activity levels it is around 3.5 °C wide (ISO7730, 1984). In order to widen this range, one has to allow for behavioural adjustments in clothing and activity. An increase in activity level will move the comfort range to lower temperatures, as will an increase in clothing insulation. E.g. an increase in metabolic rate of 20 Watts (resting levels are 100 – 160 Watts) pushes the comfort range down by approximately 1 °C, as does an increase in clothing insulation of 0.2 clo (clo is a unit for clothing insulation. For reference, a three-piece business suit is 1 clo; long trousers and short sleeved shirt around 0.6 clo). An increase in air speed will push the comfort range up (1 °C for 0.2 $m.s^{-1}$), i.e. allow for comfort at higher ambient temperatures.

5 Heat Tolerance and Individual Differences

When a group of people is exposed to a heat challenge (e.g. heat wave, or working in the heat), their body temperature will increase, but not to the same extent for all. Where some may experience extreme heat strain, others may not show any sign of strain at all. Knowledge of the mechanisms behind these differences is important for risk assessment during climatic extremes, for health screening and for selection of workers for specific stressful tasks. In order to get an insight into the relevant mechanisms, factors which may influence the response of an individual to heat exposure are discussed here and include: aerobic power, acclimation state, morphological differences, gender, use of drugs and age (Havenith, 1985, 2001 a, b, c, Havenith et al. 1995 b).

Aerobic power (fitness)

Body core temperature during work is related to the work load (metabolic rate), relative to the individuals' maximal aerobic power (usually expressed as the % of the maximal oxygen uptake per minute; $\%VO_{2\,max}$). In addition, a high aerobic power is typically associated with improved heat loss mechanisms (higher sweat rate, increased skin blood flow). The $VO_{2\,max}$ of a subject is thus inversely related to the heat strain of a subject in the heat, mainly through its beneficial effect on circulatory performance. The higher the aerobic power, the higher the circulatory reserve (the capacity for additional increase in cardiac output) when performing a certain task. Further, aerobic power is often confounded with acclimatisation as training can result in an improvement of the acclimatisation state. This is caused by the regular increase in body temperature, which normally occurs with exercise. The result is a reduction of heat strain in warm climates. This effect works both through circulation and through improved sweat response.

The only condition where a high $VO_{2\,max}$ is not beneficial is for conditions where heat loss is strongly limited (Hot, Humid), while work rate is related to the workers work capacity (same percentage of $VO_{2\,max}$). In this case, the fitter person will work harder, thus liberating more heat in his body than his unfit companion will. As he or she cannot get rid of the heat, they thus will heat up faster than the slower working unfit person will.

Acclimatisation state

A subject's state of acclimatisation appears to be of great influence on his reaction to heat stress. With increasing acclimatisation state, the heat strain of the body will be strongly reduced, resulting in lower core temperature and heart rate during a given exercise in the heat (Havenith, 2001c). This is related to improved sweat characteristics (setpoint lower and gain higher), better distribution of sweating over the body and higher efficiency of sweating (higher evaporated/produced ratio) and improved circulatory stability (better fluid distribution, faster fluid recruitment from extra-cellular space, reduced blood pressure decrease) during exercise in the heat. The individual state of acclimatisation can be changed by regular heat exposures, e.g. due to seasonal changes in the natural climate or by heat acclimation due to regular artificial (e.g. climatic chamber) heat exposure. Subjects living in the same climatic conditions can differ in acclimatisation status, however, mainly due to the above-mentioned difference in regular training activity. The combination of heat and exercise induces most optimal acclimatisation. Not all subjects acclimatise equally well: some subjects do not show acclimatisation effects at all when exposed to heat regularly.

Morphology and fat

Differences in body size and body composition between subjects affect thermoregulation through their effect on the physical process of heat exchange (insulation, surface/mass ratio) and through differences in the body weight subjects have to carry (Havenith, 1985, 2001). Body surface area determines the heat exchange area for both dry (convective and radiative) and evaporative heat and thereby affects reactions to heat stress. A high surface area is therefore usually beneficial. Body mass determines metabolic load when a subject is involved in a weight-bearing task like walking. This implies that mass correlates positively with heat production. Body mass also determines body heat storage capacity. This is relevant with passive heat exposures, or when heat loss is limited and body temperature increase is determined by storage capacity.

The effect of body fat content is somewhat confounded with that of body mass: Body fat presents a passive body mass, which affects metabolic load during weight-bearing tasks. I.e. a high fat content increases the metabolic load during activity.

In the cold, subcutaneous fat determines the physical insulation of the body (conductivity of muscle = 0.39, fat = 0.20 $Wm^{-1}\,°C^{-1}$). However, as the fat layer is well perfused by blood flowing to the skin in warm conditions, it is not expected to hamper heat loss during heat exposure. Further, as the specific heat of body fat is about half that of fat free body tissue, people with equal mass but higher fat content will heat up faster at a certain storage rate. With extreme obesity, cardiac function is reduced, which also leads to reduced heat tolerance.

Gender

When investigating the effects of subject's gender on heat stress response, investigators found that females had higher core temperatures, skin temperature, heart rates, blood pressure, and setpoints for sweating, in comparison to males. Thus on a population level, women appear to be less tolerant to heat than men, which seems to be reflected in mortality numbers during heat waves. However, on a population level females differ also in many physical characteristics from men, which may confound the gender issue in thermoregulation research. A more precise evaluation of the gender effect was described by authors, who compared gender groups which were matched in many other characteristics ($VO_{2\,max}$, %fat, size; Havenith, 1985, Havenith et al. 1995b). They observed in these matched groups that gender differences in thermoregulation are minimal, and that some of these differences are climate specific (females perform better in warm, humid; worse in hot dry climates). On a population basis however, females clearly perform worse than men and, if exercising at the same level as men run a higher risk for heat illness.

Two specific female processes do effect thermoregulation: the menstrual cycle and menopause. The effect of the menstrual cycle at rest (a higher core temperature in the postovulatory phase) is almost absent during exercise and or heat exposure, however. Others found that existing male-female differences during exercise under heat stress disappeared with acclimatization, or that they were completely absent from the beginning.

In addition, the effect of menopause on thermoregulation during heat exposure has been studied. Postmenopausal hot flashes and night sweating provide anecdotal evidence that thermoregulation is affected by oestrogen withdrawal. At equal stress levels higher core temperatures were observed in postmenopausal women compared to young females with equal aerobic power levels. Acute oestrogen replacement therapy reduces cardiovascular and thermoregulatory strain in postmenopausal women.

Hypertension

Studies of the role of hypertension in heat tolerance showed reduced circulatory performance in hypertensives compared to normotensives. Though no differences between groups were present in heart rate and core temperature response, reduced forearm blood flows were observed in hypertensives as well as reduced stroke volumes and cardiac output. This indicates a reduced heat transport capacity from body core to the skin and thus an increased risk of overheating. Sweat rate was higher in the hypertensives, however, which apparently compensated for the circulatory differences. It is difficult to estimate whether this compensation will also be effective in other climatic or work conditions. A certain increased risk seems present.

Drugs

Use of drugs such as alcohol may predispose subjects to heat illness by changes in physiological effector mechanisms and by changes in behaviour. Reviews list drugs that are potentially harmful in heat exposure. The relevant drugs have mainly effects on the body fluid balance, vasoconstrictor/dilator activity and on cardiac function. These include: alcohol, diuretics, anti-cholinergic drugs, vasodilators, anti-histamines, muscle relaxants, atropine, tranquillisers and sedatives, ß-blockers and amphetamines. Especially antihypertensive drugs deserve attention because of their widespread use.

Age

With advancing age our ability to thermoregulate tends to decrease (Havenith et al, 1995b; Inoue et al, 1999; Kenney and Havenith, 1993). This is a multi-factorial process involving many of our physiological systems with an emphasis on the cardiovascular system. The most important factor is that physical fitness tends to decrease with age (Åstrand and Rodahl, 1970; Havenith et al, 1995b) mostly due to a reduced physical activity level in the elderly (DTI, 1999). This implies that any activity performed becomes relatively more stressful with advancing age. It will put more strain on the cardiovascular system, and leave less cardiovascular reserve. The cardiovascular reserve is especially relevant to the capacity for thermoregulation as it determines the capacity to move heat for dissipation from the body core to the skin by the skin blood flow. The fitness reduction with age can work like a vicious circle as the increased strain experienced with activity may in itself promote even further activity reduction. Due to a reduced activity level, people also tend to expose themselves less to physical strain in the form of heat or cold exposure. This leads to a loss of heat and cold acclimatisation (Havenith, 1985), which will result in higher strain when the elderly are on occasion exposed to extreme climates. Typically, on a population level these and other changes lead to reduced muscle strength, reduced work capacity, a reduced sweating capacity, a reduced ability to transport heat from the body core to the skin, and a lower cardiovascular stability (blood pressure) in the elderly. These effects will put elderly people at a higher risk in extreme conditions, leading to an increase in morbidity and mortality.

Apart from the physiological changes mentioned above, the percentage of people with illnesses and disabilities increases with age as well. In the UK 41 % of people aged 65 – 74 and 52 % over 75 reported that their lifestyle was limited by an illness or disability, compared to 22 % of all age groups (DTI, 1998). This also has consequences for well being in various thermal environments.

6 Mortality and morbidity with age

Of the individual characteristics age seems to be the best predictor of mortality increases at high temperatures. As discussed above, this ageing effect is likely to be a combined effect of changes in all physiological systems with age, many of which can be avoided by staying fit. The main underlying factors with age are reducing fitness levels, which reduce the spare capacity of the physiological systems to deal with the heat, the increasing health problems in other areas (e.g. hypertension) and the concomitant increase in use of drugs.

In order to provide an overview of the problems related to ageing and temperature regulation, we will now discuss those aspects of mortality and morbidity that in the past have been associated with high or low temperatures. The main ones are heat stroke, hypothermia, cardiovascular and cerebral stroke, and in-home accidents as falls.

Statistical evidence from heat waves in American and Japanese cities shows that mortality increases dramatically with age when temperature increases. Figure 3 shows data on the number of deaths due to heat stroke by age group for the period of 1968 through 1994 in Japan (Nakai et al., 1999). It is evident that there is a sharp increase in deaths due to heat above the 6th decade. The peak daily temperatures when these heat strokes occurred were typically above 38 °C, although a relatively increased heat related mortality in the elderly compared to younger groups has also been observed at lower temperatures. Data from the United States provide a similar picture, with typically above 60 % of heat related deaths during heat waves above the 6th decade of age. While heat stroke deaths among those aged below 64 years were typically exertion induced heat strokes (outdoor sports or occupational hazards), this was not the case in the elderly. As mentioned, they have a very limited cardiovascular reserve, have a reduced sweating capacity, both putting them at higher risk to develop hyperthermia. Apart from heat stroke, the high cardiovascular strain may also set them up for cardiovascular or cerebral stroke.

The analysis of mortality and morbidity data for cold exposure is more complex (Jendritzky et al., 2000), as many cold related problems may not be attributed to the cold in statistics. Only extreme hypothermia cases may be registered as such, but mild hypothermia and in general excursions into the cold can have a severe impact on health too. Suggested causes are the raised blood pressure and the induced haemoconcentration (Donaldson, et al., 1997) which could put additional strain on the cardiovascular system and which may set the elderly up once more for cardiovascular or cerebral stroke.

➲ *Figure 3 and 4*

Another problem in the elderly is the incidence of falls. The number of falls increases dramatically with age as can be seen in figure 4. At the age of 84, the number of falls is over 60 percent higher than at the age of 65. At the same time the size of this age group is much smaller than that of 65. There is no clear statistical evidence that these falls are related to temperature however. This lack of statistical significance is most likely due to the large number of confounding factors involved in these falls. A relation with cold is however likely. Cold reduces muscle force, and increases stiffness of joints and tendons (Havenith et al., 1995a). Hence, in the elderly where fitness and muscle force are already reduced, cold will aggravate these problems. Further, people will wear more clothing in the cold, increasing the clothing stiffness and thus decreasing the freedom of movement (Havenith, 1999).

Conclusions

In summary the thermal strain of the body will be affected by climatic parameters such as temperature, the air's moisture content, wind speed, and radiation levels. In good health the body can deal well with heat

■ Fig. 3

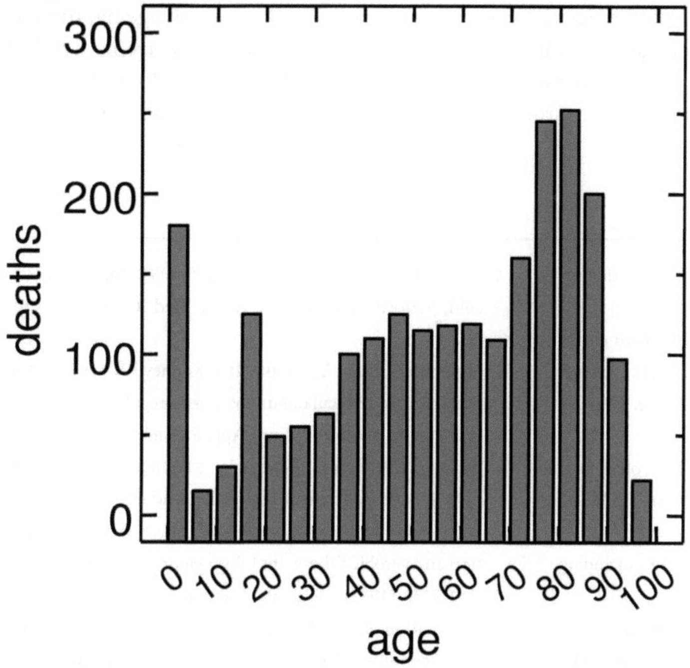

Deaths due to heat stroke by age group from 1968 to 1994 in Japan. Data from Nakai et al, 1999

■ Fig. 4

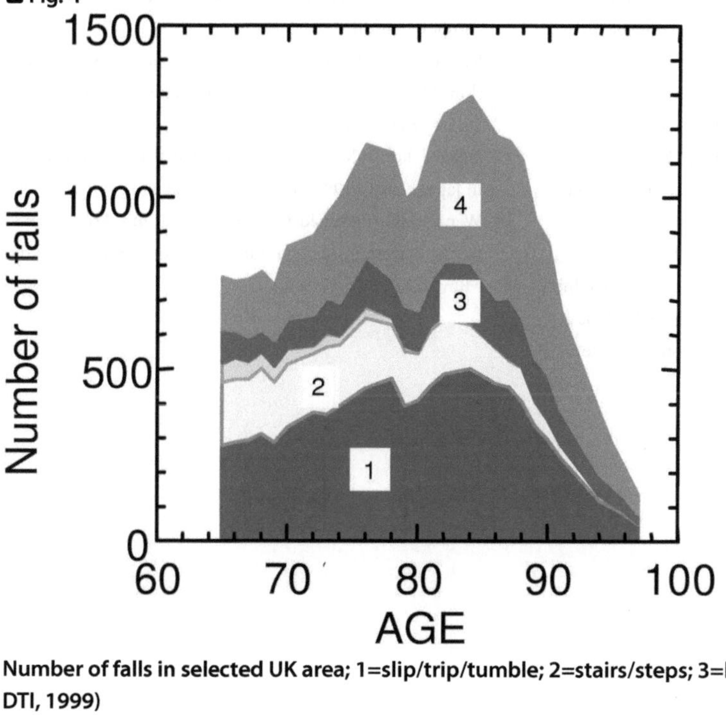

Number of falls in selected UK area; 1=slip/trip/tumble; 2=stairs/steps; 3=level; 4=unspecified (Data from DTI, 1999)

and cold stress, but when thermoregulation becomes impaired, e.g. by inappropriate clothing and activity levels, or by a reduced efficiency of the thermoregulatory system as typically occurs with ageing, the human is at risk, which is reflected in increased mortality and morbidity numbers during extreme weather events. The increased mortality with increasing age reflects a change in a number of physiological systems that are observed with age. These changes lead to a reduced capacity to deal with external stressors.

References

1. Åstrand PO, Rodahl K (1970) Textbook of work physiology. McGraw-Hill, New York
2. Donaldson, GC, Robinson D, Allaway SL (1997) An analysis of arterial disease mortality and BUPA health screening in men, in relation to outdoor temperature. Clin Sci 92:261 – 268
3. DTI (1998) Avoiding slips, trips and broken hips; older people in the population. Fact sheet Department of Trade and Industry
4. DTI (1999) Home accidents surveillance system, including leisure activities. HASS 22nd annual report
5. Havenith G (1985) Individual differences in thermoregulation; a review. Report of TNO-Institute for Perception 1985-C26
6. Havenith G (1999) Heat Balance When Wearing Protective Clothing. Annals of Occup Hygiene 43(5):289 – 296
7. Havenith G (2001a) Human surface to mass ratio and body core temperature in exercise heat stress – a concept revisited. Journal of Thermal Biology 26;4 – 5:387 – 393
8. Havenith G (2001b) Temperature regulation in the elderly – improving comfort and reducing morbidity and mortality using modern technology. Gerontechnology 1:41 – 49
9. Havenith G (2001c) Individualized model of human thermoregulation for the simulation of heat stress response. Journal of Applied Physiology 90:1943 – 1954
10. Havenith G (2004) Thermal Conditions Measurements. In: Stanton N, Hedge A, Hendrick HW, Salas E, Brookhuis K (eds) Handbook of Human Factors and Ergonomics Methods. Taylor and Francis, London
11. Havenith G, Heus R, Daanen HAM (1995a) The hand in the cold, performance and risk. Arctic Med Research 54(suppl.2)1 – 11
12. Havenith G, Inoue Y, Luttikholt V, Kenney WL (1995b) Age predicts cardiovascular, but not thermoregulatory, responses to humid heat stress. Eur J Appl Physiol 70:88 – 96
13. Inoue Y, Havenith G, Kenney WL, Loomis JL, Buskirk ER (1999) Exercise- and methylcholine-induced sweating responses in older and younger men: effect of heat acclimation and aerobic fitness. Int J Biometeor 42(4):210 – 216
14. ISO 7730 (1984) Moderate Thermal Environments – Determination of the PMV and PPD indices and specification of the conditions for thermal comfort. International Standardisation Organisation, Geneva
15. Jendritzky G, Bucher K, Laschewski G, Walther H (2000) Atmospheric heat exchange of the human being, bioclimate assessments, mortality and thermal stress. Int J of Circumpolar Health 59:222 – 227
16. Kenney WL, Havenith G (1993) Heat stress and age: skin blood flow and body temperature. J Therm Biol 18(5/6):341 – 344
17. Nakai S, Itoh T, Morimoto T (1999) Deaths from heat stroke in Japan: 1968 – 1994. Int J Biometeorol 43:124 – 127
18. Wenzel HG, Piekarski C (1984) Klima und Arbeit, Bayrisches Staatsministerium für Arbeit und Sozialordnung, Munich

Health Impact of the 2003 Heat-Wave in France

Stéphanie Vandentorren[1] · *Pascal Empereur-Bissonnet*[1]

Abstract

An unprecedented heat-wave struck France in early August 2003, associated to high levels of air pollution. The meteorological event was accompanied by an excess of mortality that started early and rose quickly. Between August 1st and 20th, the excess of deaths reached 14,802 cases in comparison to the average daily mortality in the 2000 – 2002 period. It represents +60 % of mortality for all causes. The observed excess of mortality first affected the elderly (+70 % for 75 years-old and more), but was also severe for the 45 – 74 year olds (+30 %). In all age groups, females mortality was 15 to 20 % higher than male. Almost the whole country was affected by the excess-mortality, however its intensity varied significantly from one region to another and was at a maximum in Paris and suburbs (+142 %). The excess mortality clearly increased with the duration of extreme temperatures. With regard to the location, the highest mortality rate affected nursing homes where the number of deaths observed was twice the expected number. Following the descriptive studies carried out immediately after the heat-wave, two case-control surveys were carried out. The first study was conducted to identify individual risk factors (way of life, medical history, self sufficiency) and environmental factors (housing) in elderly people living at home. The second one was conducted to identify individual risk factors (autonomy/handicap, medical condition, drug consumption) and environmental risk factors (number and quality of personnel available; facility size and characteristics; prevention plans and therapeutic protocols) for elderly residing in a nursing home. This survey was made in two parts: a "facility case-control study" and an "individual case-control study". High levels of photochemical air pollution were associated to the heat-wave. A study was conducted to estimate the fraction attributable to ozone in the excess risks of mortality jointly related to temperature and ozone, and also to identify a decrease of expected mortality in the weeks following the heat-wave. In cities having experienced the highest excess risk of mortality (Paris, Lyon) the contribution of ozone was minor relative to temperature; the relative part of this air pollutant was higher but variable in cities where the excess risk of mortality was low. The study did not show a harvesting effect within the three weeks following the heat-wave. The French Heat-Wave National Plan, developed immediately after the 2003 event, includes a Heat Health Watch Warning System operating from 2004 and covering the whole country.

1 An exceptional heat-wave

An heat-wave struck France in early August 2003 after a warm month of June with temperatures 4 to 5 °C above seasonal averages, and a month of July closer to normality except high heat during the last two weeks. Unusual periods of high temperatures were observed in France between August 4th and August 12th, 2003. Throughout the country, 2/3 of the meteorological stations recorded temperatures above

[1] for the Heat-Wave Workgroup of the French Institute of Public Health (Institut de Veille Sanitaire, InVS, France)

35 °C, sometimes above 40 °C in 15 % of the main French cities. In Paris, the temperature exceeded 35 °C for as long as 10 days, including 4 days in a row between August 8th and 11th, 2003, a situation never observed since 1873.

On August 11th and 12th, 2003, a minimal temperature of 25.5 °C was recorded in Paris during the night. This is the absolute record for minimal temperatures ever measured in Paris (source: French Weather Bureau – Meteo France). The French heat-wave was exceptional and the warmest year over the last 53 years in terms of minimal, maximal and average temperatures, and in terms of duration. In addition, the high temperatures and the sunshine causing the emission of pollutants significantly increased the atmospheric ozone level.

2 First descriptive studies

2.1 Mortality assessment during the heat-wave

The heat-wave of August 2003 had a severe impact on public health in France, with a mortality largely affecting the elderly people (1). On August 13th, 2003, data allowed the conclusion that a large-scale outbreak was to happen. An excess mortality of 3000 cases was estimated by the InVS based on the data provided by the Pompes Funèbres Générales (France's largest Funeral Parlour). A more complete analysis of the excess mortality for August 2003 was then conducted by The French Institute of Health and Medical Research (INSERM: Institut national de la santé et de la recherche médicale) (2), based on the death certificates passed on by the town councils to the county health offices (DDASS: Directions départementales des affaires sanitaires et sociales).

By comparing the number of deaths registered between August 1st and 20th, 2003 in France (41,621 reported deaths) and the expected number of deaths (i.e. estimated on the basis of the mortality in 2000, 2001 and 2002) for the same period (26,819 expected deaths), the excess mortality was estimated at 14,802 additional deaths. This is equivalent to a total mortality increase of 55 % between August 1st and 20th, 2003. The mortality increase was particularly high for the elderly (+70 % in people above 75 years old). It was different by sex (+40 % in men and +60 % in women). With regard to the location, the mortality rate was maximal in nursing homes where the number of deaths observed was twice the expected number.

Almost the whole country was affected by the excess mortality, however its intensity varied significantly from one region to another (➲ *Fig. 1*): +20 % in Languedoc-Roussillon (South), +130 % in Ile-de-France (Paris and suburbs). The excess mortality clearly increased with the duration of extreme temperatures.

2.2 Heatstroke deaths in health care facilities

A survey conducted by the InVS referred to the deaths caused by heatstroke in public and private healthcare facilities (3). This study pertained to all persons deceased from a heat stroke in a health care facility, over the whole country, between the 8th and 19th of August 2003. The case definition was: a body temperature equal to or greater than 40.6 °C before time of death, or before all cooling attempts, with reasonable exclusion of other causes of hyperthermia. A questionnaire was sent to the French medical facilities likely to care for heat stroke victims. A recall of the non-respondent facilities was undertaken on August 21st and August 25th. Overall, a response rate of 78 % was observed for state facilities and 53 % for private ones.

A total of 2851 heat stroke related deaths were reported. A body temperature ≥40.6 °C before time of death was recorded in 48 % of the cases [median: 41 °C, 25th percentile – 75th percentile: 40 – 42 °C]. More than half the deaths occurred between August 11th and 13th. The majority of deaths occurred in women (65 %), and 81 % in elderly people aged 75 and over. In this study, 16 % of the deceased lived alone,

Fig. 1

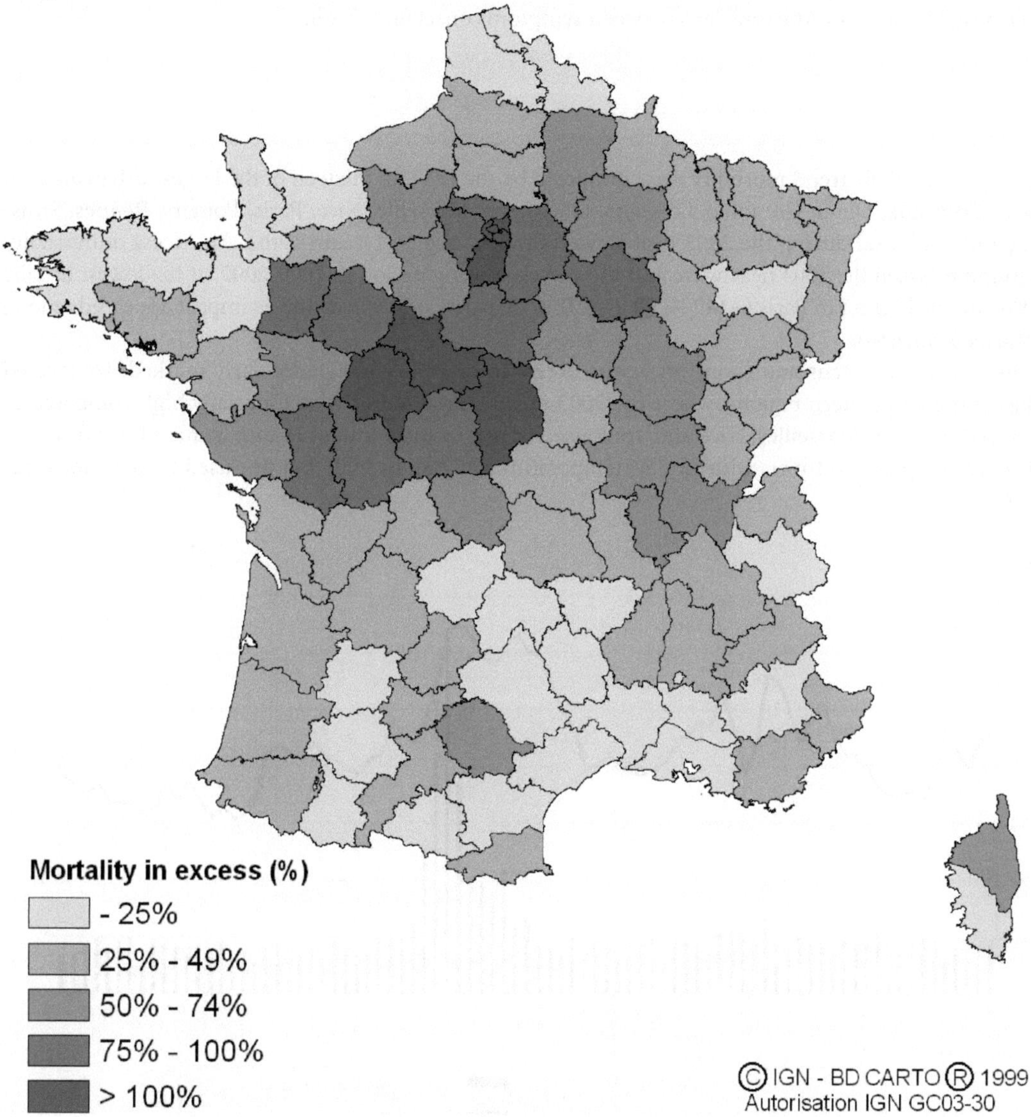

Analysis of over mortality in France August from 1st to 15th 2003, related to mean of deaths from 2000–2002 (source: InVS)

20 % in individual housing but not alone, and 63 % lived in health care facilities, mostly in retirement homes (47 %). For those aged 60 years and over (95 %), various diseases associated with heat stroke were notified, of which essentially, 30 % were mental illnesses, 23 % cardio-vascular diseases, and 12 % diabetes. For those aged under 60 (n = 146), a mental illness was notified for 41 %, excessive alcohol consumption for 20 % and a severe physical disability for 14 %. Three deaths were recorded for those less than 20 years of age. The date of onset of the symptoms, reported for 2417 deaths, corresponded to the day of or the day before death for 47 %. This time lapse was significantly shorter when the recorded body temperature was

40.6 °C or higher (stringent case definition). When restricted to this latter sub-group, the analysis shows similar results, except for the time laps between symptoms onset and death.

2.3 Mortality assessment in urban areas

A survey on the daily trend mortality was conducted by the InVS in thirteen of the largest urban areas in France (Bordeaux, Dijon, Grenoble, Le Mans, Lille, Lyon, Marseille, Nice, Paris, Poitiers, Rennes, Strasbourg and Toulouse) during the 2003 heat wave (between August 1st and 19th). An excess of mortality was found between the 2003 heat-wave and the corresponding period in 1999/2002, at the lowest in Lille (+4 %) and the highest in Paris (+142 %, ➔ *Fig. 2*). A disparity of the heat-wave's impact appeared among the thirteen cities (4).

Dijon, Paris, Le Mans and Lyon, where the excess in mortality was particularly marked, are located in the central and eastern regions, where the 2003 mean temperatures were especially high compared to the preceding years. Marseilles, Nice and Toulouse, located in the South of France, suffered less from the heat-wave compared to towns with similar temperatures in August 2003 but not used to very hot summers (5).

Fig. 2

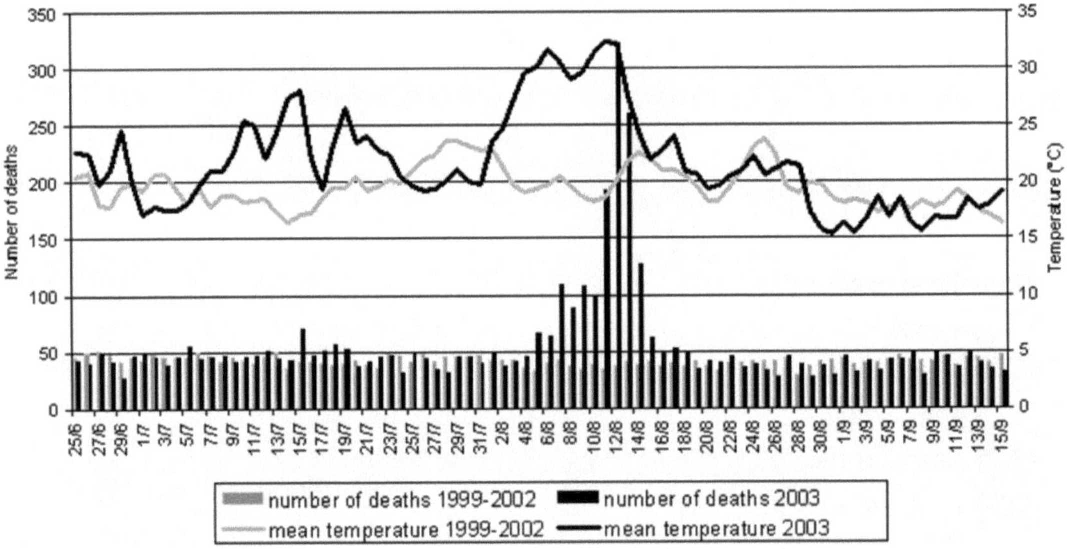

Comparison of daily mortality and mean temperature in Paris for years 2003 and 1999–2002 (source: InVS)

3 Etiologic factors studies

Following the descriptive studies carried out immediately after the heat-wave, two case-control surveys were conducted by the InVS to identify the risk factors of death for the elderly living at home and those residing in a nursing home (6).

The aim of the study for the elderly living at home was to identify individual risk factors (way of life, medical history, self sufficiency) and environmental factors (housing) in elderly people living at home (7).

A case control study was conducted in Paris, Val de Marne, Tours and Orleans. These sites were selected because they are urban areas which were affected by the heat wave and because they present different social and architectural patterns. Cases were defined as: people aged 65 and over, deceased between August 8th to 13th, living at home and with death certificate not mentioning accident, suicide or surgical complications. Controls were matched with cases on age (+/–5 years), gender and place of residence (area of 100,000 residents). A questionnaire was administered face to face or by phone to the next to kin for cases and to controls themselves or next of kin if necessary. Size of sampling was evaluated at 300 cases and 300 controls.

Data were collected using satellite pictures to assess the heat island profile of the home places. Data analysis is currently in progress, using SAS. We use matched pairs analysis to estimate odds ratios and confidence intervals for each potential risk factor. All potential risk factors derived from the questionnaire are assessed in univariate analysis. The final multivariate model will entered into a conditional, step-wise logistic regression model.

The aim of the study for the elderly residing in a nursing home was to identify individual risk factors (autonomy/handicap, medical condition, drug consumption) and environmental risk factors (number and quality of personnel available; facility size and characteristics; prevention plans and therapeutic protocols). This survey is made of two parts: a "facility case-control study" and an "individual case-control study". The target facilities are retirement homes and hospital units for the long term care of the elderly in the areas characterized by a high level of heat wave-related mortality. The selection of facilities was done on the basis of a rapid postal survey conducted in August 2003. The facilities with the higher level of mortality are the "cases" in the "facility case-control study". They are paired to controls on criteria of geographical proximity and mean level of autonomy of the residents. 200 cases and 200 controls are investigated. Within each of the "cases facilities", two "individual cases" are randomly selected among the persons deceased between 5th and 15th of August 2003 (excluding non heat-related causes). Each individual case is paired with the person who was still alive on the first of September, and whose age is the closest to the case. The data on cases and controls are gathered through a questionnaire administered during face-to-face interviews with the administrative and health care personnel. Matched pairs analysis is used to estimate odds ratios and confidence intervals for each potential risk factor. All potential risk factors derived from questionnaire are tested in univariate analysis. The final multivariate model will use a conditional, stepwise logistic regression model.

The results of theses case-control surveys have been used to define profiles of sensitive people for the 2004 Heat-Wave National Plan.

4 Contribution of air pollution by ozone and interaction with temperature

The meteorological conditions during the heat-wave contributed to a photochemical air pollution that was very unusual according to its duration and its geographical spread. Consequently, the question of the specific contribution of high levels of ozone to the mortality in excess was asked early by the Public Health Authorities. A study (8), carried out by the Heath and Surveillance Programme in 9 cities (PSAS-9), had three objectives: 1) to estimate the short-term risks of death due to ozone pollution during the heat-wave and compare them to the risk estimates observed in normal weather conditions; 2) to evaluate the excess risks of mortality jointly related to air temperature and ozone, and the relative part of each risk factor; 3) to quantify if a decrease of expected mortality in the weeks following the heat-wave (short-term harvesting effect) exists.

An interaction on mortality, between high temperatures and high ozone concentrations, was not observed. For the whole set of the studied cities (Paris, Lyon, Marseille, Lille, Bordeaux, Toulouse, Le Havre,

Rouen and Strasbourg), the pooled excess risk of mortality – for an increase of 10 μg/m^3 of the ozone air concentration – was 1,01 %. This value is slightly higher than, but not statistically different from, the excess risk usually estimated in these cities. On the other hand, the town-specific excess risks were more heterogeneous among the 9 cities than in the previous works. The health impact attributable to ozone pollution, worked out from August 3rd to 17th, was 379 deaths in excess in the 9 cities (for a total population of 11 million inhabitants).

Comparing effects of temperature and ozone, it appeared that in the cities experiencing the highest excess risk of mortality, Paris and Lyon, the relative part of ozone was minor: it represented respectively 7 and 3 % of the excess risk of mortality related to the two risk factors. For cities where the excess risk of mortality was moderate, the contribution of ozone was higher (except in Bordeaux: 2 %) but very variable, from 32 % (Rouen) to 85 % (Toulouse).

Modelling of the "heat-wave effect" on mortality, independently of any particular factor, did not show a harvesting effect within the three weeks following the heat-wave. The deaths which occurred during the meteorological event were not anticipated on only a few days, and the loss of life was probably larger for over three weeks.

5 Heat Health Watch Warning System

To prevent another terrible epidemic related to extreme temperatures during summertime, the French Public Health Authorities have developed a Heat-Wave National Plan. It includes a Heat Health Watch Warning System (HHWWS) which operates since 2004, from June to August, and covers the whole metropolitan country.

The warning system is based on thresholds of biometeorological indices, defined from historical dataset of the previous 30 years daily mortality and meteorological indicators (9). Out of seven meteorological indicators tested (related to temperature and humidity), the minimal and maximal temperatures, coupled and averaged on three consecutive days, had the highest sensitivity and specificity and have been chosen as biometeorological indices. The HHWWS aims to alert the French Authorities to the possibility of a significant outbreak: so the warning thresholds correspond to an expected daily mortality in excess of 50 or 100 %, according to the size of the city. The level 2 of the Heat-Wave National Plan is activated when the 3-day forecasting of the coupled biometeorological indices reach the thresholds.

References

1. Basu R, Samet JM (2002) Relation between elevated ambient temperature and mortality: a review of the epidemiologic evidence. Epidemiol Rev 24(2):190 – 202
2. Hemon D, Jougla E (2003) Estimation of mortality and epidemiologic characteristics. INSERM U170-IFR69 report September 25th 2003. http://www.inserm.fr
3. InVS (Institut de Veille Sanitaire) (2003) Impact sanitaire de la vague de chaleur en France survenue en août 2003. Rapport d'étape. http://www.invs.sante.fr
4. Vandentorren S, Suzan F, Medina S, Pascal M, Maulpoix A, Cohen JF, Ledrans M (2004) Mortality in thirteen French cities during the August 2003 heat-wave. Am J Public Health 94(9):1518 – 1520
5. Keatinge WR, Donaldson GC, Cordioli E, Martinelli M, Kunst AE, Mackenbach JP, Nayha S (2000) Heat related mortality in warm and cold regions of Europe: observational study. BMJ 321:670 – 672
6. Semenza JC, Rubin CH, Falter KH, Selanikio JD, Flanders WD, Howe HL et al. (1996) Heat-related deaths during the July 1995 heat wave in Chicago. N Engl J Med 335(2):84 – 90
7. Clarke JF (1972) Some effects of the urban structure on heat mortality. Environment research 5:93 – 104

8. InVS (Institut de Veille Sanitaire) (2004) Vague de chaleur de l'été 2003 : relations entre températures, pollution atmosphérique et mortalité dans neuf villes françaises. Rapport d'étude. http://www.invs.sante.fr

9. InVS (Institut de Veille Sanitaire) (2004) Système d'alerte canicule et santé (SACS) 2004. Rapport opérationnel. http://www.invs.sante.fr

Portugal, Summer 2003 Mortality: the Heat Waves Influence

Rui Manuel D. Calado · Jaime da Silveira Botelho · Judite Catarino · Mário Carreira

Introduction

In Portugal, during both June 1981 and July 1991, the air temperature rose above 32 °C during 2 or more consecutive days. Studies after these periods concluded that these heat incidents were associated with excess deaths.

This events justified the development, by the National Observatory of Health, of a heat wave vigilance and alert system, using data and information from the Meteorological Institute.

On a regular basis, since 1999, from 15th of May until the end of September, the "Civil Protection" and the General Directorate of Health receive the Ícaro Index daily. This index helps to predict the possibility of a heat occurrence which would be intense enough to affect mortality 3 days in advance.

This short article describes the effects of the summer heat-waves 2003 in Portugal.

Summer 2003 occurrence

In Summer 2003 the alert was given three times. It corresponded to the heat waves of 18 – 20 June, 29 of July to 13th August and 11 – 14 September (➲ *Fig. 1*).

Through the "Alert and Appropriate Response System" the alerts were immediately communicated to the "Health Authorities" at regional and local levels.

The alerts given on 19th of June and 29th of July led to technical recommendations sent to the "Regional Health Authorities" on 20th June and 30th July, respectively.

During the first week of August, the technical recommendations were published through the General Directorate of Health web site and were complemented with more technical data considered pertinent.

The "Civil Protection" formal reports through the media informed the population of the possibility of receiving additional information about prevention measures on the heat problem, using the Public Health Call Center. During the first two weeks of August, this phone line received more than 1400 calls. The daily frequency of the calls followed the "Civil Protection" alerts and the Ícaro Index variations (➲ *Fig. 2 and 3*).

Summer 2003 mortality

In the second week of August, the population was alerted through the media with news about the mortality increase in France and the incapacity of French authorities to identify a large number of corpses and to conduct their funerals.

Fig. 1

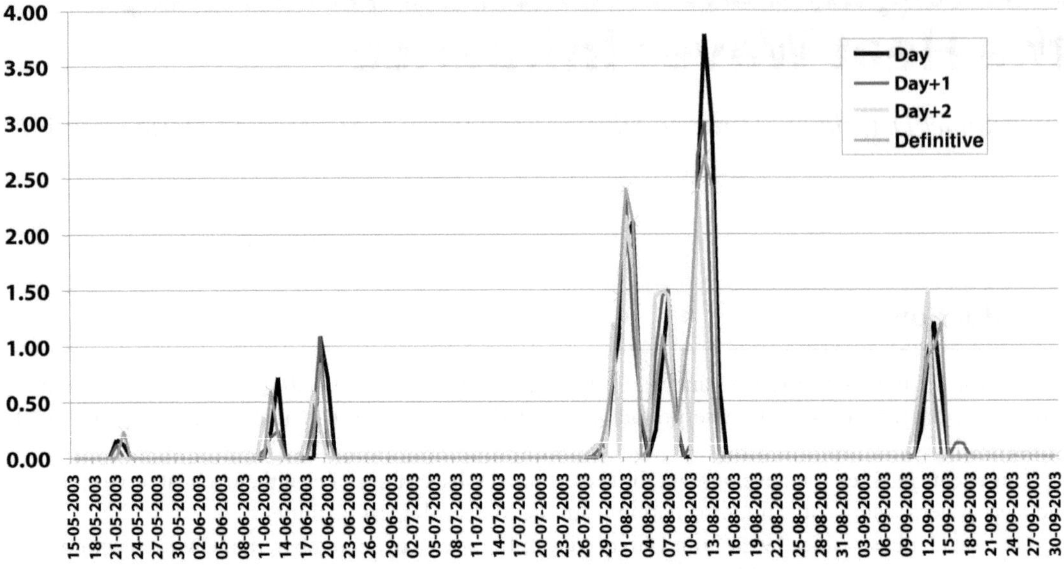

Ícaro Index between 15.05 and 30.09.2003 Source: National Observatory of Health (ONSA)

Fig. 2

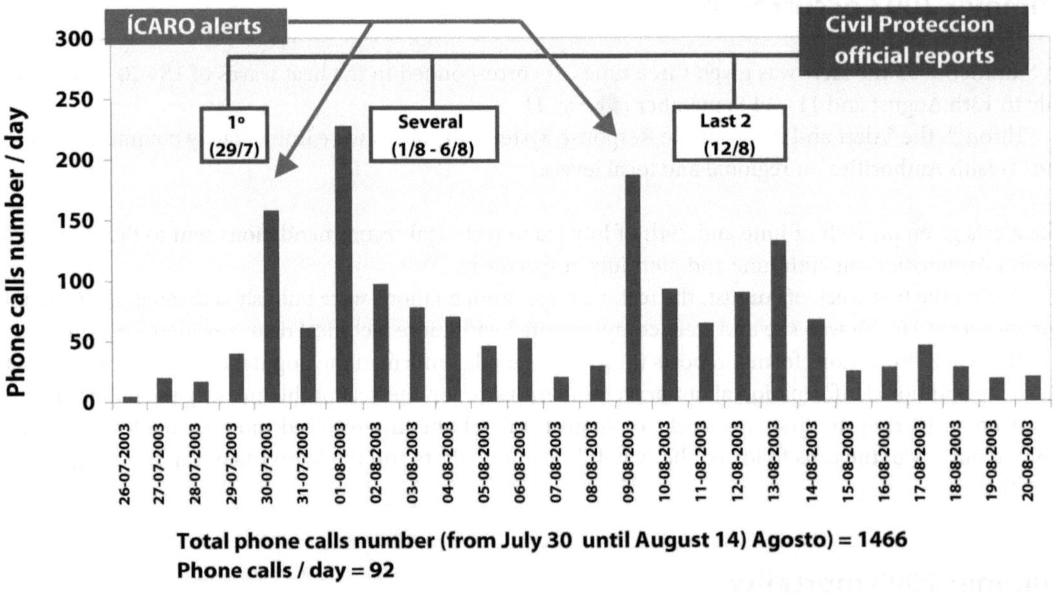

Heat Wave 2: ÍCARO alerts, Civil Protection official reports and Public Health Call Center activity. Source: General Directorate of Health (DGS)

Fig. 3

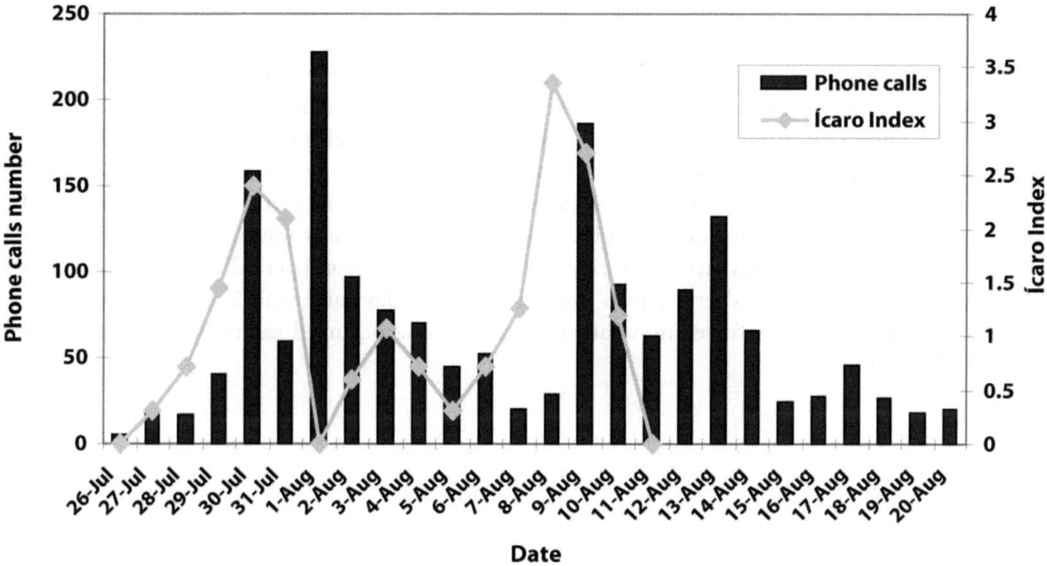

Heat Wave 2: relationship between Ícaro Índex and use of the Public Health Call Center. Source: General Directorate of Health (DGS)

In Portugal between 30 July and 14 August, our central department received no formal or informal communication about hospital problems related to the heat waves. But the analysis of "people movement" at the Hospital Emergency Services between 1 June and 31 August revealed an increase in its use of 11,6 % in relation to the "people movement" average verified in the previous 2 years. There is an increase in percentage to 27,2 % for people over 74 years of age and this percentage increases to 40,5 % if we consider the search for health care by aged people in the South of Portugal alone (*Tab. 1*).

Tab. 1

Hospital emergency consultations between June 1 and August 31. Source: General Directorate of Health (DGS)

Regions	Age group	2001/02	2003	Variation	Δ %
Portugal	All ages	1289916	1439875	149959	11,6
	75E+	99445	126506	27062	27,2
North	All ages	527603	553414	25812	4.9
	75E+	40735	48089	7355	18.1
Centre	All ages	300765	306494	5729	1.9
	75E+	14090	15777	1687	12.0
South	All ages	461549	579967	118419	25.7
	75E+	44620	62640	18020	40.4

At the end of August, a "National Observatory of Health" study alerted to an eventual increase of deaths during the first two weeks of August related to the "heat wave." The estimation predicted an excess of

1316 deaths, distributed over "all ages" but with a predominance in people of over 74 years (58 % of the total excess).

It was thus necessary to confirm the conclusions of that study. The General Directorate of Health requested of the "Death Registration Department" immediate access to the death certificate copies of the Summer months registers, to compare its number with equivalent periods (it was used the 5 years average of 1997 – 2001), and to study not only the cause of death described, but also the co-morbidities.

The data was obtained from 32778 death certificates, of which 16563 pertained to males (50,5 %).

During the 4 months analyzed (1 June to 30 September) the daily mortality incidence represented a strong increase in 2 periods, coincident with the first and the second heat waves. But whilst in the first period we can see the anomalous increase during just 4 days, in the second period the increased number of deaths is much higher, the adverse effects present during 17 days and with 3 acute moments (2, 8, 13 August), immediately after worsening weather conditions that occurred respectively in the 1, 6 and 12 August (*Fig. 1 and 4*).

Fig. 4

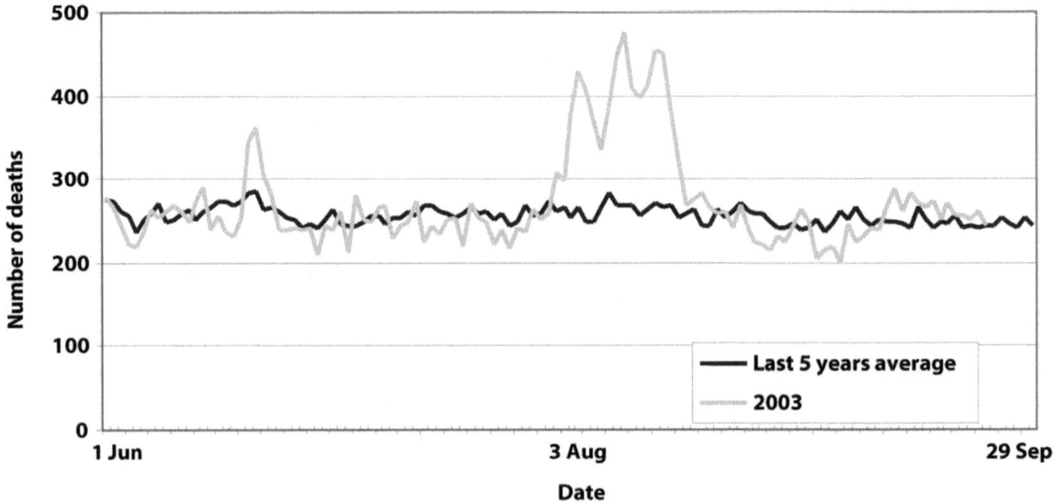

Heat Waves: daily deaths between June 1 and September 30. Source: General Directorate of Health (DGS)

The increase of mortality during the first heat wave occurred among both genders, but during the second it was predominant in females (69 % of the death excess over the first two weeks of August). During the third heat wave, the mortality numbers didn't show any important change, with the numbers by gender maintaining them usual pattern, which means higher values for males (*Fig. 5*).

The analysis of weekly incidence reveal that only the effects of the second heat wave were of enough magnitude to influence in a significant way this indicator, during the week 9th,10th and 11th of the considered period (*Fig. 6*).

The comparison of death by age group registered between 1997 – 2001 and 2003, for equivalent periods during Summer time, revealed that the "heat wave" adverse effects affect in particular persons over 74 years of age. During August 2003 the excess was of 2131 deaths, with 2055 (96,6 %) belonging to this age group (*Tab. 2*).

Fig. 5

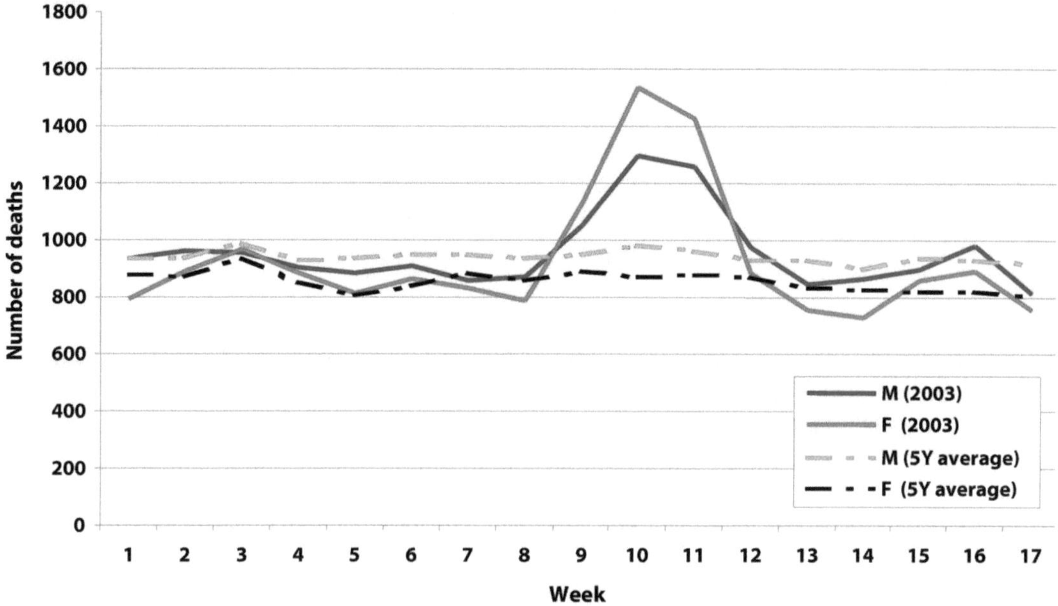

Heat Waves: weekly deaths by sex, between June 1 and September 30. Note: The week 1 begin in 01.06.2003 and the week 17 end in 27.09.2003. Source: General Directorate of Health (DGS)

Fig. 6

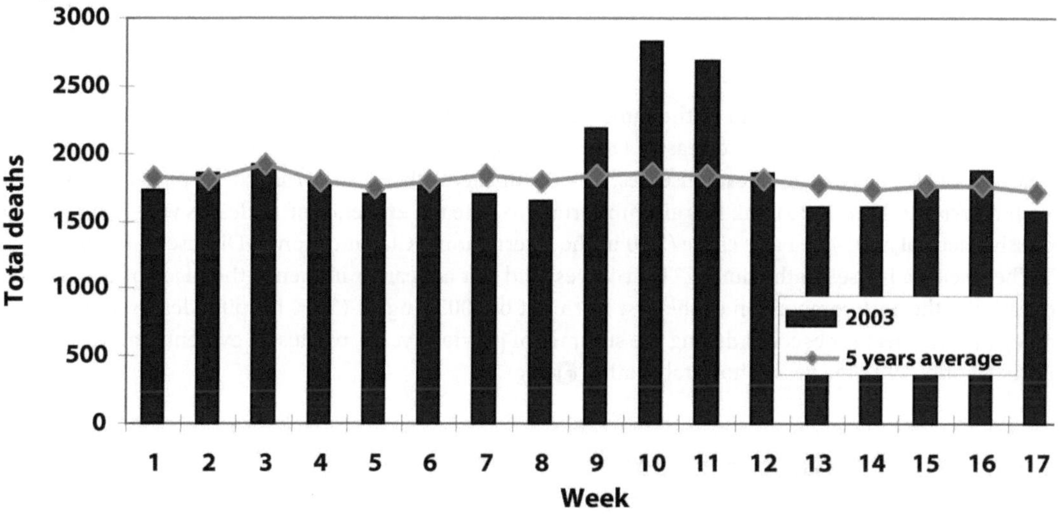

Heat Waves: weekly deaths, between June 1 and September 30 and last 5 years average. Source: General Directorate of Health (DGS)

Tab. 2
Heat Waves: total deaths by age group, during August 2003. Source: General Directorate of Health (DGS)

Age group	Last 5 years average	2003	Difference	%-difference
< 15 years	103	94	-9	-8 %
15–64 years	1849	1795	-54	-3 %
65–74 years	1688	1827	+139	+8 %
75E+	4338	6393	+2055	+47 %
Total	7978	10109	+2131	+27 %

Curiously, in the other Summer 2003 months (June, July and September) the death number was always lower than that expected for this period. So the deaths excess for the period between 1 June and 30 September 2003 was 1802 (*Tab. 3).*

Tab. 3
Heat Waves: total deaths by age group, between June 1 and September 30. Source: General Directorate of Health (DGS)

Age group	Last 5 years average	2003	Difference	%-difference
< 15 years	385	314	-71	-18 %
15–64 years	7026	6393	-633	-9 %
65–74 years	6607	6330	-277	-5 %
75E+	16962	19746	+2784	+16 %
Total	30981	32783	+1802	+6 %

In the first 2 weeks of August and in the same period for 1997–2001, the percentage difference between deaths was +159 % for endocrine diseases, +83 % for respiratory diseases, +51 % for cerebrovascular diseases and +47 % for ischemic heart diseases. The death excess that result from these 4 causes was 1029, which represents 45 % of its total. It is also important to note the existence of 14 deaths with "exposure to excessive natural heat" like basic cause (X30 in the International Classification of Diseases – ICD10).

The rise in adverse deaths during "heat waves" did not appear to influence the place of occurrence of death. So the data comparison of the first fortnight of 2003 August (54 % hospital deaths and 32 % at home) with the deaths observed during the summer of previous years, reveals an evident maintenance of the distribution of home versus hospital deaths (*Fig. 7).*

Identification of risk population

Given current demographics, it is easy to understand that the major risk group is elderly women, with some diseases that make them more vulnerable to heat adverse effects.

The ageing of the Portuguese population, where between 1960 and 2000 there was a gain of 12,2 years in life expectancy, in clear convergence with the EU, has strong implications for the population health situation. The process has created favourable conditions in the increase in vulnerability of the aged population, which is becoming more and more numerous.

Fig. 7

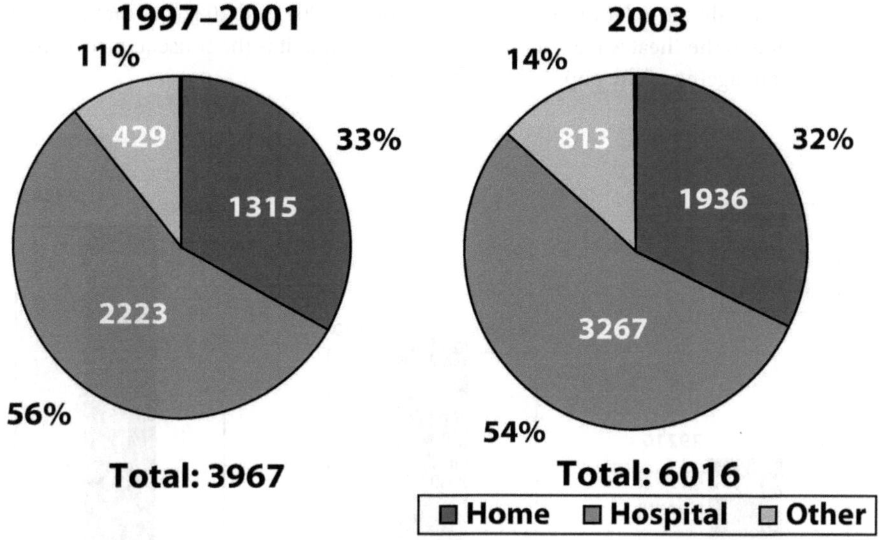

Deaths during the first fortnight in August: occurrence place. Source: General Directorate of Health (DGS)

The increase of the susceptible population could also be the consequence of a small number of winter deaths in Portugal between November 2002 and March 2003. The number was lower than that expected (comparing with the homologous 1997 – 2001 period), with the enormous difference of –4746 deaths (*Fig. 8*).

Fig. 8

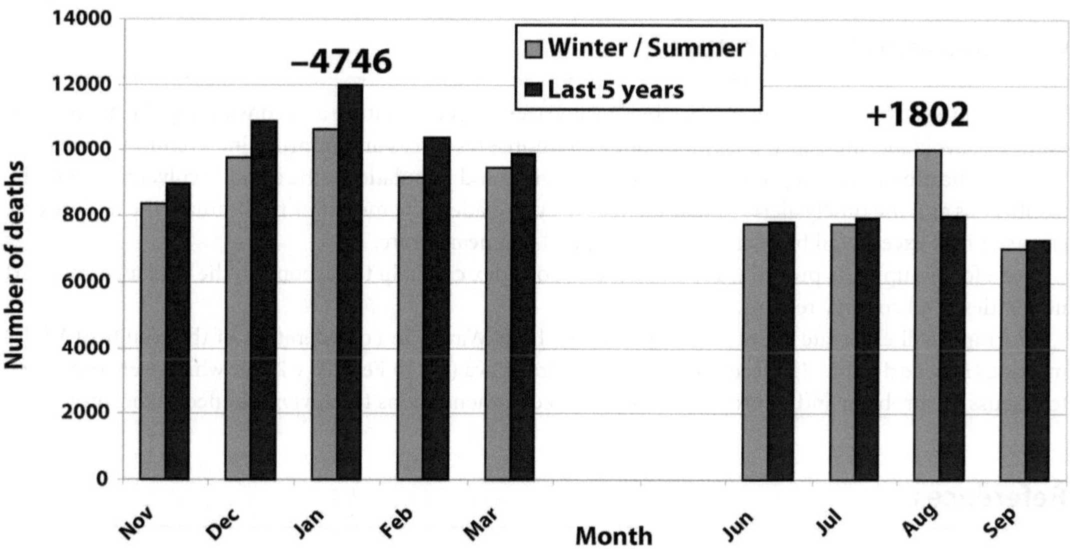

Heat Waves: deaths in the last winter and summer months, and in the same periods of the last 5 years. Source: General Directorate of Health (DGS) and National Statistics Institute (INE)

It is also possible to observe that, between January and September 2003, the global number of deaths correspond to an intermediate value in relation to the same period of 2001 and 2002. As that number was influenced by excess related to the "heat waves", it is possible to think that it is the consequence of mortality decrease and a concomitant ageing of the population (Fig. 9).

Fig. 9

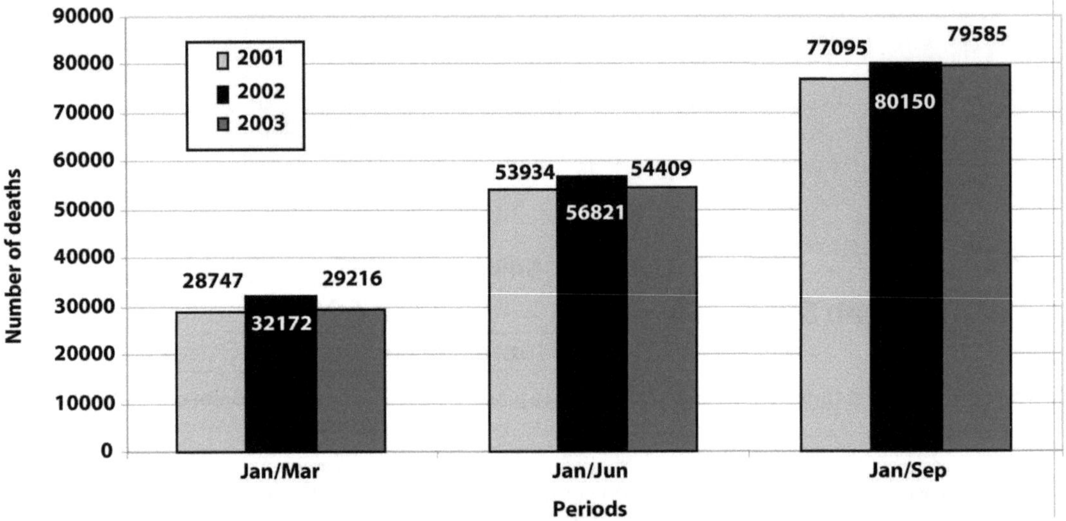

Heat Waves: 3 periods 2003 total deaths and last 2 years similar periods. Source: National Institute of Statistics (INE)

Civil measures for the future

To combat the adverse effects of predictable "heat waves" effects in future and considering the necessity to minimize avoidable mortality, it is important to reinforce resources and improve intervention strategy.

The education of health professionals, social services and populations, the active involvement of social institutions and the timely alerts to take all necessary and adequate measures to eliminate the undesirable effects of heat excess, will be considered and applied in a near future.

As a follow-up of its proved usefulness, the Ícaro Index can help to respond to the vigilance and alert necessities of all country regions.

Portugal will elaborate a Contingency Plan for Heat Waves, in consideration of the results of WHO meetings in Madrid (14 – 15 December 2003) and Bratislava (9 – 10 February 2004) which were organized to discuss the problem and to propose all suitable recommendations to prevent its effects efficiently.

References

Calado R, Nogueira PJ, Catarino J, Paixão EJ, Botelho J, Carreira M, Falcão JM (2003) Onda de calor de Agosto de 2003: os seus efeitos sobre a mortalidade da população portuguesa (aceite para publicação na Revista Portuguesa de Saúde Pública)

Falcão JM, Nogueira PJ, Contreiras MT, Paixão EJ, Brandão J, Batista I (2003) Onda de calor de Agosto de 2003: efeito sobre a mortalidade da população. Estimativas provisórias (até 12.08.2003). Observatório Nacional de Saúde

Garcia CP, Nogueira P, Falcão JM (1999) Onda de calor de Junho de 1981: efeitos na mortalidade. Clima e Saúde. Revista Portuguesa de Saúde Pública. Volume temático:I

Nogueira P (2003) Como fazemos as estimativas do excesso de óbitos numa onda de calor. Observações Nº 20; Observatório Nacional de Saúde

Paixão EJ, Nogueira PJ (2003) Efeitos de uma onda de calor na mortalidade. Revista Portuguesa de Saúde Pública 21;1:41–54

The Effect of Temperature and Heat Waves on Daily Mortality in Budapest, Hungary, 1970 – 2000

A Páldy · J Bobvos · A Vámos · RS Kovats · S Hajat

Keywords

Heat waves, temperature, respiratory, cardiovascular, mortality

Summary

We investigated the association of weather on daily mortality in Budapest, 1970 – 2000, with special regard to heat waves. Budapest has a continental climate and experiences extreme heat episodes. In the past 30 years, the minimum and maximum daily temperatures in Budapest has significantly increased, as well as daily variability in summer. A 5 °C increase in daily mean temperature above 18 °C increases the risk of total mortality by 10.6 % (95 % CI 9.7, 14.0). The effect of hot weather on cardiovascular mortality is even greater. Six heat episodes were identified from 1993 to 2000 using standardized methods. During each episode, a short term excess in mortality occurred. During the early June heatwave in 2000, excess mortality was greater than 50 % over the three day period. We conclude that temperature, especially heat waves, represent an important environmental burden on mortality in the residents of Budapest. Heat waves that occur early in the summer are particularly dangerous. There is a need to improve public health advice in order to reduce the burden of heat waves on human health in Hungary.

Introduction

Global climate change is one of the most important environmental problems facing the twenty-first century. In 1862, John Tyndall, the natural philosopher, made the prediction that anthropogenic emissions of carbon dioxide would trap the radiative energy of the sun within the earths's atmosphere and raise surface temperature (Tyndall 1862). Global warming is accelerating at a rate far greater than that predicted a century ago. Scientists now agree that the balance of evidence indicates that recent warming is due in large part to the combustion of fossil fuels (IPCC 2001). The global mean surface temperature has increased by 0.4 °C in the past 25 years (IPCC 2001).

Climate change threatens our health because of adverse affects to environment, ecosystem, economy and society (Haines et al. 2000, Patz and Khaliq 2002). Moreover, climate change may affect food safety, changes in vector-borne diseases patterns both in space and time, and increases the frequency of extreme weather events, such as heat waves. Global mean temperature is projected to rise by 2.5 to 5.8 °C by the end of this century (IPCC 2001). In the Carpathian basin, a 0.5 to 1.0 °C increase in temperature can be expected by the 2050s, with increases of 0.8 °C and 1.0 – 2.5 °C in mean summer and winter temperatures, respectively. In addition, a 10 % increase in the solar radiation and 20 – 100 mm decrease in precipitation is also projected (Mika 1988).

In response to these changing risks, the Third Ministerial Conference on Environment and Health in London in 1999 recommended developing the capacity to undertake national assessments of the potential health effects of climate variability and change, with the goal of identifying: 1) vulnerable populations and subgroups and 2) interventions that could be implemented to reduce the current and future burden of disease. Hungary has recently completed its National Environmental Health Action Program (1997). As a part of a national health impact assessment in Hungary, the effect of temperature and heat waves on daily mortality in Budapest was investigated. Ambient thermal conditions are an important type of environmental exposure and are responsible for a quantifiable burden of mortality and morbidity. A range of epidemiological methods has been used to estimate the effect of the thermal environment on mortality and morbidity and thus estimate temperature-attributable mortality. Using data on mortality from Budapest, we illustrate two of these methods: time series analysis, and episode analysis.

Analysis of time series data

Daily time series studies are considered the most robust method for quantifying the effects of temperature on mortality. Such studies have shown that the temperature-mortality relationship in temperate countries is consistently non-linear across the temperature range. Mortality increases at both low and high temperatures. The majority of studies report an approximately linear relationship above and below a minimum mortality temperature (or range of temperatures). Thus the temperature mortality relationship in temperate countries is often described as v-shaped or u-shaped (Kunst et al. 1993). The optimum or threshold temperature varies between populations and is assumed to be a function of the population adaptation to the local climate. That is, the warmer the climate, the higher the threshold temperature above which heat-related mortality is detectable.

Heat waves

Heat-waves are rare events that vary in character and impact even in the same location. Arriving at a standardised definition of a heat-wave is difficult; the World Meteorological Organisation (WMO) has not yet defined the term, and many countries do not have operational definitions for health warnings or other purposes. The Hungarian weather service does not have an operational defiinition of a heat wave, and Budapest does not currently have a heat health warning system.

The essential components of a heat wave definition should include high temperatures in the area of interest and some component of duration. A heat wave is a sustained exposure to high temperatures over several days. A heat-wave can be defined based on an absolute or a relative threshold of weather variables or as a combination of both. However, an absolute threshold fails to address the differences between populations in response to temperature, and also within a single population over time.

Studies of heat waves in urban areas have shown an association between increases in mortality and increases in heat, measured by maximum or minimum temperature, heat index (a measure of temperature and humidity), or air mass (Kalkstein, Green 1997). An infamous example is the effect of the 5-day heat wave in Chicago in 1995, in which maximum temperature reached 40 °C and deaths increased by 85 % (CDC 1995). Increases in temperature have a direct and substantial impact on excess mortality for elderly individuals and individuals with pre-existing illnesses. Much of the mortality attributable to heat waves is a result of cardiovascular, cerebrovascular, and respiratory diseases (Koppe et al. 2004).

Material and methods

Daily mortality data for Budapest (permanent residents) from 1970 to 2000 were obtained from the Central Statistical Office. Mortality was divided into three series by cause of death: total mortality except external causes, TM (ICD9 < 800); cardiovascular, CM (ICD9: 430–438); and respiratory: RM (ICD9: 460–519). In 2000, the population of Budapest was approximately 1.7 million. During the 31-year data period, the number and age distribution of inhabitants in Budapest has changed and therefore mortality data were standardised for the average of the whole period. Observed daily mortality for each day was weighted by the ratio of the average mortality of the respective year to the average mortality over all 31 years.

The National Meteorological Service provided meteorological data: 24 hour average values of maximum and minimum temperature, barometric pressure and relative humidity. Data were from the "Pestlörinc" meteorological station which is situated at the Southern outskirts of Budapest, 12 km from the centre (World Meteorological Organisation station number 12843).

The effect of high temperatures on mortality was investigated using two methods. First, the relationship between mortality and temperature was investigated for the whole temperature range using regression models adapted for time series data. Second, individual heat wave episodes were identified, and the associated excess mortality during the defined heat wave periods was quantified.

The time series analysis was carried out using generalised additive models (GAM). Non-parametric smoothing techniques were used to control for season and trend. We investigated the effect on mortality of same day and previous day's temperature and humidity, as previous studies have shown that the effect of high temperatures is short lived. The effect of temperature was investigated separately for the summer months (April to September inclusive). Influenza is known to affect daily mortality in winter, and so days with respiratory mortality over the 98 percentile (over 10 cases per day) were excluded. Previous sensitivity analyses have shown the confounding effects of influenza epidemics on the short term effect of air pollution on total mortality (Touloumi et al 2004). In this analysis, no adjustment was made for air pollution as data were not available for the whole time period.

A range of definitions has been used to define heat wave events (Koppe et al. 2004). We used a relative definition based on the temperature record. Events were identified as a minimum three-day period of days with mean temperature (lags 0–2) above the 99th centile of daily mean temperature (26.6 °C) between 1990 and 2000. Baseline (expected) mortality was estimated for each heat episode with a regression model that included day of week, time of year, ozone, PM_{10}, an influenza indicator, and smoothed functions of temperature and humidity to describe the underlying seasonal pattern. The heat wave attributable mortality was estimated by subtracting the "expected" mortality from the observed mortality during the predefined periods.

Results

Climate changes in Budapest

Budapest has experienced a significant warming trend since 1970. The minimum and maximum daily temperatures have both increased. Further, the number of hot days and the variance of daily temperature by year have increased. The variability of temperature changed during the 31 years. Days with mean temperature different from the mean of the previous five, ten or fifteen days became more frequent, especially during the summer period. These differences in temperature were not so common in wintertime.

Eight of the ten hottest years in the series were in the 1990s. The cause of the warming may be due, in part, to urbanisation, and, in part, to global warming, as Western and Central Europe has warmed by

0.3 °C per decade since the 1970s (Klein Tank et al. 2002). The number of hot days (daily max temperature over 32 °C) became more frequent in the nineties (Fig. 1).

Fig. 1

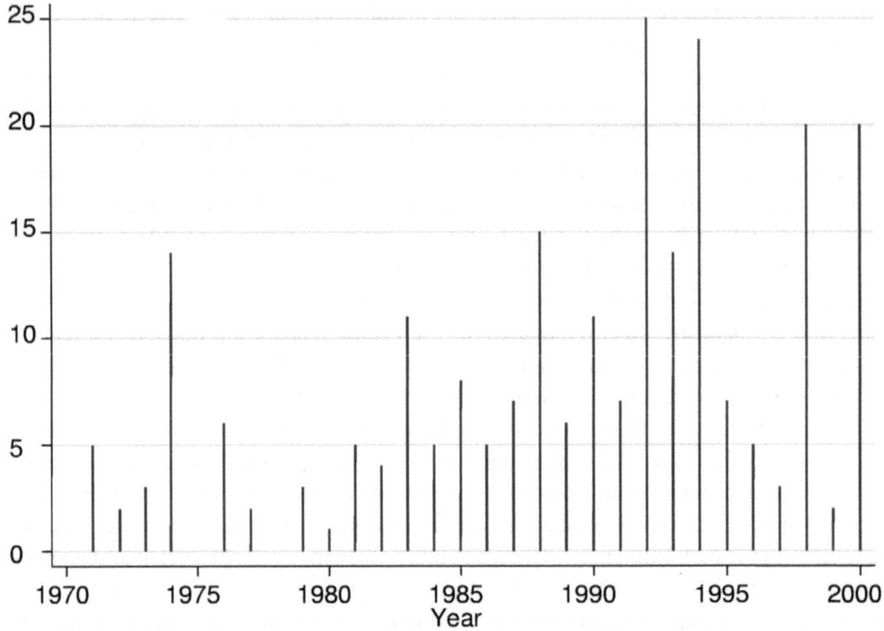

Number of hot days per year (days with maximum temperature ≥ 32 °C), 1970–2000

Relationship between meteorological factors and mortality

Mortality in Budapest shows a seasonal pattern, with deaths highest during the winter months. Seasonal respiratory infections have an important role in winter mortality, and their association with meteorological factors is uncertain. Within the summer period, the unadjusted (crude) relationship between temperature and mortality is approximately linear $\beta = 1.704$ (95 % CI 1.569; 1.839) for 1 °C change in temperature). During the winter period, the association between temperature on same day and total mortality is negative and not so strong $\beta = -0.055$ (95 % CI −0.108; −0.003), as the effect of cold is known to last for up to two weeks. There is no corresponding change in the slope for the effect for very cold days. The acute effect of weather variables on mortality was more pronounced during summer than winter.

Figure 3 shows the relationship between temperature and total mortality, adjusted for season and trend, and day of week. As has been shown for other cities in temperate regions, the relationship is u-shaped. Mortality is lowest when daily mean temperature (lag 0,1) is approximately 18 °C. Above 22 °C daily mean temperature, the linear association has a steeper slope.

Assuming a linear relationship within the summer months, we quantified the slope of the relationship for each mortality series (*Fig. 4*). A 5 °C increase in temperature increased the risk of total mortality by 10.6 % (96 % CI 0.97, 14.0); cardiovascular disease mortality by 18 % (96 % CI 11, 29); and respiratory disease mortality by 8.8 % (96 % CI 5.4, 23). A 5 °C increase from the previous 15-day moving average temperature also had a significant impact on mortality in both seasons.

Fig. 2

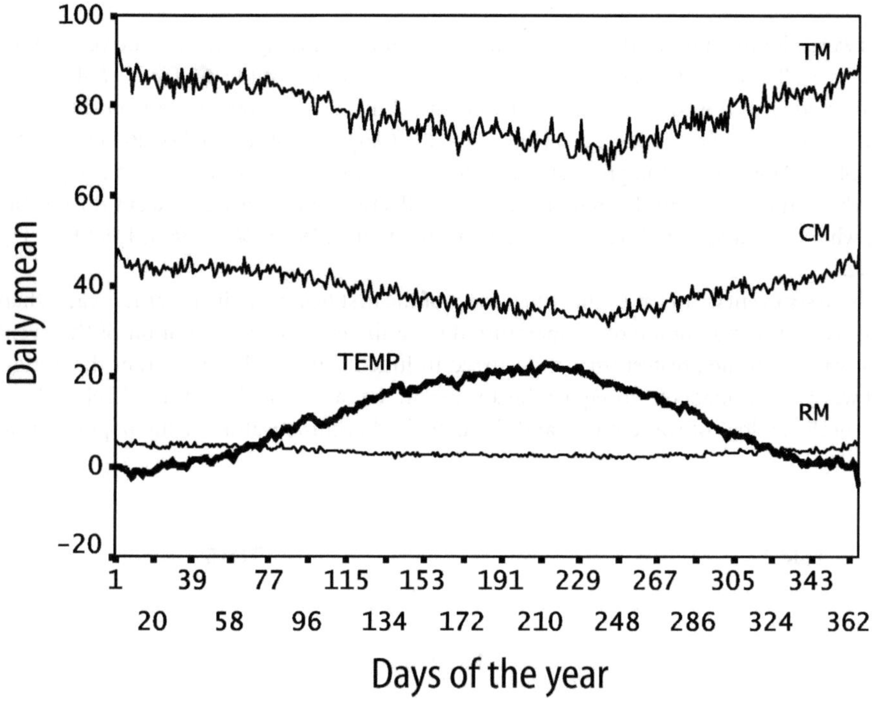

Seasonal pattern of daily total, cardiovascular and respiratory mortality (average of years 1970–2000)

Fig. 3

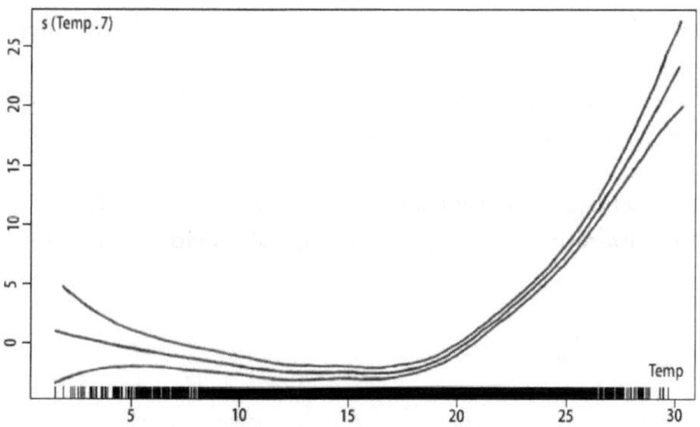

Effect of temperature on daily total, cardiovascular and respiratory mortality in winter and summer (fully adjusted models). Relationship between temperature and total mortality (model adjusted for season, influenza activity, day of week, trend).

Heat waves

Six heat wave events were identified in the data series, with two events occurring in each of the years 1994, 1998 and 2000 (➔ *Tab. 1*). The episodes varied in duration, timing and magnitude. ➔ *Figure 4* illustrates mortality peaks associated with the two heat waves that occurred in the year 2000. Excess mortality (all cause) was observed during all the episodes: 22 % (Jun 94), 12 % (Aug 94), 24 % (Jul 98), 26 % (Aug 98), 52 % (Jun 00), and 14 % (Aug 2000). The greatest excess for each heat wave is in the oldest age groups (75+). Mortality in the adult age group did not appear to be affected by heat waves, except during the August 2000 event, which was associated with an excess mortality of 72 % (95 % CI 36.6, 116.6) during the three day episode.

Evidence from these six events, indicatest that the impact of the first heat wave in a year is greater than the second, irrespective of the magnitude of temperature during the event or the duration of the event (➔ *Fig. 5*). The heat wave with the greatest impact occurred in June 2000, which was neither the hottest nor the longest heat wave. The impact of subsequent heat waves in the same year, however, is likely to have been diminished by both the loss of susceptibles and the increased acclimatisation of the population to hot weather.

Fig. 4 a and b

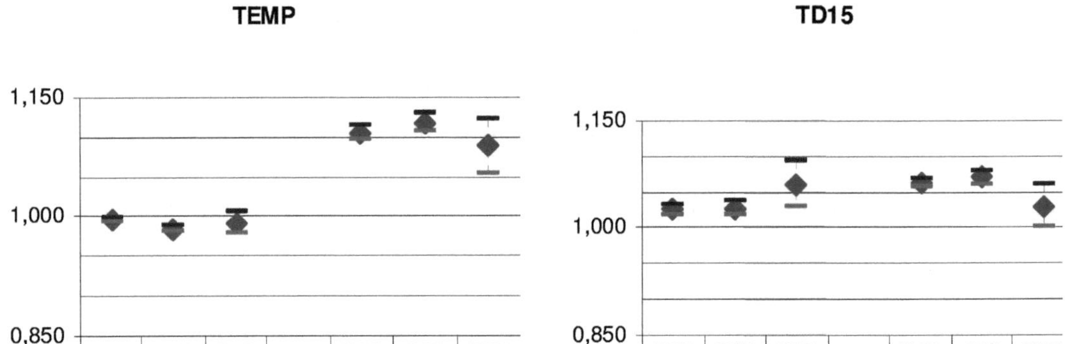

TEMP	TMST	CMST	RMST	TMST	CMST	RMST
◆ RR	0,997	0,984	0,992	1,106	1,118	1,088
‒ RRMIN	0,993	0,980	0,978	1,097	1,106	1,054
‒ RRMAX	1,000	0,989	1,007	1,114	1,129	1,123

TD15	TMST	CMST	RMST	TMST	CMST	RMST
◆ RR	1,027	1,028	1,061	1,063	1,071	1,030
‒ RRMIN	1,020	1,018	1,029	1,056	1,062	1,000
‒ RRMAX	1,033	1,038	1,093	1,070	1,081	1,059

Effect of temperature on daily total, cardiovascular and respiratory mortality in winter and summer, for a) daily mean temperature (no lag) and b) temperature difference compared to average of previous 15 days.

Tab. 1
Heat wave events

Heat wave event	Duration	Mean temp (average)	Max temp (maximum)
28 June to 1 July 1994	4 days	27.0	36.3
30 July to 8 August 1994	10 days	27.5	36.3
22 July to 25 July 1988	4 days	27.4	34.6
3 August to 5 August 1998	3 days	27.6	36.7
13 June to 15 June 2000	3 days	27.5	36.2
20 August to 22 August 2000	3 days	28.1	37.9

Fig. 5

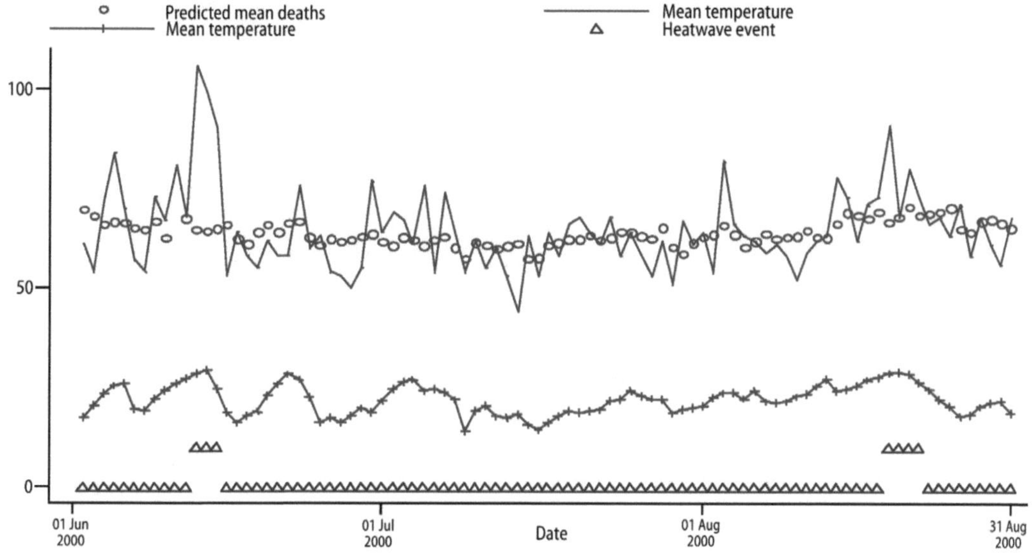

Mortality and temperature series, Summer 2000

Discussion

The association between mortality and temperature in Budapest is U-shaped, as has been previously shown in other cities in Europe (Kunst et al. 1993). The effect of temperature on total mortality is greater than that observed for London (mortality increased by 1.2 % (95 % CI 1.0 to 1.3) for each degree above 18 °C), but similar to that observed in Sofia: 2.2 % (95 % CI 1.6, 2.9) (Pattenden et al. 2003). Between the two ends of the temperature range, a physioclimatic optimum exists where mortality is at the minimum. For example, mortality is estimated to be lowest at 14.3 – 17.3 °C in north Finland but at 22.7 – 25.7 °C in Athens (Keatinge et al. 2000). In Budapest, this optimum is at approximately a daily mean temperature of 18 °C.

As has been found in other studies in Europe and the US, the most sensitive age group to high temperatures was the over 75s (Paldy et al. 2001). The highest risk estimate was estimated for cardiovascular mortality, followed by total and respiratory mortality (although the differences are not statistically significant). Studies in other cities in Europe have often shown the greatest effects of high temperatures on respiratory mortality (Koppe et al. 2004). However, there are likely to important differences in the coding of cardio-respiratory deaths between countries that may explain this result.

In ecological studies, problems could arise from the fact that different populations are at risk in summertime and wintertime. This problem however is of less importance in Budapest, because the most vulnerable part of the population does not leave the city for holidays during the summertime.

Another concern that has been raised is the possibility that air pollution is a confounder of the relationship between weather and mortality, as the same meteorological conditions that cause heat waves also trap photochemical pollutants. A number of studies have proved the independent effect of air pollutants on health using different approaches for weather adjustment (Pope, Kalkstein 1996, Samet et al. 1998). Within the APHEA2 project, the effect of air pollution on short term mortality was studied and temperature as a confounder was considered. The effects of air pollutants on mortality in Budapest were weaker than that of temperature (Paldy et al. 2000). However, the combined effect of weather and tropospheric ozone should be further studied in this population.

Fig. 6

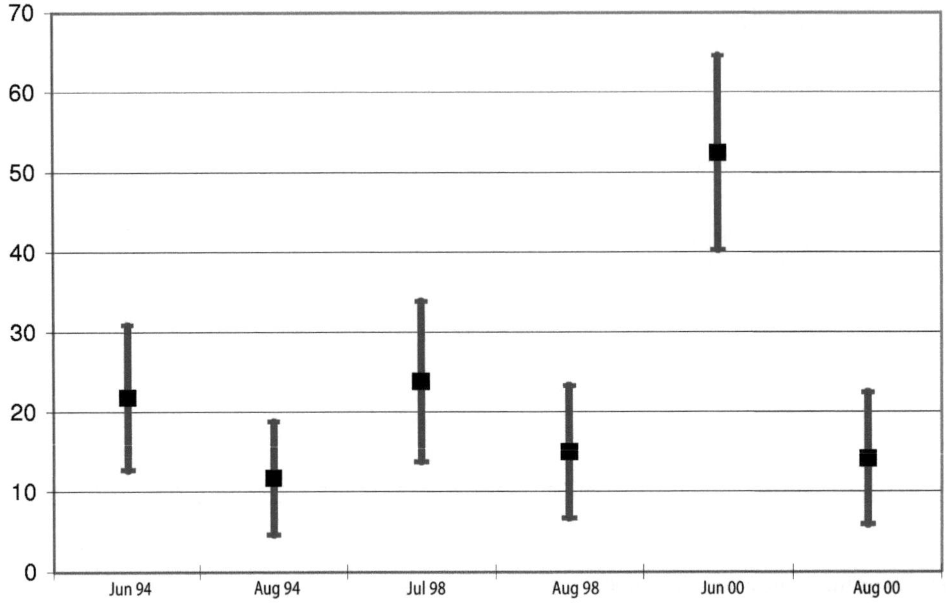

Excess mortality associated with each heat wave episode (%) with 95 % confidence intervals. Expected mortality was estimated using a regression model. Excess total mortality (%) for six heat wave episodes, 1994 to 2000.

Conclusion

Heat waves and high summer temperatures contribute to significant burden on mortality in Budapest, particularly heat wave events that occur early in summer. The elderly are most at risk from extreme hot weather, but mortality in all ages is affected.

Studies of weather-mortality relationships indicate that populations in north-eastern and Midwestern U.S. cities may experience the greatest number of heat-related illnesses and deaths in response to infrequent extremes of summer temperature (Kalkstein, Green 1997). Keatinge et al. (2000) have also reported that deaths in mid- and high latitude countries occur most frequently during conditions of extreme cold or extreme heat. Mortality in Budapest is certainly sensitive to high temperatures (above 18 °C), although these do not occur infrequently. Therefore, there may be non-climate factors that are important determinants of the temperature-mortality relationship such as a relatively high prevalence of chronic diseases (Széles et al 2003).

The August 2003 heat wave affected the population of Budapest, although temperatures were not so extreme as those experienced in France. An increase in ambulance call outs was reported (Paldy, pers com) but the impact on mortality and morbidity has not yet been reported. In response to this extreme event, the National Institute of Environmental Health distributed 15,000 leaflets with advice on how to avoid heat related illness. There is a need to further improve public health advice in order to reduce the burden of heat waves on human health in Hungary.

References

CDC (1995) Heat related mortality – Chicago, July 1995. MMWR 44:577 – 579

Haines A, McMichael AJ, Epstein PR (2000) Environment and health: 2. Global Climate change and health. CMAJ 163(6):729 – 734

Hajat S, Kovats S, Atkinson R Haines A (2002) The impact of hot temperatures on death in London: a time-series approach. JECH 56:367 – 72

Intergovernmental Panel on Climate Change (IPCC). Houghton J, Ding Y, Griggs M et al. (eds) (2001) Climate Change 2001. The Scientific Basis: Contribution of Working Group I to the 3rd Assessment Report of the IPCC. Cambridge University Press, Cambridge

Kalkstein LS (2000) Biometeorology – looking at the links between weather, climate and health. Biometeorology Bull 5:9 – 18

Kalkstein LS, Green JS (1997) An evaluation of climate/mortality relationships in large US cities and the possible impacts of climate change. Environ Health Perspect 105(1)84 – 93

Keatinge WR, Donaldson GC, Cordioli E, Martinelli M, Kunst AE et al. (2000) Heat related mortality in warm and cold regions of Europe: observational study. BMJ 81:795 – 800

Klein Tank A, Wijngaard J, van Engelen A (2002) Climate of Europe: Assessment of observed daily temperature and precipitation extremes. KNMI, De Bilt, the Netherlands, pp 36ff

Koppe C, Kovats RS, Jendritzky G, Menne B (2004) Heat-waves: impacts and responses. In: Health and Global Environmental Change Series, No. 2. WHO Regional Office for Europe, Copenhagen

Kunst AE, Looman CW, Mackenbach JP (1993) Outdoor air temperature and mortality in the Netherlands: a time series analysis: American Journal of Epidemiology 137:331 – 341

Mika J (1988) Characteristics of global warming in the Carpathian Basin (in Hungarian) Időjárás 92:178 – 189

National Environmental Health Action Program (1997) http://www.antsz.hu/oki/nekap/

Paldy A, Bobvos J, Erdei, Vámos A, Károssy Cs (2001) Short-term effects of weather and synoptic weather conditions on daily mortality in Budapest Epidemiology 12:(Suppl.)S57

Paldy A,. Bobvos J, Erdei E, Károssy Cs, Vamos A (2000) Short-term effects of classical and biological air pollutants – Synoptic weather conditions on daily mortality in Budapest: APHEA2 project Epidemiology 11:(4)S123 – S123

Pattenden S, Nikiforov B, Armstrong BG (2003) Mortality and temperature in Sofia and London. J Epidemiol Community Health 57:628 – 633

Patz JA, Khaliq M (2002) Global Climate change and Health: Challenges for Future Practitioners. JAMA 287:2283 – 2284

Pope CA, Kalkstein LS (1996) Synoptic weather modelling and estimation of exposure-response relationship between daily mortality and particulate air pollution. Environ Health Perspect 104:414 – 420

Samet JM, Zeger S, Kelsall J et al. (1998) Does weather confound or modify the association of particulate air pollution with mortality? An analysis of the Philadelphia data, 1973 – 1980. Environ Res 77:9 – 19

Széles G, Fülöp I, Bordás I, Ádány R (2003) Characteristics of chronic non-communicable diseases in Hungary based on the data of GP Morbidity data Collection Program. In: Ádány R (ed) The Health State of the Hungarian Population at the Turn of the Millenium (in Hungarian). Medicina Press, Budapest, pp 27 – 42

Touloumi G, Samoli E, Quenel P, Paldy A, Anderson HR, Zmirou D, Galan LI, Forsberg B, Schindler C, Schwartz J, Katsouyanni K (2004) Confounding effects of influenza epidemics on the short-term effects of air pollution on total and cardiovascular mortality: a sensitivity analysis Epidemiology: In press

Tyndall J (1862) On the absorption and radiation of heat by gases and vapours, and on the physical connection of radiation, absorption and conduction. Philosophicsal Magazine 122-169-194:273 – 285

Epidemiologic Study of Mortality During Summer 2003 in Italian Regional Capitals: Results of a Rapid Survey

Susanna Conti · Paola Meli · Giada Minelli · Renata Solimini · Virgilia Toccaceli · Monica Vichi · M. Carmen Beltrano · Luigi Perini

Abstract

Following the unusually hot summer in 2003 and the dramatic news from neighbouring countries such as France, the Italian Minister of Health commissioned an epidemiologic mortality study to investigate whether there had been an excess of deaths in Italy, particularly for the elderly population. Communal offices, which provide vital statistics, were asked to provide data on the number of deaths among residents between June 1 and August 31, for the years 2003 and 2002, for the 21 capital cities of Italy's regions. A mortality increase of 3,134 deaths was observed, most of which (92 %) occurred among persons aged 75 years and older. The highest increases were observed in northwestern cities (Turin, Milan, Genoa). A clear correlation was observed between mortality and climatic indexes (maximum temperature, Humidex).

Key Words

Heat Wave, Elderly, Urban Island

Introduction

It has been widely recognized that extreme weather conditions constitute a major public health threat [1–6]. Extreme weather events are, by definition, rare stochastic events, yet recent assessments indicate that, as global temperatures continue to increase due to climatic changes, the number and intensity of extreme events may increase [7–9].

Moreover, the Urban Heat Island (UHI) effect creates an additional risk for persons living in cities, which have higher temperatures than suburban and rural areas [10, 12]. Furthermore, the larger the city, the more pronounced the UHI and the higher the risk of heat stress during summer [13]. Heat waves present special problems in urban areas also because of the retention of heat by buildings; during heat waves, inhabitants of urban areas may experience sustained thermal stress also during the night, whereas inhabitants of rural areas often obtain some relief from thermal stress at night [15]. Additionally, heat-related mortality is higher in those areas with more temperate climates, as shown in the United States, where a north-south gradient in heat-related mortality has been observed [16, 17].

Studies on heat-wave related mortality have also demonstrated that the greatest increases in mortality occur in the elderly. In some studies, persons over 60 years of age have been reported to be at highest risk of mortality following heat waves [18]; other investigations have indicated persons different minimum ages: over 65 years of age [6, 19], 70 years of age [20], and 75 years of age [21]. Vulnerability to heat in old age occurs because of intrinsic changes in the regulatory system or the presence of drugs that interfere with normal homeostasis [22–25]. As homeostasis is impaired, an elderly person may not be aware that he/she is becoming ill due to high temperatures and therefore may not take action to reduce the exposure. Another factor that makes the elderly susceptible to heat-related morbidity and mortality is the relatively high percentage of persons with illnesses and disabilities. The physical and social isolation of elderly persons further increases the risk of death during a heat wave [26].

In Italy, the highest temperatures of the last decades, or even centuries, were recorded in the summer of 2003. For example, in Turin (which is located in northern Italy, near the French border), the mean temperature was the highest since temperatures began to be recorded (1752); in Milan, in the first half of August, temperatures reached nearly 40 °C, breaking the 37 °C record reached twenty years earlier. In Rome, the highest temperature, nearly 40 °C, was recorded in June, breaking the former record of 37 °C in 1965.

During August 2003, to cope with the social alarm caused by the unusual weather conditions, and following the dramatic news from France (thousands of deaths among the elderly were reported), the Italian Minister of Health commissioned the Istituto Superiore di Sanità (Italy's National Institute of Health) to complete a rapid epidemiologic survey by the end of September 2003. The Bureau of Statistics was in charge of this survey. The objective of the survey was to investigate whether there had been an excess of deaths in Italy's major cities, with a particular focus on the elderly population. The results of the survey are reported herein.

Materials and methods

Various epidemiological methods have been used to estimate the effects of excess heat on mortality [27]; in particular, the episode analysis method may be used to estimate excess mortality following heat waves. In many studies, attributable or excess mortality is estimated by subtracting the expected mortality from the observed mortality; the expected mortality is calculated using a variety of measures, including moving averages and averages of similar time periods in previous years; thus estimates are sensitive to the methods used [28].

One of the methods generally used compares mortality counts or rates during the heat wave with figures observed during the same period in the previous year [25].

As it was necessary to obtain mortality data quickly from many municipalities (whose archives were not always computerized) and to conduct the analysis as soon as possible, the latter method was adopted and the comparison was limited to the year 2002, given that the data from this year were readily available.

Italy is divided into 21 administrative Regions, each with a capital city; our study concerned all 21 regional capitals: Turin, Aosta, Genoa, and Milan (northwestern Italy); Trento, Bolzano, Venice, Trieste, Bologna (northeastern Italy), Ancona, Florence, Perugia, and Rome (central Italy) ; and Naples, L'Aquila, Campobasso, Bari, Potenza, Catanzaro, Palermo, and Cagliari (southern Italy). All Italian cities with more than 1 million inhabitants are included among these 21 cities.

Mortality data

The Communal Office of each city, which records vital statistics, was asked to provide the individual records of deaths among residents for the period June 1–August 31, 2003 and during the same period of 2002. Since the number of individuals actually present in a given city varies from one year to another (especially during the summer time, when many Italians go on vacation), only data regarding residents who died in their own city were taken into consideration.

Meteorological data

Within the framework of a collaboration with the Italian Central Office for Agriecology (UCEA), daily meteorological data (minimum/maximum temperatures and humidity) were obtained for the observed periods of 2002 and 2003, allowing us to calculate discomfort indexes, such as Humidex, a measure of perceived heat that results from the combined effects of excessive humidity and high temperature. The Humidex formula is based on the work of Masterton and Richardson [29].

For those cities that recorded the highest excess mortality, the daily Humidex values and the distribution regarding the categories of comfort were calculated for the three summer months.

Epidemiologic methods

The summer period was divided into six 15 day periods (1 – 15 June, 16 – 30 June, 1 – 15 July, 16 – 31 July, 1 – 15 August and 16 – 31 August), in order to analyze the phenomenon more thoroughly.

Mortality counts among residents of all ages in the 21 capital cities during the three summer months of 2003 were compared with the mortality counts recorded in the same months of 2002.

The 21 cities were also grouped into three large areas (Northern Capitals, Central Capitals, and Southern Capitals). As it was not possible to hypothesise the distribution of the deaths, Wilcoxon Matched Pairs Signed Rank Test was used to determine the statistical significance of the difference in daily death counts, comparing 2002 to 2003.

The analysis was performed by gender and for all age-groups, focusing on individuals of 75 years of age and older, since the excess mortality mainly regarded this group. Standardized mortality rates by age were calculated, using the 1991 Italian census population as standard. The correlation between mortality and climate was investigated by comparing daily distributions of deaths and of the Humidex values and applying Spearman's Coefficient to measure this correlation.

The so called "lag time" (i.e. time elapsed between exposure to the heat wave and the outcome, in this case death) was investigated. Lag time is known to have an important impact on public health [30,31]

Analyses were carried out with the SPSS package.

Results

An overall increase in mortality was observed in the summer of 2003 (23,698 deaths, compared to 20,564 deaths in 2002; difference of 3134). The greatest increase concerned persons aged 75 years and older, among whom 16,393 deaths occurred in 2003, compared to 13,517 deaths in 2002. Of the total 3134 excess deaths in 2003, 2876 (92.0 %) occurred among the elderly. The extent of the increase in mortality varied by city and month . As shown in ❯ *Table 1*, the highest increases (considering the entire three-month period) were observed in northwestern cities. The highest increases for the elderly occurred in Turin (44.9 % increase), Trento (35.2 %), Milan (30.6 %) and Genoa (22.2 %).

Tab. 1

Mortality observed in 2002 and 2003 in the 21 capitals of Region 1 June – 31 August

All Ages					75 yrs and older			
Cities	2002	2003	Differences 2003 – 2002	Differences %	2002	2003	Differences 2003 – 2002	Differences %
Turin	1780	2341	561	31.5	1134	1643	509	44.9**
Aosta	96	101	5	5.2	59	70	11	18.6
Genoa	1829	2136	307	16.8	1295	1575	280	22.2**
Milan	2438	2953	515	21.1	1612	2105	493	30.6**
Trento	168	223	55	32.7	122	165	43	35.2**
Bolzano	196	251	55	28.1	135	156	21	15.6
Venice	706	763	57	8.1	491	541	50	10.2
Trieste	795	835	40	5.0	571	606	35	6.1
Bologna	968	1144	176	18.2	698	880	182	26.1**
Ancona	271	309	38	14.0	187	227	40	21.4
Florence	941	1015	74	7.9	707	790	83	11.7**
Perugia	332	368	36	10.8	229	268	39	17.03
Rome	5246	5849	603	11.5	3334	3899	565	16.9**
Naples	2033	2339	306	15.1	1231	1458	227	18.4**
L'Aquila	125	138	13	10.4	77	96	19	24.7
Campobasso	71	78	7	9.9	42	54	12	28.6
Bari	535	675	140	26.2	340	455	115	33.8**
Potenza	109	122	13	11.9	63	79	16	25.4
Catanzaro	135	142	7	5.2	86	76	-10	-11.6
Palermo	1469	1558	89	6.1	896	1010	114	12.7*
Cagliari	321	358	37	11.5	208	240	32	15.4
ITALY	20564	23698	3134	15.2	13517	16393	2876	21.3**

** $p < 0.01$; * $p < 0.05$ (Wilcoxon Matched Pairs Signed Rank Test)

Also of note are the apparent increases in two southern cities, where the weather is usually cool, such as L'Aquila (24.7 %) and Potenza (25.4 %). In Bari, an increase of 33.8 % was observed, and most of this increase was due to the extremely high mortality increase in August.

The smallest increases were observed in Trieste (6.1 %) and Florence (11.7 %). Catanzaro showed a decrease (–11.6 %). For Rome, a mid-level increase (16.9 %) occurred, although it must be considered that in absolute terms Rome is Italy's largest city and a large proportion of the population is old or very old.

The results obtained using the standardized mortality rates were consistent with those for mortality counts. As observed for mortality counts, the greatest increase among the elderly occurred in Turin (49.2 %), followed by Milan (34.1 %), Bologna (36.3 %) and Trento (33.0 %) (*Tab. 2*). In Bari, the increase among the elderly was quite large (40.2 %) when using standardized rates, which can be attributed to the fact that the city's population is relatively young. No significant differences were observed between males and females, as observed in other studies [32,33].

Tab. 2

Age – standardized mortality rates × 1000 persons aged 75 years and older. 1 June – 31 August 2002 and 2003.

	2002		2003		
Cities	Rate	ES	Rate	ES	Diff. %
Turin	15.6	0.4	23.3	0.5	49.2
Aosta	18.3	2.2	21.6	2.4	17.9
Genoa	19.5	0.5	23.8	0.5	22.2
Milan	14.1	0.3	18.9	0.4	34.1
Trento	15.9	1.2	21.1	1.4	33.0
Bolzano	17.6	1.3	19.1	1.4	8.3
Venice	19.6	0.8	21.1	0.8	7.4
Trieste	21.8	0.9	23.6	0.9	8.3
Bologna	15.8	0.5	21.5	0.6	36.3
Ancona	19.0	1.3	24.7	1.4	30.2
Florence	17.6	0.6	20.2	0.6	15.0
Perugia	20.2	1.1	22.5	1.2	11.4
Rome	19.1	0.3	22.7	0.3	18.8
Naples	24.0	0.6	29.5	0.6	22.9
L'Aquila	16.0	1.6	19.6	1.8	22.9
Campobasso	15.9	1.9	17.2	2.1	8.5
Bari	17.9	0.8	25.0	1.0	40.2
Potenza	20.7	2.0	26.9	2.2	29.5
Catanzaro	20.4	1.7	16.3	1.6	−19.8
Palermo	31.3	0.8	34.7	0.8	10.9
Cagliari	20.4	1.2	22.5	1.3	9.9

The level of excesses in mortality varied by period (*Tab. 3*). In northwestern cities, such as Turin, Genoa and Milan, the highest increases were observed during the first fifteen days of August (increases among individuals 75 years of age and older were, respectively: 181.7 %, 110.8 % and 110.2 %). In Rome the increases were more equally distributed over the summer (1 – 15 June, 18.6 %; 16 – 31 July, 65.2 %; 1 – 15 August, 37.4 %; and 16 – 31 August, 34.2 %). For Bari and Campobasso, the entire increase occurred during the last fifteen days of August: 186.2 % and 450 %, respectively. It should be considered that whereas Bari, given its population size, recorded around one hundred deaths, in Campobasso only 11 deaths occurred.

Taking into account the cities with the highest increase in mortality (Turin, Milan, Genoa and Bari), together with Rome and its relatively old population, comparisons were made between mortality levels and Humidex values. *Figures 1, 2, 3, 4 and 5* show, for the 5 selected cities, the trend in the daily number of deaths and Humidex, observed during the summer of 2003 and 2002.

Tab. 3
Mortality in the sub-periods of 15 days – 75 years and older

Cities	1–15 June			16–30 June			1–15 July			16–31 July			1–15 August			16–31 August		
	2002	2003	Δ %*	2002	2003	Δ %*	2002	2003	Δ %*	2002	2003	Δ %*	2002	2003	Δ %*	2002	2003	Δ %*
Turin	191	264	38.2	277	265	–4.3	174	213	22.4	171	243	42.1	153	431	181.7	168	227	35.1
Aosta	12	14	16.7	8	8	0.0	6	16	166.7	13	7	–46.2	12	14	16.7	8	11	37.5
Genoa	211	229	8.5	263	231	–12.2	201	184	–8.5	194	285	46.9	204	430	110.8	222	216	–2.7
Milan	260	365	40.4	415	347	–16.4	249	260	4.4	218	278	27.5	235	494	110.2	235	361	53.6
Trento	19	27	42.1	23	27	17.4	17	21	23.5	25	29	16.0	21	34	61.9	17	27	58.8
Bolzano	31	28	–9.7	29	22	–24.1	21	23	9.5	24	29	20.8	18	30	66.7	12	24	100.0
Venezia	78	91	16.7	102	82	–19.6	71	68	–4.2	76	90	18.4	89	117	31.5	75	93	24.0
Trieste	84	92	9.5	106	93	–12.3	92	97	5.4	104	100	–3.8	86	117	36.0	99	107	8.1
Bologna	113	139	23.0	180	161	–10.6	101	119	17.8	95	133	40.0	94	159	69.1	115	169	47.0
Ancona	36	45	25.0	38	45	18.4	22	30	36.4	27	32	18.5	30	28	–6.7	34	47	38.2
Florence	116	132	13.8	164	127	–22.6	109	103	–5.5	101	155	53.5	103	134	30.1	114	139	21.9
Perugia	36	34	–5.6	53	38	–28.3	42	40	–4.8	37	52	40.5	20	48	140.0	41	56	36.6
Rome	531	630	18.6	850	708	–16.7	566	543	–4.1	460	760	65.2	447	614	37.4	480	644	34.2
Naples	135	194	43.7	319	307	–3.8	146	172	17.8	277	310	11.9	118	172	45.8	236	303	28.4
L'Aquila	15	9	–40.0	18	19	5.6	13	9	–30.8	9	18	100.0	10	19	90.0	12	22	83.3
Campobasso	6	13	116.7	10	9	–10.0	8	10	25.0	8	7	–12.5	8	4	–50.0	2	11	450.0
Bari	56	63	12.5	54	82	51.9	66	69	4.5	64	91	42.2	71	67	–5.6	29	83	186.2
Potenza	11	18	63.6	13	13	0.0	8	8	0.0	9	13	44.4	9	11	22.2	13	16	23.1
Catanzaro	21	13	–38.1	9	13	44.4	13	12	–7.7	20	13	–35.0	9	10	11.1	14	15	7.1
Palermo	140	167	19.3	154	167	8.4	148	151	2.0	167	230	37.7	162	142	–12.3	125	153	22.4
Cagliari	27	34	25.9	33	42	27.3	42	40	–4.8	33	55	66.7	36	31	–13.9	37	38	2.7

* Differences 2003–2002

Fig. 1

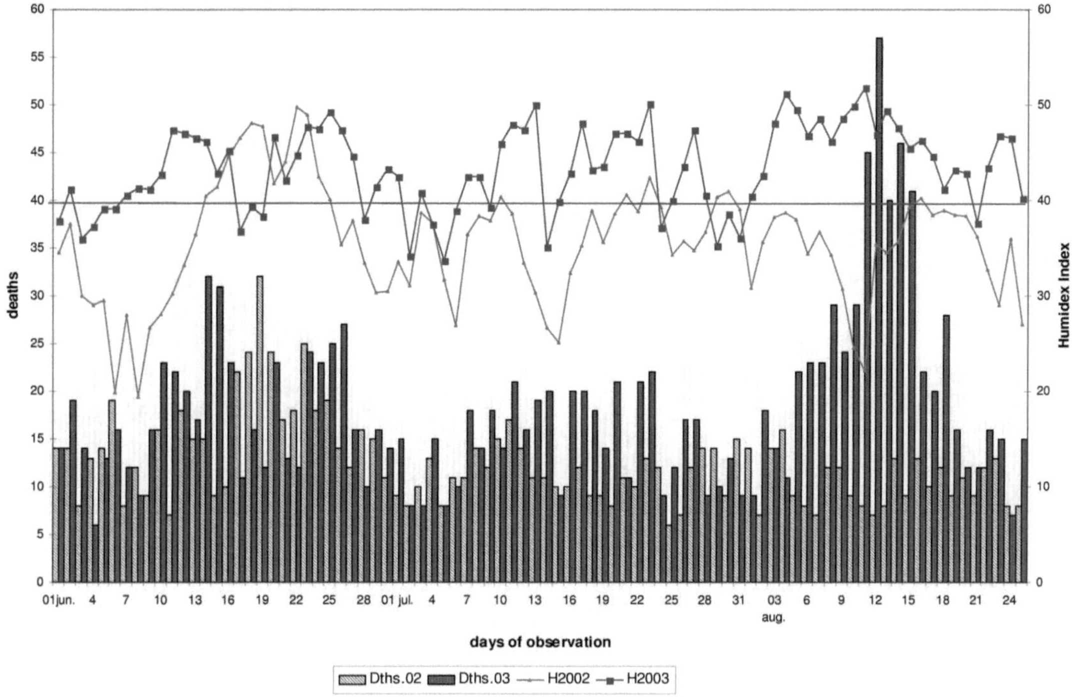

Turin

Fig. 2

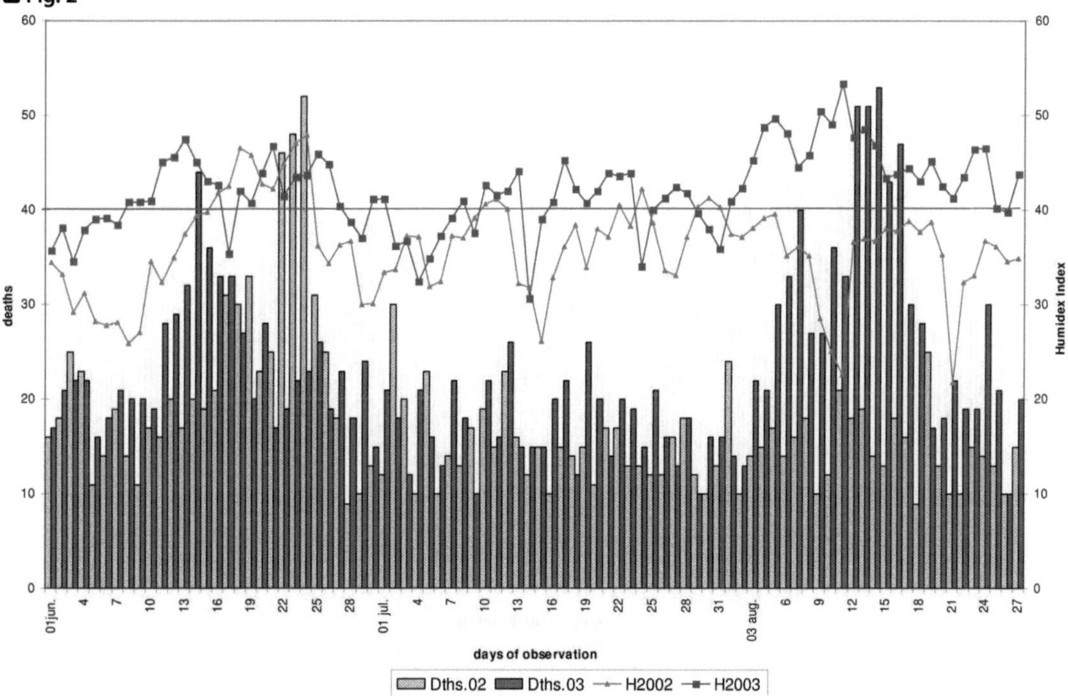

Milan

Fig. 3

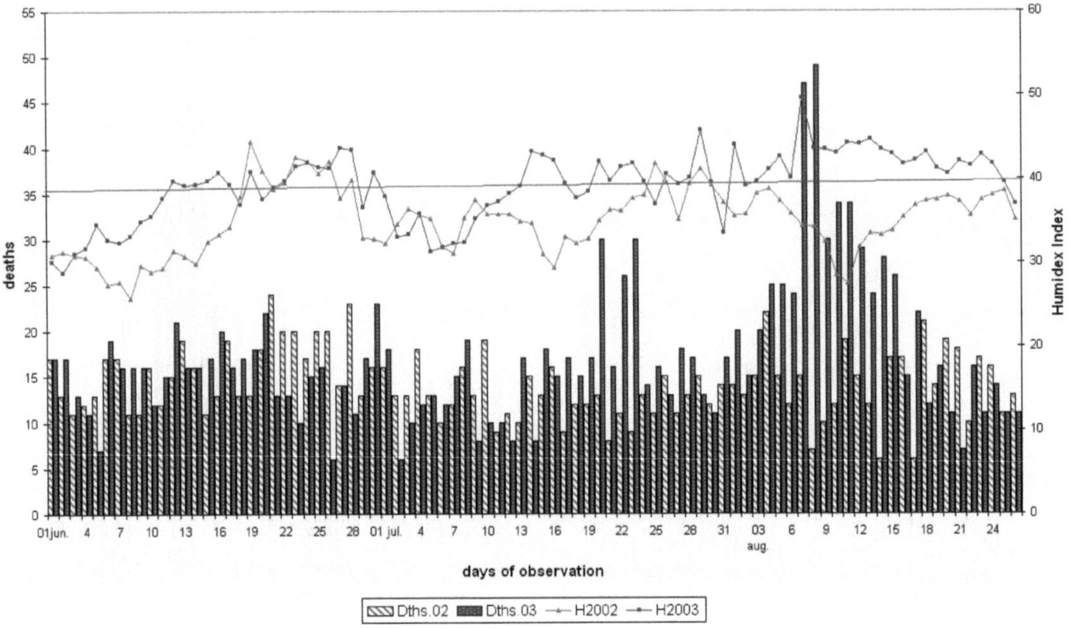

Genoa

Fig. 4

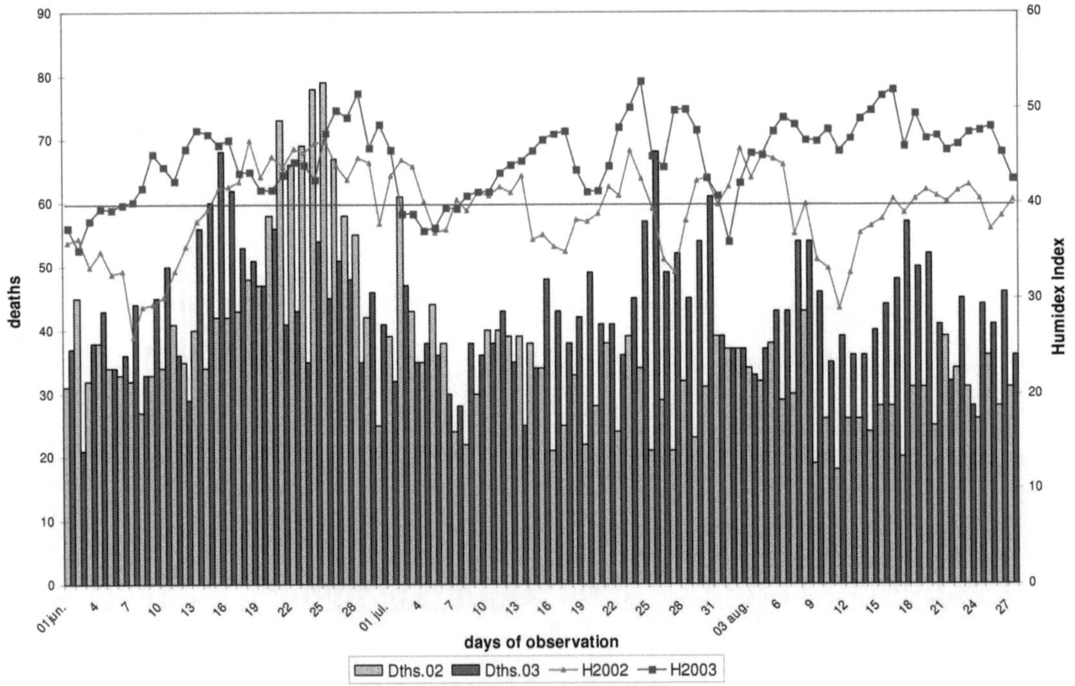

Rome

Fig. 5

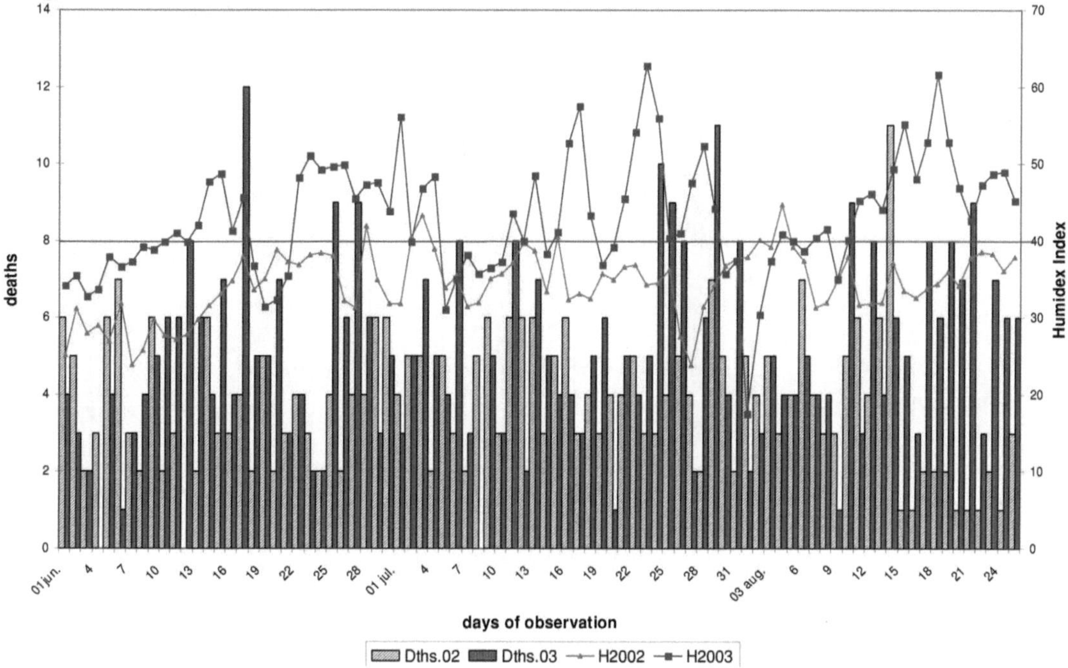

Bari

The correlation between the two series is consistently high, in particular for Turin (ρ=0.639); the values for the correlation index ρ for Milan, Genoa, Rome and Bari are, respectively: 0.520, 0.338, 0314 and 0.241; all indexes were associated with a p-value of less than 0.05.

In Turin, in 2003, there were no "comfortable days", and there were three times as many "dangerous days" (Humidex index > 40 C°) compared to the previous year (81.4 % vs. 31.4 %). The same situation was observed in Milan (79.5 % "dangerous days" in 2003 compared to 28.4 % in 2002) and Genoa, which in 2003 had no "comfortable days" and three times as many "dangerous days" (65.1 % in 2003 and 20.9 % in 2002). In Rome, there were neither "comfortable" nor "some discomfort days" in 2003; only "great discomfort" and above all "dangerous days" were recorded (90.9 % in 2003 compared to 61.4 % in 2002). In Bari, in 2003, more than 57 % of the days were "dangerous days", compared to 14.0 % in 2002); 6.5 % of the days were "very dangerous" (data not shown).

Figure 6 summarizes, for the selected cities, the percentage of "very dangerous days" during the entire two summers, clearly reflecting the heat wave recorded in 2003.

Table 4 shows, for the selected cities, the results of the "lag time" analysis: the maximum correlation was shown for the few days before the deaths: 2 days for Rome; 3 days for Bari, Turin and Genoa; and 4 days for Milan, confirming that the lag time was quite brief.

Fig. 6

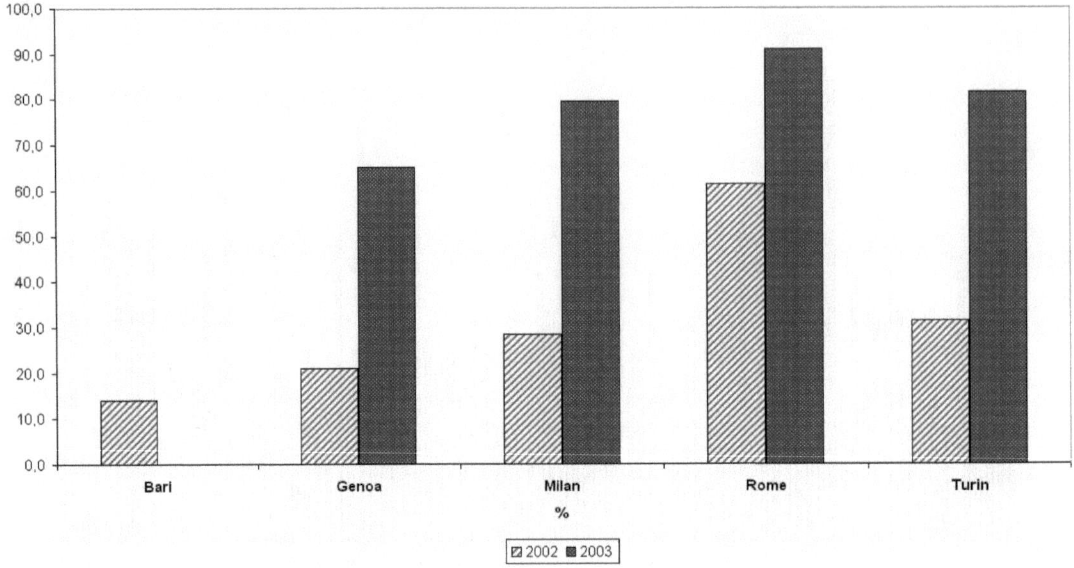

Level of "danger" according to Humidex in some selected cities. 1st June – 31 August, years 2003 and 2002

Tab. 4

Lag-Time: maximum correlation between deaths observed in each day and average of Humidex calculated for a few days before. People aged 75 years and older.

Cities	Period	Correlation (ρ Spearman)
Turin	3 days before	0.652
Milan	4 days before	0.661
Genoa	3 days before	0.355
Rome	2 days before	0.449
Bari	3 days before	0.396

Finally, to have an idea of the magnitude of the excess in mortality among the elderly population in Italy, an empiric estimate was calculated based on the mortality data recorded in the 21 regional capitals and on demographic figures. In particular, based on the subdivision by population size of cities and towns, performed by the Italian Census Bureau (fewer than or equal to 100,000, from 100,001 to 500,000, and more than 500,000 inhabitants), the mortality force observed in the 21 capitals was applied to the entire Italian population, divided into these groups. An excess of more than 7000 deaths (7659) was estimated for the period from 16 July to 31 August 2003, compared to the same period in 2002, among individuals 65 years of age and older. The mean increase was 19.1 %. The percent increases were higher for more populated cities (39.8 % for those with more than 500,000 inhabitants) and lower for smaller towns (13.8 % for towns with a population equal to or less than 100,000 inhabitants and 29.2 % for those with a population of 100,001 to 500,000 inhabitants).

Discussion and conclusion

During the summer of 2003, heat waves affected large areas of Western Europe, with France, Germany, Italy, Portugal, Spain, and the United Kingdom experiencing unprecedented heat and humidity. The effects of heat waves on health are well known, and elderly persons, who have impaired homeostasis, are the most severely affected, with dramatic increases in mortality. In fact, the excesses in mortality mostly affected elderly persons living in cities, also because of the UHI phenomenon and what has been referred to as "the urban loneliness island" [26].

As expected based on reports in the international literature, those cities that normally have a fairly cool climate reported the highest excess mortality, specifically, northern cities, especially those in the northwest, which are close to France and have similar climates, and a few towns in the south, such as L'Aquila, which is located more than 700 m above see level, and Potenza, located more than 800 m above see level.

In the second half of August, an high excess in mortality was recorded even in a few southern towns with normally hot weather, such as Bari and Campobasso. In these cases, it is very likely that, as previously observed in other epidemiological studies on heat waves, the greatest excess occurred after prolonged exposure to high temperatures.

Although a time-series approach or a comparison between 2003 and several previous years would have been more methodologically correct, mortality data needed to be obtained quickly from many cities, whose archives were not always computerized. We thus chose to limit the comparison to the year 2002, given that the data were readily available. However, our results for Turin, Milan, Genoa, Venice, Bologna and Rome are similar to those obtained by local studies conducted later in these cities, which used mortality data directly provided by their own municipalities for a number of previous years.

Our results are also consistent with those of studies conducted in other parts of the world, also with regard to the lag time between the onset of extreme climatic conditions and death, which was found to be quite brief. This finding, together with the correlation between mortality and discomfort climate conditions, provide clear indications for public-health interventions that need to be performed; specifically, social and healthcare support must be given to elderly and frail persons, so as to prevent the occurrence of such a great number of deaths.

References

1. Mackenbach JP, Borst V, Schols JM (1997) Heat-related mortality among nursing-home patients. Lancet 349:1297–98
2. Faunt JD, Wilkinson TJ, Aplin P et al. (1995) The effect of the heat: heat-related hospital presentations during a ten day heat wave. Aust N Z J Med 25:117–21
3. (2002) Heat-related deaths – four states, July-August 2001, and United States, 1979–1999, MMWR 51:567–60
4. Bridger CA, Ellis FP, Taylor HL (1976) Mortality in St. Luis, Missouri, during heat waves in 1936, 1953, 1954, 1955, and 1966. Environ Res 12:38–48
5. Kunst AE, Looman CW, Mackenbach JP (1993) Outdoor air temperature and mortality in the Netherland: a time-series analysis. Am J Epidemiol 137:331–41
6. Mc Farlane A (1978) Daily mortality and environment in English conurbations. Deaths during summer hot spells in Greater London. Environ Res 15:332–41
7. National Research Council (2000) Reconciling observations of global temperature change. National Academy Press, Washington DC
8. Yoganathan D, Rom WN (2001) Medical aspects of global warming. Am J Int Med 40:199–210
9. Meehl GA, Zwiers F, Evans J et al. (2001) Trends in extreme weather and climate events: issues related to modeling extremes in projections of future climate change. Bull Am Met Soc 81:427–36
10. Landsberg HE (1981) The Urban Climate. Academic Press Inc, New York
11. Buehley RW, Van Bruggen J, Truppi LE (1972) Heat Islands equals Death Island? Environ Res 5:85–92
12. Kalkstein LS (1993) Health and climate change: direct impacts in cities. Lancet 342:1397–9

13. Oke TR (1973) City size and the urban heat island. Atmospheric Environment 7:769 – 79
14. Yannas S (2001) Towards more sustainable cities. Solar Energy 70(3):281 – 94
15. Clarke JF (1972) Some effects of the urban structure on heat mortality. Environ Res 5:93 – 104
16. Kalkstein LS, Davis RE (1989) Weather and human mortality: an evaluation of demographic and interregional responses in the Unites States. Ann Assoc Am Geogr 79:44 – 64
17. Kalkstein LS, Greene JS (1997) An evaluation of climate/mortality relationships in large U.S. cities and possible impacts of a climate change. Environ Health Perspect 105:84 – 93
18. Applegate WB, Runyan JW, Brasfield L et al. (1981) Analysis of the 1980 heat wave in Memphis. J Am Geriatr Soc 29:337 – 42
19. Saez M, Sunyer J, Castellsague J et al. (1995) Relationship between weather temperature and mortality: a time series analysis approach in Barcelona. Int J Epidemiol 24:576 – 82
20. Ballester F, Corella D, Perez-Hoyos S. et al. (1997) Mortality as a function of temperature: a study in Valencia, Spain, 1991 – 1993. Int J Epidemiol 26:551 – 61
21. (1995) Heat-related mortality – Chicago, July 1995. MMWR Morb Mortal Wkly Rep 44:577 – 9
22. Kenney WL, Hodgson JL (1987) Heat intolerance, thermoregulation and ageing. Sports Med 4:446 – 56
23. Drinkwater BL, Horvath SM (1979) Heat Tolerance and aging. Med Sci Sports 11(1):49 – 55
24. Havenith G (2001) Temperature regulation and technology. Gerontecnology 1:41 – 49
25. Basu R, Samet J (2002) An exposure assessment study of ambient heat exposure in the elderly population in Baltimore, Maryland. Env Health Pers 110(12):1213 – 24
26. Klinenberg E (2002) Heat wave : a social autopsy of disasters in Chicago. University of Chicago Press, Chicago
27. Basu R, Samet JM (2002) Relation between Elevated Ambient Temperature and Mortality: A Review of the Epidemiologic Evidence. Epidemiological Reviews, Department of Epidemiology, School of Public Health at Johns Hopkins University, Baltimore, 24:2, pp 190 – 202
28. (2003) Briefing note for the fifty-third session of the WHO Regional Commitee for Europe. Vienna, 8 – 11 September 2003
29. Masterton JM, Richardson FA (1979) Humidex, a method of quantifying human discomfort due to excessive heat and humidity. CLI 1-79, Environment Canada, Atmospheric Environment Service, Downsview, Ontario, 45 pp
30. Oechsli FW, Buechley RW (1970) Excess mortality associated with three Los Angeles September hot spells. Environ Res 3:277 – 84
31. Wyndham CH, Fellingham SA (1978) Climate and disease. S Afr Med J 53:1051 – 61
32. Ellis FP, Nelson F (1978) Mortality in the elderly in a heat wave in New york City, August 1975. Environ Res 15:504 – 12
33. Ellis FP, Nelson F, Pincus L (1975) Mortality during heat waves in New York City July 1972 and August and September 1973. Environ Res 10:1 – 13

Heat Waves in Italy: Cause Specific Mortality and the Role of Educational Level and Socio-Economic Conditions

P. Michelozzi · F. de'Donato · L. Bisanti · A. Russo · E. Cadum · M. DeMaria · M. D'Ovidio · G. Costa · C. A. Perucci

Introduction

Record temperatures were observed across Europe during the summer of 2003. There is debate among experts as to whether the high temperatures observed in recent years are a normal fluctuation in the climate or a sign of global warming characterized by wider temperature variations and an increase in climate extremes. The full impact of climate change on health still remains unclear, and an accurate analysis and quantification of the possible effects, both in the short and long term, still has to be carried out (1,2). Although interest on the impact of heat on mortality is increasing, it is clear from summer 2003 that most European countries were unprepared to cope with this emergency.

The relationship between weather, temperature and health has been well documented throughout the literature, both during the winter and summer seasons. The relationship between mortality and temperature graphically presents a "U" or "V" shape, with mortality rates lowest when the average temperature ranges between 15 – 25 °C and rising progressively as temperatures increase or decrease. In relation to hot weather and the effects of high temperatures on mortality, the literature has concentrated on the effect of extreme temperatures, often denoted as "heat waves", which are known to enhance deaths from cardiovascular, cerebrovascular, and respiratory conditions (3,4). The increased frequency and intensity of heat waves may lead to an increase in heat-related deaths with the greatest impact on urban populations, particularly the elderly and the ill.

The report presents an evaluation of the health impact of heat waves recorded during the summer of 2003 (June 1st – August 31st) in three major Italian cities (Rome, Milan and Turin). The analysis aims to analyse the impact of heat waves on cause-specific mortality and to analyse the role of demographic characteristics and socio-economic conditions that may increase the risk of mortality.

Data and Methods

The dataset for each city is comprised of daily mortality counts among the resident population by age, sex, and cause of death. Temperatures (mean, maximum, minimum), and maximum apparent temperature (Tappmax an index of human discomfort based on air temperature and dew point temperature) ($T_{app\text{-}max} = -2.653 + 0.994Ta + 0.0153(Td^2)$) (5) during the summer period are also considered. The latter index combines two meteorological variables that have been shown to have an impact on human health, namely temperature and humidity.

Expected daily mortality was computed as the mean daily value from a specific reference period, respectively 1995 – 2002 for Rome and Milan, and for the period 1998 – 2002 for Turin. The daily mean expected value was smoothed using a smoothing spline. Daily excess mortality was calculated as the difference between the number of deaths observed on a given day and the smoothed daily average. Confidence limits were determined assuming a Poisson distribution.

In Rome, excess mortality by socio-economic level was evaluated for the census tract of residence using a deprivation index based on a series of components, namely education, occupation, unemployment, the number of family members, overcrowding and household ownership data (6). In Turin the level of education was considered as socio-economic indicator.

Results

During the summer of 2003 Tappmax was higher than the mean for the reference period in all cities; the greatest increase was observed in Milan with +16 %, with peak values in August that recorded the highest monthly mean since 1838. In Rome temperatures registered a 13 % increase, while Turin a +11 % increase was observed during summer 2003 (*Tab. 1*).

The definition of a heat wave has to be site specific in order to reflect local conditions; in fact an international definition of heat waves currently does not exit. In this article heat waves are defined as days with Tappmax >90th annual percentile and for the first day an increase of 2 °C compared with the previous day. This definition was elaborated on the basis of the literature reviewed and on the relationship observed between temperature and mortality. Three major heat wave periods occurred in Rome, while in the north of Italy two heat waves occurred; a minor one at the beginning of the summer (mid June) and a major one in August (2nd – 18th).

The results of the analyses indicate record excess mortality during the summer 2003 heat waves. A strong association between daily mortality and temperature was observed; with peaks in mortality corresponding to peaks in temperature or with a lag of 1 – 2 days following the peaks (*Fig. 1*). The heat waves recorded between June and August 2003 are associated with significant health effects; a total of 944 excess deaths were observed in Rome (+19 %), while in Milan and Turin 559 (+23 %) and 577 (+33 %) excess deaths were recorded respectively (*Tab. 1*). In Rome, excess mortality was observed throughout the summer, but predominantly during the three heat waves. The first heat wave (June 9 – July 2) was associated with an increase in mortality of 352, a total of 319 excess deaths occurred during the second heat wave period (July 10 – 30), and 180 excess deaths during the third (August 3 – 13) (*Fig. 1*). In the northern cities, excess mortality was mainly concentrated in the first part of August, when weather conditions were more extreme (*Tab. 2 and 3*). In Milan, 141 excess deaths were recorded during the first heat wave (9 – 20 June) and 380 excess deaths during the second heat wave (5 – 18 August) (*Fig. 1*). In Turin, 76 excess deaths were recorded during the first heat wave (20 – 27 June) and 257 excess deaths during the second heat wave (3 – 14 August) (*Fig. 1*).

When subdividing by age group, excess mortality increased dramatically with age, with the greatest impact observed in the old (75 – 84 years) and very old (85+ years) age groups. In the latter group, mortality increased by 65 % in Turin, 46 % in Rome and 40 % in Milan (*Tab. 2*). When stratifying by age group, there probably is some residual confounding related to gender, in that a larger proportion of females are in the older age groups. In fact, when stratifying by gender, the increase in mortality seems to be greater among females, probably reflecting both the higher proportion of females in the elderly population (*Tab. 2*) and their possible higher susceptibility.

The analyses of cause-specific mortality illustrate how the greatest excess in mortality was observed for the central nervous system, cardiovascular, respiratory diseases and metabolic\endocrine gland and psychological illnesses (*Tab. 3*). In Rome, the highest excess was registered for disorders of the central nervous system (+85 %) and respiratory diseases (+39 %). When subdividing by age group, the excess, for both causes, was greatest in the old (+123 %, +52 %) and very old (+100 % and +45 %) age groups. In Turin the greatest increase in mortality was observed for metabolic\endocrine gland disorders (+143 %), central nervous system (+122 %) and respiratory diseases (+57 %). Milan registered the greatest excess in mortality for disorders of the central nervous system (+118 %), followed by respiratory diseases (+82 %)

and metabolic\endocrine gland disorders (+68 %). Cardiovascular disease registered an increase only in Rome (+24 %) and Turin (+41 %) (Tab. 3).

Results show that the socio-economic level is an important risk factor since the greatest excess was observed in the lower levels in both Turin (+43 % low education level) and Rome (+17.8 % low socio-economic level).

Tab. 1

Maximum apparent temperature and % variation in mortality for Rome, Milan and Turin during summer 2003

	Rome	Milan	Turin
mortality			
2003	6009	2968	2332
reference period	5065 (1995–2002)	2409 (1995–2002)	1755 (1998–2002)
variation	19 %	23 %	33 %
max. apparent temp (°C)			
2003	35.2	32.7	31.7
reference period	31.1 (1995–2002)	28.3 (1995–2002)	28.6 (1990–1999)
variation	13 %	16 %	11 %

Tab. 2

Total and excess mortality by age group and sex in Rome, Milan and Turin during summer 2003

Mortality reference period	Rome 1995–2002				Milan 1995–2002				Turin 1998–2002			
	observed	expected	excess	%	observed	expected	excess	%	observed	expected	excess	%
0–64	915	973	–58	**–6**	372	407	–35	**–9**	307	286	21	**7**
65–74	1163	1112	51	**5**	480	503	–23	**–5**	416	358	58	**16**
75–84	1938	1541	397	**26**	1020	715	305	**43**	752	539	213	**40**
85	1993	1439	554	**38**	1096	784	312	**40**	857	572	285	**50**
Total												
All ages	6009	5065	944	**19**	2968	2409	559	**23**	2332	1755	577	**33**
Gender												
Male	2768	2522	246	**10**	1299	1158	141	**12**	1074	859	215	**25**
Female	3241	2543	698	**27**	1669	1251	418	**33**	1258	896	362	**40**

Fig. 1

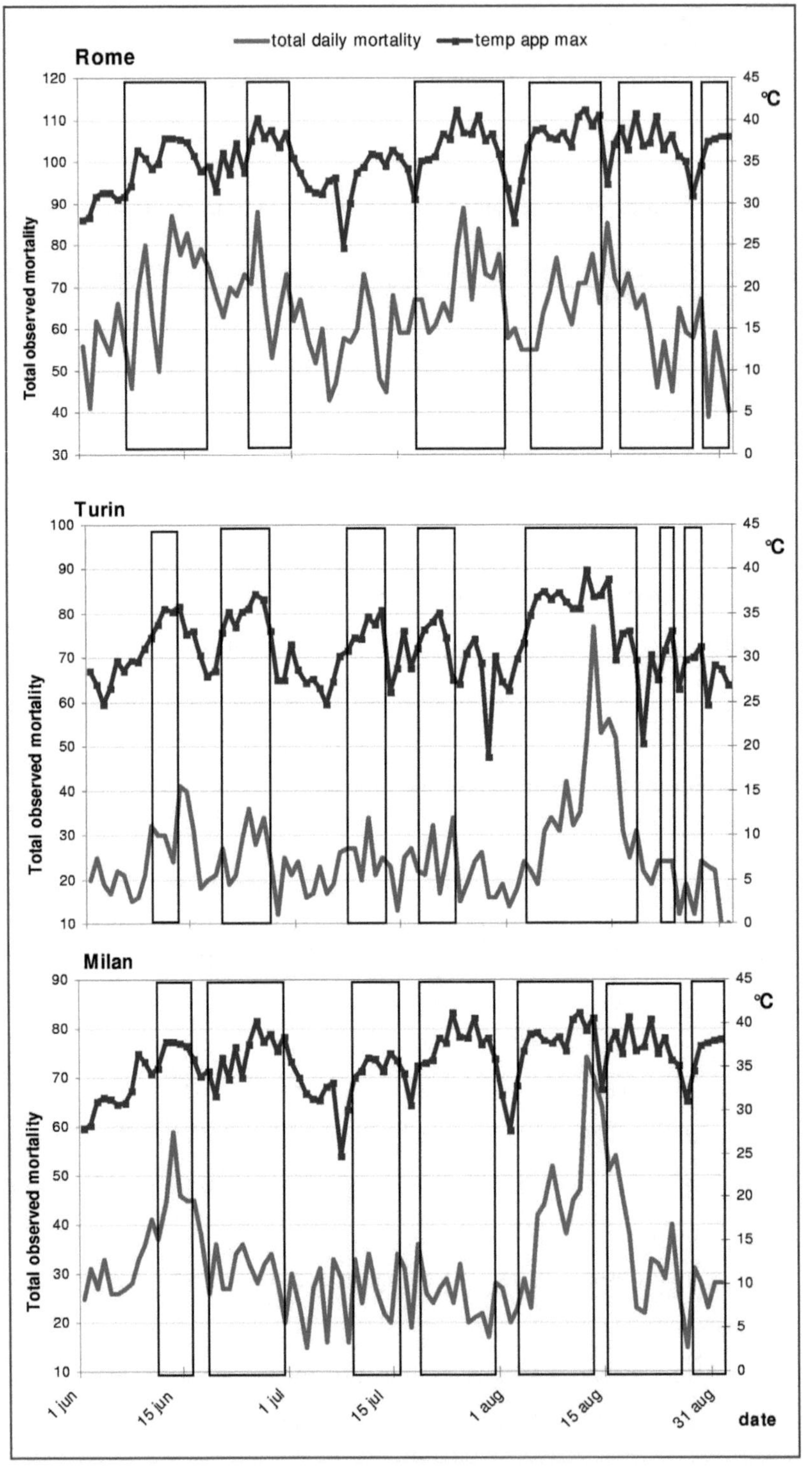

Total daily mortality and maximum apparent temperature, June 1st – 31st August 2003

Tab. 3

Total and excess mortality by cause of death in Rome, Milan and Turin during summer 2003

Causes of death	Rome				Milan				Turin			
	observed	expected	excess	%	observed	expected	excess	%	observed	expected	excess	%
Tumours	1921	1779	142	**8**	926	935	−9	**−1**	656	639	17	**3**
Circulatory	2328	1876	452	**24**	1044	832	212	**25**	892	631	261	**41**
Respiratory	327	236	91	**38**	282	155	127	**82**	201	128	73	**57**
Digestive system	227	253	−26	**−10**	121	103	18	**17**	97	85	12	**14**
Genito-urinary	81	63	18	**29**	57	41	16	**39**	40	27	13	**48**
Metabolic/endocrine gland disorders	307	247	60	**24**	111	66	45	**68**	103	42	61	**145**
Psycho-logical illnesses	96	57	39	**70**	38	34	4	**12**	70	42	28	**67**
Central nervous system	254	137	117	**86**	133	61	72	**118**	85	38	47	**124**
Total												
All causes	6009	5065	944	**19**	2968	2409	559	**23**	2332	1755	577	**33**

Discussion

The exceptional heat waves of summer 2003 had a strong impact on the population in terms of mortality, especially in the northwest where peak temperatures reached record values. The high temperatures and the persistence of these conditions were a strong determinant of the increase in observed mortality.

The study considers a limited time window of three months as a more complete time series was not available. However, throughout the literature, the time series approach is considered the most valid tool for analysing heath impact of climate change as it accounts for short-term variation and allows for an accurate estimate of long-term effects. The reference period for the quantification of excess mortality is a matter of discussion as by using different reference periods, different estimates of excess mortality are produced. The reference period has to be sufficiently long to be representative of the variability of exposure and of the observed effect but, on the other hand, not too long in order to account for long-term variation of mortality due to variations in the denominator and of mortality rates.

Daily mortality trends and peaks in mortality show a temporal variation associated to temperature trends (7,8). However, prolonged periods of high temperature may have a diverse effect on health compared to periods with extreme peak values but a lower mean. In summer 2003, persistent high temperatures maintained above-average levels of mortality. The limited time window analysis did not permit an evaluation of the harvesting effect (displacement of mortality), but lower excess mortality during the third heat wave period in Rome for example, could be attributed to a reduction in the susceptible population, as observed in other cities (7).

It is equally important when comparing results for different cities and regions to consider whether heat waves are a recent phenomenon or have been occurring on a regular basis, as the estimate of expected deaths and the associated excess might vary considerably.

During the summer months many Italian cities are typified by seasonal migration, hence populations in urban areas are reduced (eg. Milan, Rome). Although this aspect may be accounted for in the time series, it is important to note that the migratory pattern is not equally distributed among the population (9). Susceptible groups, such as the elderly and ill of lower socio-economic status, often remain in the city, creating a bias in predicted excess death. The high number of excess deaths in these subgroups might reflect the higher proportion of elderly people of low socio-economic status who remain in the city during summer. Other socio-economic factors that might have an impact on health include poor housing quality, lack of air conditioning, lack of access to social and health services, and individual behaviours (e.g., alcohol consumption and taking medication).

Analysis of mortality by cause (➋ *Tab. 3)* gave a valid insight on the effects of heat waves on health in medical terms, confirming previous results of increases in heat-related mortality by respiratory and cardiovascular diseases (4,10,11) and showing that extreme heat can worsen the conditions of people suffering from chronic disease. These results may be an important in the identification of susceptible populations and the development of effective warning systems and prevention programs. In Italy, as in other countries, the possible effects of global warming could make susceptible subgroups more vulnerable (1,2) and together with the increasing proportion of elderly people may enhance heat-related mortality. It is important to recall the heterogeneous nature of the health impact of heat waves in terms of its characteristics, such as the intensity and temporal variation in relation to the meteorological conditions between the different cities. Results from 2003 highlight the necessity of implementing further preventive actions targeting the groups of susceptible people involved (over 75+ especially females) as well as disadvantaged areas of the city and low income populations.

Demographic and social factors, as well as the level of urbanisation, air pollution and the efficiency of social services and health-care units represent important local modifiers of the impact of heat waves on health. Further analysis to observe the lag effect between heat exposure and health effects as well as the cumulative effect linked to time of exposure between extreme and normal conditions.

Concerning the latency between the peak in temperature and the increase in the mortality trends, the data showed that peaks in mortality was observed 1 – 2 days following the heat wave. These results are coherent with results of previous time series studies that reported temperature lags at 0 – 3 days as having the maximum effect on mortality and demonstrate that heat related-mortality is a very acute event requiring timely intervention.

The evaluation of the heat waves during 2003 stress the importance of introducing further preventive measures, for both the general population and susceptible groups, to reduce heat-related deaths during summer. Heat stress conditions may be predictable, and appropriate prevention measures may reduce heat-related mortality. This is achievable if efficient and effective warning systems are introduced to alert the residents of urban areas of the oppressive weather conditions.

In 2004, the Italian Department for Civil Protection implemented a national program for the evaluation and prevention of the health effects of heat waves during the summer period. Heat\Health Watch\ Warning System (HHWWS) (12,13) and city-specific prevention programs are activated during the summer period in order to reduce the heat-related deaths.

The implementation of warning systems integrated with prevention and response programmes at the national level are a valid tool for the monitoring and surveillance of mortality and the reduction of heat-related deaths during heat waves. Furthermore, the national project includes the identification of susceptible subgroups, such as the elderly aged 75+ and people with specific illnesses who are at higher risk during heat waves. Health guidelines, developed by the Ministry of Health, have been put in place for the implementation of appropriate local prevention programs.

References

Yoganathan D, Rom WN (2001) Medical aspects of global warming. Am J Ind Med40:199–210

Patz JA, McGeehin MA, Bernard SM, et al. (2000) The potential health impacts of climate variability and change for the United States: executive summary of the report of the health sector of the US National Assessment. Environ Health Perspectives 108:367–76

Albertoni F, Arcà M, Borgia P, Perucci CA, Tasco C (1984) Heat-related mortality Latium Region summer 1983. MMWR 33(37):518–521

Semenza J et al. (1996) Heat-Related Deaths during the July 1995 Heat Wave in Chicago. The NE J of Med Jul 11;335(2):84–90

Kalkstein LS, Valimont KM (1986) An Evaluation of summer discomfort in the United States using a relative climatological index. Bulletin of the American Meteorological Society 67(7):842–48

Michelozzi P, Perucci CA, Forastiere F, Fusco D, Ancona C, Dell'Orco V (1999) Inequality in health: socio-economic differentials in mortality in Rome, 1990–95. Journal Epidemiology and Community Health 11:687–93

Braga AL, Zanobetti A, Schwartz J (2001) The time course of weather-related deaths. Epidemiology 1;12:662–7

Hajat S, Kovats R, Atkinson R, Haines A (2002) Impact of hot temperatures on death in London: a time series approach. J Epid Comm Health 56:367–372

Michelozzi P, Fano V, Forastiere F, Barca A, Kalkstein LS, Perucci CA (2000) Weather conditions and elderly mortality in Rome during summer. Bulletin of the World Meteorological Organization 49(4):348–355

Semenza JC et al. (1999) Excess hospital admissions during the July 1995 heat wave in Chicago. Am J of Prev Med 16:269–277

Weisskopf M et al. (2002) Heat wave morbidity and mortality, Milwaukee, Wisconsin, 1999 vs 1995: an improved response? Am J of Public Health 92(5):830–833

Kalkstein LS, Nichols MC, Barthel CD, Greene JS (1996) A new spatial synoptic classification: application to air mass analysis. International Journal of Climatology 16:983–1004

Sheridan SC (2002) The re-development of a weather type classification scheme for North America. International Journal of Climatology 22:51–68

Response to Temperature Extremes

Lessons of the 2003 Heat-Wave in France and Action Taken to Limit the Effects of Future Heat-Waves

T Michelon · P Magne · F Simon-Delavelle

Abstract

In August 2003, France was hit by a severe heat-wave, with catastrophic health consequences (an imputed 14,800 deaths). This health crisis was unforeseen, was only detected belatedly and brought to the fore several deficiencies in the French public health system: a limited number of experts working in the sphere; poor exchange of information between several public organizations which were understrength because of the summer holidays and whose responsibilities were not clearly defined in this particular area; health authorities overwhelmed by the influx of patients; crematoria/cemeteries unable to deal with the influx of bodies; nursing homes underequipped with air-conditioning and in manpower crisis; and a large number of elderly people living alone without a support system and without proper guidelines to protect themselves from the heat.

This health crisis, without precedent since the Second World War, has had serious repercussions and has led the French government to take various steps to limit the effects on public health of any future heat-waves. Firstly, a number of studies are looking at the risk factors associated with the heat-wave. These should lead, in particular, to action thresholds being defined for given meteorological parameters. Secondly, a health surveillance (checks on the number of admissions to emergency wards) and environmental surveillance (meteorological data) mechanism is to be put in place. Finally, national and local action plans are to be drawn up by June 2004. These will clearly identify the public organizations with responsibility for heat-wave issues, their roles and the action to be taken at each level.

Introduction

In summer 2003, France was struck by a severe heat-wave, exceptional in both its intensity and its duration. These extreme conditions led to very high excess mortality that highlighted the absolute need to anticipate and better manage such extreme climate events. Against this background, the ministers of health and social affairs asked their services to identify short and medium term action to be taken to cope with heat-waves in the future.

In December 2003, the General Directorate of Health (DGS) was asked to lead a heat-wave report drafting committee. The aim of the report is to look at the lessons of past heat-waves to justify and formulate short and medium term actions in the area of crisis prevention and management in order to reduce the health effects of any future heat-waves. In particular, the report explains the national environment and health surveillance mechanism that has been established, gives a list of national bodies concerned by heat-waves with an action list for each of them, and describes the national alert organization programme, a standard departmental plan to be offered in a variety of forms, standard information messages for the various population groups at risk, and methodologies and procedures to be implemented by the préfet (head civil servant of each département, administrative division).

Lessons from past heat-waves

France was not prepared to cope with a heat-wave in any organized way. The health effects of the unprecedented heat-wave were only detected and quantified belatedly. In addition to this, management of the crisis was hampered by the lack of preparation, the shortage of cooling equipment in nursing homes and hospital facilities, and the lack of any clearly defined roles for the agencies involved.

I.1 The need for foresight

Build up expertise

As was found by the report of the parliamentary commission of inquiry, very few studies had been made prior to summer 2003 of the consequences of heat-waves. Few research teams had taken any real or long-term interest in the subject. Because of the absence of any strong signals from research, the shortage of funds, the very broad range of issues they were required to deal with and the absence of any prior discussion of the matter, the French health and safety agencies (the Institute for health surveillance (InVS) and the French agency for environmental health safety (AFSSE) set up or reformed since the law of 1 July 1998 and mandated to identify threats, coordinate expertise and assess health risks related to the environment had not included heat-waves in their programmes of work.

These findings highlighted the need for strong international surveillance of emerging risks by the above agencies, for more research into the health consequences of heat-waves (such as the climate change management and impact (GICC) programme) and for better coordination between the expert agencies and research bodies (research units, the Higher council of meteorology (CSM), the National institute for health and medical research (INSERM), etc.), so that the signals get through to the authorities more clearly and at an earlier stage.

However, the shortcomings of French expertise in heat-waves should be set in context. The international meetings which followed the August 2003 heat-wave (Centers for Disease Control (CDC), Paris, August 2003; World Health Organization (WHO), Bratislava, February 2004) showed the extent of scientific uncertainty that still exists at global level: what will the climate be like in 2100? What synergy is there between atmospheric pollution and heat? What are the characteristic meteorological parameters of a heat-wave and what levels show a significant risk? There are no clear answers to these questions.

Plan action to prevent and manage the consequences of a heat-wave

In contrast to the United States, where the geoclimatic configuration makes it possible to anticipate a heat-wave some weeks in advance, in Europe, the meteorological services can only forecast between three and seven days ahead, in the best case (source: the French meteorological service, Météo France). This leaves the authorities little time for preparation.

A heat-wave action plan – something that France did not have in August 2003 – needs to be drawn up well before the next heat-wave arrives, its content could be inspired by local action in some towns and the action plans used in other countries and towns (Chicago, Toronto, Montreal, Portugal, etc.). However, the lack of preparation in France should also be set in perspective. The CDC recently conducted a survey on the level of preparation for a heat-wave in some American towns, finding that very few of them have any real heat-wave action plan and that "they are not adequately prepared" (meeting with the CDC at the Ministry of Health, Paris, August 2003).

I.2 Build up a reliable surveillance system

Optimize environmental surveillance and pass on the information produced

There is no universally accepted definition of a heat-wave. Opinions differ on maximum, minimum and mean temperatures, relative and absolute humidity, indeed, on all the meteorological parameters that could be used to characterize one. There are several competing models: the ICARO system used in Portugal, the National Weather Service (NWS) system used by the city of Chicago, the University of Delaware system used in Toronto, the European Union "Assessment and prevention of acute health effects of weather conditions in Europe" (PHEWE) project, and others.

Despite this lack of scientific agreement, Météo France did, on 1 August 2003, note the exceptional intensity and duration of the heat-wave which was to hit France between 6 and 14 August. However, the warnings issued by Météo France between 1 and 7 August 2003 were not passed on, because, as extreme as the weather event was, the state of scientific knowledge at the beginning of August 2003 meant it was impossible to have any idea of the scale of the health catastrophe about to occur, and the messages were issued to the press rather than to anyone likely to take action. There is no clearly established dose/effect relationship, and the alert levels set by NWS for the town of Chicago cannot simply be applied even to another American town (different climate, inhabitants accustomed to heat). At the above-mentioned meeting, the CDC stated that only post-event analysis of the heat-waves that had affected France in the previous thirty years (1976, 1990 and 2003) could help set the meteorological parameters and alert levels for France. The latter must also be made specific to the principal climatic zones in the country.

Establish a surveillance mechanism to provide rapid measurements of the health consequences of a heat-wave

The August 2003 heat-wave highlighted the fact that data on the activities of the main emergency and medical services and from death certificates are not used for health surveillance purposes at national level. The InVS health surveillance mechanism was based on the notifiable diseases system and the detection of very localized infectious disease epidemics. While it may seem unbelievable that the cumulative deaths – 4000 by 8 August, 8000 by 11 August and 12 000 by 13 August, the date on which the government first announced an estimated 3000 deaths – were not detected in "real" time, it must be remembered that they were scattered through most of the country. Hence the 2000 deaths in nursing homes corresponded on average to one death per nursing home in August instead of the expected 0.5 deaths.

The phenomenon was thus practically undetectable within individual establishments or at departmental level. Only by aggregating the figures at regional or national level would it have become obvious. However, the following three points should be added to this analysis of the situation:

- the Paris region should be considered separately from the rest of the country. Given the density of both population and hospitals, an analysis of hospital and funeral figures for this zone alone would have made it possible to detect the health catastrophe occurring;
- "gross" figures for numbers of deaths or the activities of the emergency services are not informative since these data fluctuate significantly from year to year without being linked to any major health catastrophe. These figures must therefore be accompanied by a qualitative analysis (type of medical treatment, etc.);
- "the mortality curve rose from the beginning of the heat-wave, culminating 24 or 48 hours after the first temperature peak": any alert triggered on the basis of health surveillance would come too late to

warn or protect those at risk, unless the heat-wave continued over a long period. Environmental surveillance and the setting of intervention thresholds are therefore fundamental.

I.3 Better coordination of national crisis management

How should the August 2003 heat-wave be classified? Can it be considered as a natural catastrophe "like an earthquake"? Is it a health catastrophe, like an epidemic of severe acute respiratory syndrome (SARS) or legionellosis? Is it a social catastrophe, as declared by Mr Besançenot ("the hecatombs caused by heat-waves are less and less natural (…) but should increasingly be seen as the price of new poverty")?

The ministries with main responsibility for managing crisis varies depending on which way heat-waves are classified: the ministry of the interior in the first case (natural catastrophe), the ministry of health in the second (health catastrophe) and the ministry of social affairs in the third (social catastrophe).

However, traditional descriptions need to be set aside and heat-waves considered as a new type of catastrophe which demands that the divisions between the social affairs, health and civil security administrations be removed. Heat-waves should be seen as the first symptom of an underlying trend, the ageing of the population, which means that the administrations must better coordinate their own actions with those of others involved, to ensure that these particularly vulnerable sections of the population are cared for.

In any case, following the advice of the CDC, if future heat-waves are to be better managed, the roles of all those involved must be clearly defined, together with the working procedure and exchange of information at national and local level. A steering body must be identified to coordinate the action taken. The DGS could do this, but it would require more funding. The InVS should also be given a stronger position as the public organization in France responsible for receiving, detecting and dealing with all alerts with potential public health implications. This would require proper organization in the institute (24-hour on-call unit; emerging risk surveillance unit). The new public health policy act (8/9/2004) allows for such changes to be made.

Response strategy

II.1 Prevention

II.1.1 Information to those at risk

A brochure on heat-wave risk prevention aimed at the general public and particularly elderly people, parents of young children, athletes and manual workers is to be distributed nationwide from June onwards through three main channels: pharmacies, federations of home-helpers' associations, and building workers' federations (Fédération française du bâtiment). The brochure will be available on the health ministry website (www.sante.gouv.fr) in an easily downloadable version which can be printed together with technical leaflets produced by the DGS giving more specific recommendations for different population groups. These publications will be promoted to health professionals.

An information campaign will begin before the summer season in magazines for the elderly. Specific information will be given in a brochure insert containing recommendations to be followed in the case of extreme heat. A general press communiqué will be issued in April giving advice for the summer including, amongst other issues, basic advice on preventing risks related to extreme heat.

II.1.2 Cooling for susceptible persons in health establishments for the elderly

The beneficial effect of air-conditioning during heat-waves is recognized on the basis of a few, solely American, scientific studies of only cooling centralized systems. It is often stated in publications that cooling for susceptible persons, particularly the elderly, for a period of two to three hours each day would significantly reduce the risk of excess mortality. However, this statement is based on no clinical or epidemiological studies, but rather on an empirical estimate of the time needed for the human body to return to a normal temperature. According to the AFSSE, a target temperature of 25 °C or 26 °C would seem reasonable.

Although no scientific publication has yet demonstrated the usefulness of such an approach, given the number of deaths noted during the August 2003 heat-wave, the ministries of health and social affairs have strongly recommended that establishments for the elderly should have an air-conditioned area where the residents could spend a few hours each day in a cooler environment during heat-waves. Those ministries have allocated around 60 millions euros to health establishments for the elderly to buy such installations. The AFSSE has proposed recommendations for such establishments on the type of air-conditioning equipment, its installation and use.

II.1.3 Measures concerning isolated elderly or handicapped people

In order to alert isolated and/or frail elderly or handicapped people, who are the most exposed and the most vulnerable to heat-waves, a new act (6/30/2004) demands that the mayor of each 36 000administrative areas organizes a survey in which these people are identified, at their own request or that of their family or friends. The mayor will need to use all possible ways of encouraging elderly and handicapped people living in their own homes to register on the list of people to be contacted systematically as a priority to check whether they need assistance, support, visits or help, when an alert is issued. The list should be updated on a continual basis. It should be checked to ensure that the mechanism for identifying those at risk is working properly whenever the alert plan is triggered or brought into seasonal use.

II.1.4 Measures for nursing homes and health facilities

The introduction of a heat-wave "blue plan" is recommended for all collective residential establishments for elderly people: retirement homes, sheltered accommodation and long-term health care units. This plan specifies general organizational measures to be taken by each institution, whether public, private, association-run or commercial, if a crisis is announced and the alert mechanism triggered by the préfet. The plan should be discussed with organizations representing the establishments and professionals involved in the social, public health and hospital sectors. It should be regularly updated and assessed after the crisis mechanism is deactivated.

It should include in particular:

- the appointment of a specific contact person to be referred to in the case of crisis;
- the establishment of an agreement with a nearby health facility and the introduction of good therapeutic practices to prevent hospitalization and to facilitate transfers to a hospital environment;
- the installation of a cooled room by June 2004;
- the introduction of rules of procedure for organization of the establishment in case of crisis and for the issuing of an alert (mobilization of staff and possible recall of staff from leave, temporary arrangement of premises to limit the effects of the heat-wave, etc.).

Health establishments should also have an organizational plan ("white plan") to cope with a crisis, including measures to be taken to accommodate large numbers of victims. These measures would cover general organization as well as adapted measures for taking care of new patients and protecting staff (circular of 3 May 2002). The new public health act (8/9/2004) provides a legal basis for these "white plans". Each préfet should also draft a "white plan" programme to help coordinate planning at regional level and to assist with crisis management. A guide giving recommendations on the drafting of these plans (both the "white plans" and the departmental "white plans" programme) is to be published in May 2004.

At last, thanks to special funding of the ministries of health and social affairs, people were hired in nursing homes (26 millions euros in 2004) and health facilities (150 millions euros in 2004, 2000 people hired in the emergency services and the services in charge of old people).

II.1.5 Research and expertise

The french national environment and health action plan (NEHAP) recently released (6/21/2004) recommends to develop research programmes in the field of exceptional climate events health impacts. Several research programmes are already looking at heat-waves, as extreme climate events, particularly in terms of their health consequences. They are being run by the InVS, INSERM and the Ministry of the Environment. A project to set up a one million euro heat-wave and health research fund has also been planned in the framework of A new call has just been made for "climate change management and impact" research proposals. This covers five main themes, one of which concerns the health impacts of climate change. The research projects should consider:

- assessment of excess mortality caused by heat-waves or cold-waves and the definition of risk prevention and management strategies and mechanisms;
- the emergence or re-emergence of particular human and animal diseases involving a combination of various biological, environmental and anthropogenic factors linked to climate warming.

II.2 Managing a heat-wave

The national and local heat-wave management mechanism is based on levels of alert, the identification of the agencies involved and measures to be taken by those agencies when the different levels are reached. "Decision support forms" appended to the report (cf. introduction) lay out the measures that could be taken by the main national agencies concerned at each of the alert levels. They constitute the national heat-wave management plan (PGCN). Similar work has been done at local level, and a departmental heat-wave management plan (PGCD) has been drawn up for adaptation and approval by the préfets, taking local specificities into account.

II.2.1 Alert levels

The national and local heat-wave management mechanisms have four graded levels of response succeeding each other in time.

- Level 1 corresponds to seasonal surveillance being activated. This happens on 1 June each year to allow each service concerned at national, departmental and local council level to check that the system is working, including the mechanisms for identifying people at risk, and that the measures in the higher levels of the plan can be put into operation. This level is deactivated on 1 October each year.

The three other levels (level 2, level 3 and level 4) involve graduated responses. They are triggered or deactivated by regional biometeorological thresholds (cf. II.2.2).

- Level 2 corresponds to mobilization of local and national public services, primarily in the health and social sectors, when Météo France forecasts three or more days in advance that biometeorological thresholds will be exceeded in at least one region.
- Level 3 is activated primarily when an InVS alert bulletin announces that daily biometeorological indicators will exceed thresholds in at least one region, or on the basis of other available elements (excess human or animal mortality linked to high heat levels). The local and national public services take mainly health and social measures, with particular emphasis on publishing information and helping to cool down the people at risk identified.
- Level 4 is activated primarily when it is forecast that biometeorological indicators will exceed thresholds in several regions in the next 24 hours and that this will persist over a long period with the appearance of collateral effects (power cuts, drought, hospitals overwhelmed, etc.). The heat-wave causes a crisis with consequences that go beyond the health and social fields. Exceptional measures are taken to cope with the event.

II.2.2 Health and environmental surveillance mechanism

Provisional heat-wave alert system based on meteorological data

On the basis of meteorological and mortality data recorded between 1973 and 2003 in 14 french towns throughout the country with contrasting climatic situations, the InVS has tested several biometeorological indicators and retrospectively calculated the number of alerts detected as true or false and the number of missed alerts. The choice of indicators and thresholds beyond which a fairly large-scale epidemic phenomenon (daily mortality multiplied by 1.5 or 2) can be expected during a summer heat-wave event, at the same time minimizing the number of missed and false alerts, has been made. The indicator chosen is the three days average of the minimum temperature and the three days average of the maximum temperature. 96 thresholds (couple of figures) have been determined: one per département (cf. annex).

In June 2004, Météo France has introduced a mechanism which allows it to specifically monitor the biometeorological indicators set by the InVS and to inform it whenever these indicators or the three-day or seven-day forecast are exceeded in one or more climatic regions of France; the InVS will then be responsible for alerting the national and local agencies concerned (cf. II.2.3.).

In the case of a 24-hour forecast that exceeds the abovementioned thresholds set by the InVS, Météo France will issue a map of France with orange or red zones, together with regular bulletins describing the development of the phenomenon, its trajectory, intensity and end, updated twice a day at 6.00 hrs and 16.00 hrs, and made available on the Météo France website (www.meteo.fr).

Surveillance and alert network based on health data

By June 2004, the InVS has introduced a surveillance and alert network based on computerized feedback from emergency services in 20 health facilities. It is intended to expand the network after its evaluation phase to include other establishments; it will also include some departmental fire and emergency services (SDIS) and associations such as SOS médecins.

The InVS is also to conduct a feasibility study in 2004 on the implementation of a health surveillance system based on a daily mortality report from computerized local council data. This system should allow for early detection of any significant change in mortality figures and make it possible to generate an alert where necessary. The system will be part of the wider epidemiological surveillance programme described above.

II.2.3 National heat-wave management organization

"Decision support forms" have been drawn up for the national heat-wave management plan (PGCN). These lay out the measures that could be taken by the main national agencies concerned by the heat-wave for each of the alert levels.

Measures for each level

- Level 1. An interministerial heat-wave committee, chaired by the DGS, checks at the beginning of the summer is to ensure that the structural measures have been taken in nursing homes and hospitals and that the PGCN and local plans are operational. It also ensures that national information campaigns to publicise heat-wave recommendations to the various population groups at risk have been properly implemented. At the end of the season, it will be responsible for assessing the effectiveness of measures taken during the summer.
- Level 2. The InVS issues a health alert bulletin to the health minister, as well as to local agencies concerned. A ministerial crisis unit will be set up. Its responsibilities will be: to guide and coordinate general action, to process information continuously submitted by the different ministries and decisions taken by the authorities responding to the crisis, and to channel announcements and the supply of information to the press and the general public. The crisis unit will implement the action planned in the decision support forms: information messages to the public and health professionals, updating of the heat-wave information on the health ministry web-site, preparation for the activation of a low-cost hotline of information, information to health facilities and professionals to ensure care for frail or dependent people. At a local level, the préfets and their services alert nursing homes, health facilities and the mayors and ask them to be ready to activate their "blue" or "white" plan in case of level 3.
- Level 3. The InVS issues a health alert bulletin to the health minister, as well as to local agencies concerned. The crisis unit mentioned at level 2 above ensures it is able to work 24 hours a day if necessary. It has the same general tasks as under level 2. From the data collected from the services throughout the country, it makes daily checks to ensure that the measures implemented are adequate. The ministerial crisis unit implements the action provided for in the decision support sheets, particularly those mentioned in the previous paragraph for level 2. It issues warnings, alerts and recommendations as planned in the appendix to the PGCN. Heat-wave zones are labelled orange and/or red on the weather watch map. A low-cost hotline of information is activated. At a local level, the "blue" or "white" plans of nursing homes and health facilities are activated and the mayors contact isolated elderly people at risk already registered (cf. II.1.3) and check whether they need assistance.
- Level 4. The ministers of health and the interior propose implementation of level 4 to the prime minister, who gives national responsibility for heat-wave management to the minister of the interior. The crisis unit mentioned in level 3 above makes itself available. The heat-wave zones are labelled red on the weather watch map. The minister of the interior takes all useful measures to cope with the event, including policing and requisition measures.

II.2.4. Recommendations in the case of extreme heat

The health consequences of a heat-wave require recommendations consisting of simple and operational health messages to the general public; vulnerable people, such as the elderly, children and infants, people with chronic pathologies, people taking certain medicines, people with mental problems and their carers; specific population groups (athletes, manual workers, vulnerable people); and health and social professionals responsible for those groups.

These messages and recommendations for individuals and groups are adapted to the four levels of the heat-wave plan. They are explained in the form of directly usable brochures which may be adapted according to the level and the target group, and can be summarized as follows:

- Before the extreme heat begins, generally at the beginning of the summer, action should include: arranging premises, particularly living areas; logistics (supplies of water and medicines, staff management); identification of those at risk; information to the public on the risks and recommendations for protection against the consequences of heat.
- When a heat-wave is forecast, the principle is to remind the public of ways of protecting themselves from the heat (arrangement of living area, individual protection, advice on hygiene), to organize surveillance of people at risk, to check that those concerned are well aware of the problem and are familiar with the measures to take to protect themselves against the health consequences of heat, and to recommend to people at major risk that they should consult their doctor to ensure that their care is properly adapted.
- When a heat-wave occurs, the principle is to implement the accompanying measures, surveillance and protection of those at risk, to ensure that mutual aid networks function, to organize individual and group protection and to identify and give warning of the signs of pathologies related to excess heat as soon as possible.

Conclusion

All the action suggested above is intended to reduce the impact of any future heat-wave which, according to the weather experts, is a strong probability in France as well as the rest of Europe because of climate change. The health consequences of the August 2003 heat-wave in France were particularly serious, but could have been exacerbated by related and highly probable events: power cuts, drought and others. This justifies the authorities' devoting funding to the issue and developing organizational programmes to cope with such situations.

➔ *Figure 1*

Fig. 1

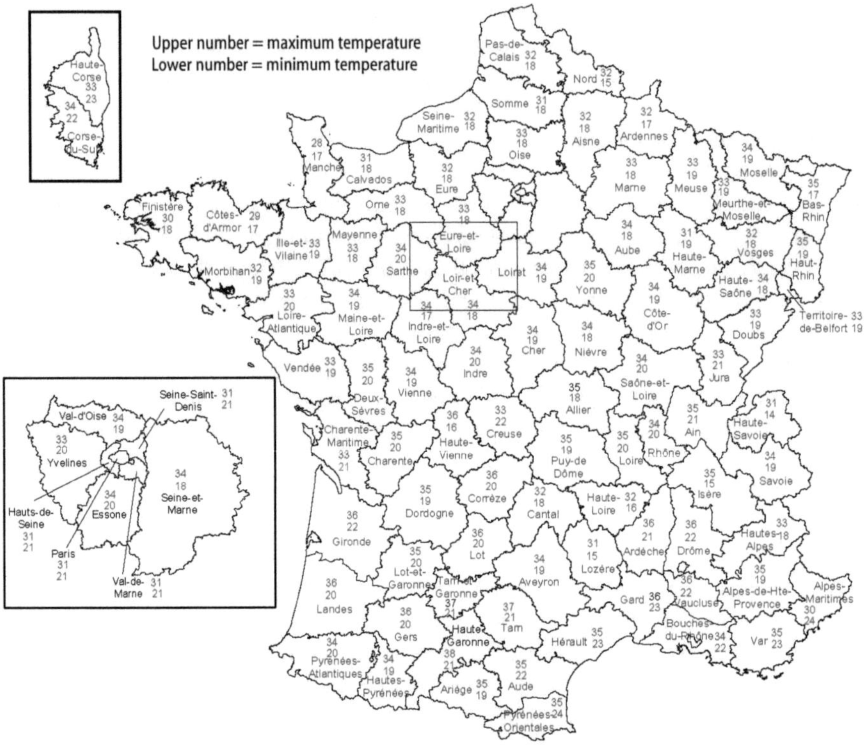

Annex: Cutoff Points for every Département (copyright Fond cartographique©IGN – BDCARTO® 1999)

Examples of Heat Health Warning Systems: Lisbon's ÍCARO's Surveillance System, Summer of 2003

Paulo Jorge Nogueira

Abstract

During the summer of 2003 Portugal was under unusual heat stress, particularly in the period from 27th July to 15th August, when almost all Portuguese districts had weekly maximum temperatures above 32 °C.

In Portugal an operational Heat Health Warning System has existed since the summer of 1999. This system is based on meteorological data and gives three days advanced heat wave predictions. This 2003 summer had several unusual heat periods that were extremely well predicted.

This article aims at presenting the Heat Health Warning System, detailing its background, methods and its five years experience. Beyond that a particular review of the summer of 2003 is done.

The 2003 summer July–August 17 day heat wave seems to have generated about 2200 excess deaths. When age, sex and district population adjustments are made the excess mortality is evaluated at 1953 heat related deceased.

Heat related mortality affected mainly elder and female individuals.

The surveillance partners had difficulties in conveying out messages to the population, using the media, late in the heat stress period.

Our 2003 summer experience lead to the conclusion that active ways must be sought to convey information to the population, when such a silent disaster is predicted. Passive systems, such as using the media to spread messages of interest during heat stress periods, are not reliable especially in a very long heat wave.

Introduction

During the summer of 2003 Portugal was under unusual heat stress, particularly in the period from 27th July to 15th August, when almost all Portuguese districts had weekly maximum temperatures above 32 °C. In fact, 13 of 18 districts had weekly mean maximum temperatures above 35 °C during that period.

The impact was detected early by the ÍCARO surveillance system, on the 30th July high ÍCARO-index was observed and a special warning was issued. In fact, from 28th July to 14th August the predicted Index for the following 3 days was always above zero, indicating presence of heat-mortality risk. On 11 of those 17 days, the ÍCARO index was above the warning threshold 0.93 (➋ *Tab. 1*). The higher indexes were predicted on 1st August with 2.44 and on 12th and 13th August with indexes above 3.

Tab. 1
Pre-established ÍCARO surveillance system warning levels

ÍCARO-INDEX Value	Warning Level
ÍCARO-Index = 0	No Effect – No Warning
0 <= ÍCARO-Index = 0,31	Non statistically significant effects on daily mortality
0,31<= ÍCARO-Index <0,93	Possible effect on mortality
0,93 <= ÍCARO-Index <1.55	Heat wave alert in analysis
ÍCARO-Index >=1.55	Heat wave alert – serious consequences on health and mortality expected

Note: since these warning levels are based on predicted excess mortality the Index cannot have negative values. Observed mortality when converted to ÍCARO-index can have negative values.

A posterior look at mortality data showed that predicted peaks of the ÍCARO index, which reflected also variation on daily predicted temperatures, related suitably with overall observed mortality. In 2003, the major summer heat wave was responsible for about 2000 excess deaths.

This article is divided in three sections. The first section is called Lisbon's ÍCARO Surveillance System and describes the ÍCARO surveillance system background, origin, implementation, organics and past experience up to the year 2002. The second section is called ÍCARO's summer of 2003 experience and reports last summer events from systems' perspective, describing observed summer temperatures, collected mortality data, summer predicted ÍCARO'S indexes and their interrelation; overall estimates of observed heat related excess deaths is made using both 2003 and past summer mortality data; a brief discussion of the effective system's intervention is also made. The third section is a conclusion and discussion section where several strong and weak points of the work herein are presented and discussed.

Lisbon's ÍCARO Surveillance System

1 Genesis

Ícaro is the Portuguese version of the name Icarus, the Greek mythological character that died because of the sun's heat. In 1998 when a name was sought for the study of the importance of heat on mortality, the name ÍCARO sprung naturally as an acronym for the Portuguese title "Importância do Calor e a sua Repercussão nos Óbitos" [importance of heat and its repercussion on deaths].

The "ÍCARO Project" stands for a research line within the Portuguese National Health Observatory. And the ÍCARO Surveillance System (of heat waves with probable impact on human health) stands for the Portuguese Heat Health Warning System. On a romantic note, the ÍCARO Surveillance System aims at transforming all potential Icarus' into wiser Deadalus' (Deadalus was Icarus' father, who while using the same wings did not die of the sun's heat).

The ÍCARO surveillance system of heat waves with probable impact on human health and mortality was optimised to its current version in 1998 and implemented in the summer of 1999.

Its roots are well defined in the events that occurred in the heat wave of June 1981 in Portugal. This was a notorious event, where official heat related deaths, along with deaths of animals as chicken and rabbits, lack of drinking water in several districts of the country, concerns about exhausted stocks of bottled refrigerants and beer were cited on the daily newspapers during the 10 day heat wave itself. At the time, ruptures at health services levels were felt, as well as excess of patients and lack of place to hold the deceased. Evidence of a heat wave effect on weekly mortality for the district of Lisbon was shown in Falcão et al. (1988).

In July 1991 another noteworthy heat wave occurred, with similar length, apparently not as hot, and with milder effects (Paixão & Nogueira 2002 and 2003).

By 1998, a full study on the effects of the 1981 heat wave was available, estimating an outstanding overall excess of 1906 deaths in what is called continental Portugal (Portugal minus the autonomous regions Madeira and Azores). This effect was statistically significant in 16 of 18 districts, for both sexes, and for all age groups except females below 15 years old (Garcia, Nogueira & Falcão 1999).

By then further studies were underway which showed that it was possible to model the heat-mortality relationship in a suitable simple way. Nunes e Castro (1998) used a time series approach that could be simplified when using only summer mortality and temperature data instead of full year data.

Therefore, a simple way of predicting heat waves with impact on mortality was available. Such knowledge could generate life and health gains and prevent premature avoidable deaths. The idea of a surveillance system of heat waves sprung up naturally and a partnership with the Portuguese Meteorology Institute was established allowing daily summer surveillance.

The creation of the surveillance system only within the two institutions was not ideal, since these institutions do not have an active intervening role near the health system and the population in general. In order to have a full organic surveillance system, the Portuguese National Health Directorate and the National Service of Firemen and Civil Protection, which have the mandatory ability to act within the health system and the population in general, were contacted and were manifestly interested in being part of the system.

Since the modelled heat-mortality relation is based on predicted number of deaths, and this is just an epidemiological indicator, an alternative indicator was created instead to parlay information between the surveillance system partners – the ÍCARO-Index (Nogueira et al., 1999).

2 Organic flow

The ÍCARO surveillance system information flow starts in early May every year. The system's exchange of information starts between the Meteorology Institute (IM) and the National Health Observatory (ONSA), with the former sending forecasted weather conditions for 3 days (current day plus 2 days). Based on that information the ÍCARO-Index for the 3 days period is calculated daily at ONSA and sent back. On the 15th May this procedure is broadened to all system partners – by this date a full report can be produced (➔ *Fig. 1).*

The system has 4 warning levels (➔ *Tab. 1)* based on the 95 % confidence interval for Lisbon's district mortality when no abnormal weather conditions occur. Converting this mortality to ÍCARO-indexes, daily mortality varies within the values –0,31 and 0,31. The several warning levels were successively built adding this interval range (see Nogueira 1999) for technical details).

The several warning levels imply different direct interactions among the various system partners. Higher levels of risk, which imply heat wave alerts, lead to great interaction and systematic re-evaluation of observed temperatures and predicted indexes and weather conditions. When any established risk level is reached, personal messaging is made and intervening institutions issue their pre-established prevention measures.

Fig. 1

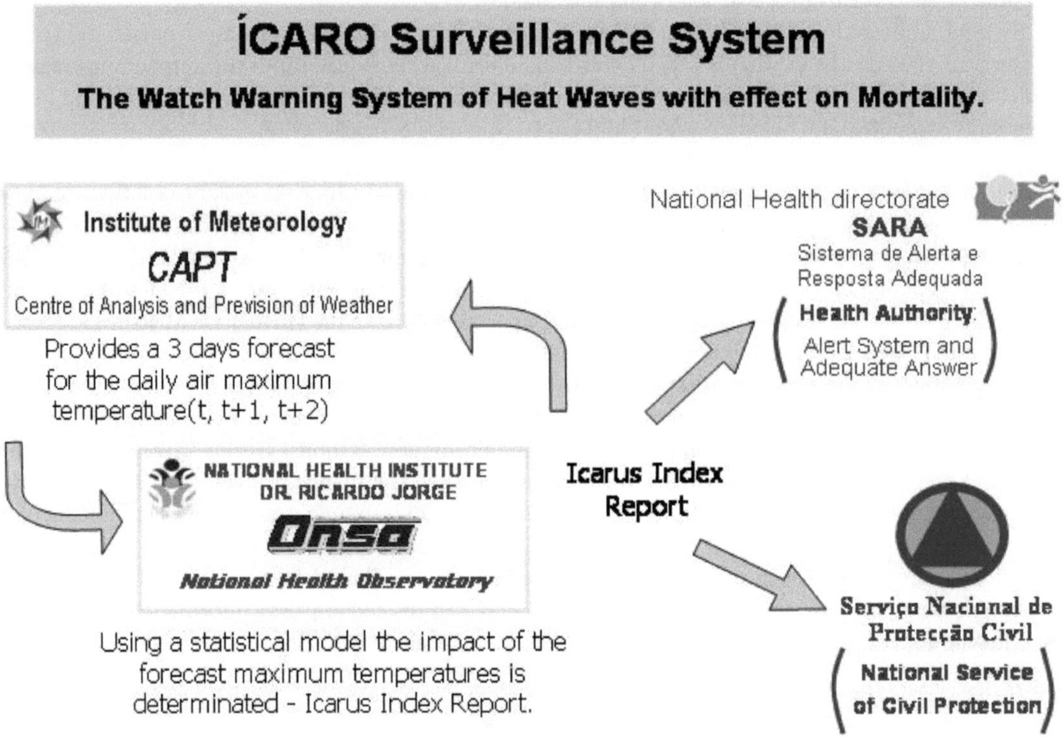

Lisbon ÍCARO Surveillance system (source: Onsa: Observatório Nacional de Saúde)

3 System scope and its past experience

The system is based on the information for the district of Lisbon and is expected to be a reasonable indicator for the rest of the country's situation. The existent information on 1981 and 1991 supports this assumption (Garcia, Nogueira & Falcão 1999 and Paixão & Nogueira 2003). Although it is a city-based system, it was mainly built on the information for its district level and is meant as a good indicator for the national level.

Since 1999 there have been several different episodes of heat waves' alerts. In 1999, the system's first year, no warning or alert was issued. The ÍCARO-Index rose up to 0.7 for one day in the beginning of July and although the observed daily mortality is not statistically different from the expected, an alteration on the randomness of the mortality process was visible (Nogueira 2001; WHO 2004). This new information gave the surveillance system an insight beyond what was previously modelled and known.

In 2000, on 15th July a first alert was issued with the 3-day predicted index with similar severity of the observed heat wave of 1991. Efforts were done by the system partners to put the information out to the health system and the population. Alert response was probably not as prompt as was required and collaboration from social media was disappointing. But oddly, two days later no such indexes were expected due to sudden changes in the weather forecast for Lisbon. Later analysis, when mortality data was available, showed that ÍCARO was absolutely correct for the district of Lisbon. But curiously, a national overall daily excess mortality was observed for the period from 15th July to 17th July just as predicted by the surveillance system on the 15th July (Nogueira 2001) and illustrated in ❯ *Figure 6.*

In 2001, a remarkable situation occurred in late May where temperatures rose from around 18 °C on the first fortnight to seven consecutive days above 30 °C. At the time awareness existed that the system was not calibrated for such occurrence.

Though heat was felt and a relative awareness of a heat wave was developing, the ÍCARO index issued only very mild warning levels. This happened because temperatures never really rose above 32 °C for more than two days, which is the departure level of the system.

A first estimation on this early heat wave was evaluated while preparing Paixão and Nogueira (2002), where an excess of 397 deaths was presented for the period of 27th May to 1st June. Future re-evaluations of this occurrence with different heat wave definitions might lead to a slightly wider estimate. In 2002 several warnings were issued but an evaluation is not yet possible because mortality data is not available yet.

ÍCARO's summer of 2003 experience

4 Temperature

This summer had several hot periods and was characterised by the following point:

- Hot temperatures occurred in late May, with temperatures equal or above 30 °C for 6 days between 21st and 31st May;
- In week 25 (➲ *Tab. 2*), maximum air temperature raised from about twenty degrees up to 39 °C in Lisbon on the 19th June. Although it only lasted about three days it resulted in a week with excess temperature profile;
- A long hot period occurred late July – August, it lasted for about a fortnight from the 29th July up to 14th August (weeks 31 to 33). As it is shown in table 2, 16 out of the 18 continental districts of Portugal had mean temperatures above 32 °C for two full weeks, two of those had even temperatures above 40 °C. This event had no known precedent, but Lisbon's maximum air temperature from June 1981 was not reached.
- Another hot period was felt on early September (weeks 37 and 38) which was also an unexpected late summer hot period, with no presently known equivalent in the past.

In ➲ *Figure 2*, daily temperatures for Lisbon and the daily mean temperature of the 18 districts are presented. They show that the temperature of Lisbon is very closely related to the overall temperature of the mainland. Differences only arise in the relative minimum and relative maximum temperatures occurrences. The most noteworthy is the difference on 4th August where the daily maximum temperature for Lisbon goes clearly below 32 °C while the overall mean stays above 32 °C.

5 Mortality data

Mortality data presented here was only possible to assess due to the unusual occurrence of this summer heat wave. In fact, due to the awareness of this heat wave problem, the Portuguese General Health Directorate and the Portuguese National Registrar Directorate were able to optimize the regular system flow of the death certificate registration and coding. This change consisted in the registrars' offices actively sending facsimiled death certificate copies to the Mortality Coding Unit of the General Health Directorate. This resulted in years advanced data (as an example at the time this paper is being written 2002 data is still not available for Health scientific purposes).

Tab. 2

Weekly mean maximum daily air temperatures for 18 districts of Portugal (districts Ordered from North to South)

Week	Viana do castelo	Porto	Braga	Vila Real	Bragança	Viseu	Aveiro	Coimbra	Guarda	Castelo Branco	Leiria	Lisboa	Santarém	Portalegre	Évora	Setúbal	Beja	Faro
18	18,6	17,5	18,6	18,4	18,4	17,5	19,2	19,7	16,1	20,7	21,4	20,3	22,6	20,0	21,7	19,4	21,7	21,0
19	21,6	18,6	22,1	21,3	21,1	20,3	18,4	21,0	15,8	23,8	20,9	21,8	23,8	22,7	24,8	23,7	24,4	22,1
20	21,7	19,7	22,3	22,5	23,4	21,9	19,5	22,8	20,2	26,5	23,1	23,7	25,4	25,9	27,9	25,9	28,1	26,5
21	25,7	23,6	26,1	24,0	24,1	23,6	22,8	26,7	20,7	27,1	27,6	28,0	**31,9**	26,8	29,7	29,4	**30,8**	27,6
22	25,1	23,0	26,4	26,1	27,2	24,9	23,7	27,0	23,3	28,1	27,2	26,7	28,0	27,0	29,4	29,0	**30,5**	25,4
23	22,7	20,4	23,5	24,4	26,4	23,1	20,9	23,8	21,8	28,7	24,3	25,1	27,5	27,7	**30,4**	27,3	**31,3**	26,8
24	27,5	25,5	28,5	29,4	**31,4**	28,3	24,7	29,4	28,2	**32,7**	28,2	28,6	**31,8**	**31,3**	**32,7**	**31,2**	**32,7**	29,5
25	**30,6**	28,4	**31,0**	**31,7**	**32,9**	30,0	27,8	**31,4**	29,1	**33,7**	**31,3**	**30,1**	**34,0**	**33,0**	**34,7**	**33,0**	**35,1**	28,7
26	21,9	20,5	22,6	22,9	26,0	21,1	20,6	23,1	25,7	27,7	24,1	24,0	26,3	26,1	29,0	26,3	29,4	26,5
27	25,1	21,9	24,0	24,1	25,6	23,4	22,7	24,8	22,6	28,6	24,9	25,5	27,9	27,4	**30,1**	28,3	**30,8**	27,0
28	25,4	21,9	26,1	**30,2**	**33,4**	28,0	22,2	24,3	29,1	**32,7**	24,9	26,4	28,5	**30,3**	**32,5**	29,7	**33,2**	28,5
29	23,1	21,7	22,6	24,1	25,3	22,5	22,9	24,9	22,1	28,2	26,1	26,2	28,1	27,2	29,8	28,4	**30,4**	29,1
30	24,0	22,6	24,6	25,9	28,6	25,0	23,2	26,0	25,1	**32,3**	27,2	27,6	**30,3**	**30,8**	**32,5**	30,0	**33,3**	**31,8**
31	**32,6**	**30,6**	**33,5**	**35,4**	**35,8**	**34,5**	28,7	**36,6**	**33,1**	**38,8**	**36,4**	**36,9**	**41,0**	**39,0**	**41,3**	**40,3**	**41,7**	**33,5**
32	**34,4**	**31,1**	**36,3**	**37,2**	**38,1**	**36,9**	27,7	**36,4**	**34,1**	**39,2**	**36,1**	**35,9**	**40,4**	**39,0**	**40,2**	**35,7**	**40,5**	**33,5**
33	28,3	25,3	28,9	**30,1**	**32,5**	29,0	24,9	28,5	29,0	**33,7**	28,9	**30,3**	**32,2**	**32,7**	**34,8**	**32,4**	**35,0**	**32,6**
34	28,1	24,4	29,4	29,9	**30,8**	29,5	23,5	28,8	28,0	**32,5**	28,8	27,5	**31,2**	**31,4**	**32,9**	29,5	**33,0**	27,8
35	23,9	22,9	23,5	22,3	22,4	21,2	24,1	25,0	19,2	25,6	26,8	25,3	28,3	24,8	28,6	26,3	29,2	25,8
36	23,8	22,4	25,3	23,5	23,7	24,1	22,7	23,7	20,4	25,3	25,5	23,7	27,3	26,1	27,6	26,9	29,1	26,8
37	**30,5**	29,9	**32,1**	**30,2**	29,5	28,3	28,8	**32,6**	26,4	**32,1**	**33,7**	**32,6**	**35,8**	**31,2**	**34,2**	**35,0**	**35,2**	29,5
38	26,3	25,7	28,3	27,9	28,3	29,3	26,6	28,9	23,3	28,3	**30,8**	28,9	**31,7**	**30,1**	**30,3**	**30,3**	**30,5**	27,0
39	23,1	22,7	25,2	25,3	25,6	25,4	23,0	24,7	23,7	25,2	26,7	24,7	29,1	27,4	29,3	28,0	28,8	25,7

Source: Instituto de Meteorologia, Portugal

Fig. 2

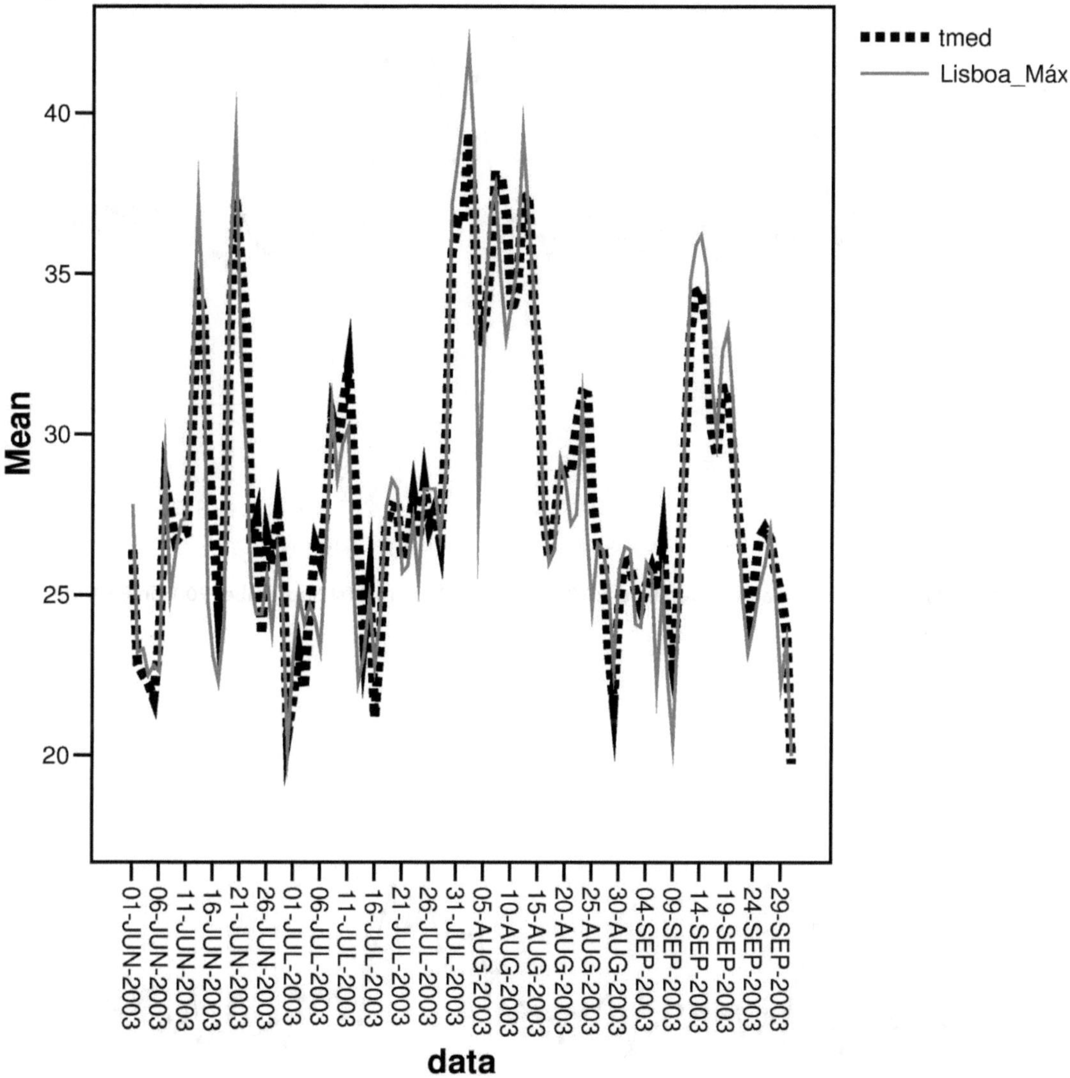

Lisbon Daily maximum temperature and Mean maximum daily temperature for the 18 districts of mainland Portugal (source: Instituto de Meteorologia, Portugal)

Fig. 3

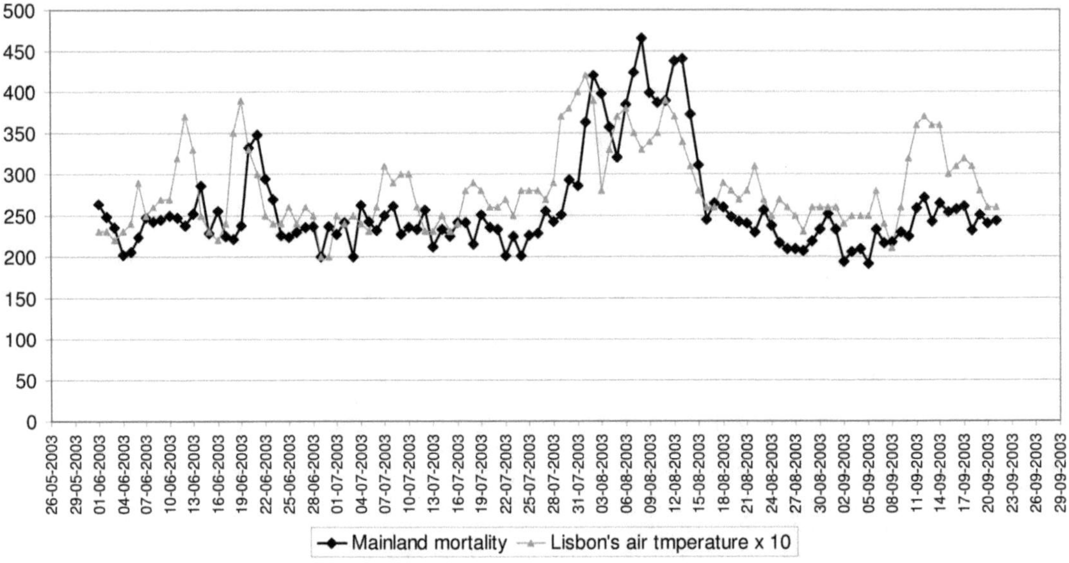

Daily Mortality and Maximum air temperature × 10 °C from 1st June to 21st September 2003 in Portugal mainland

Fig. 4

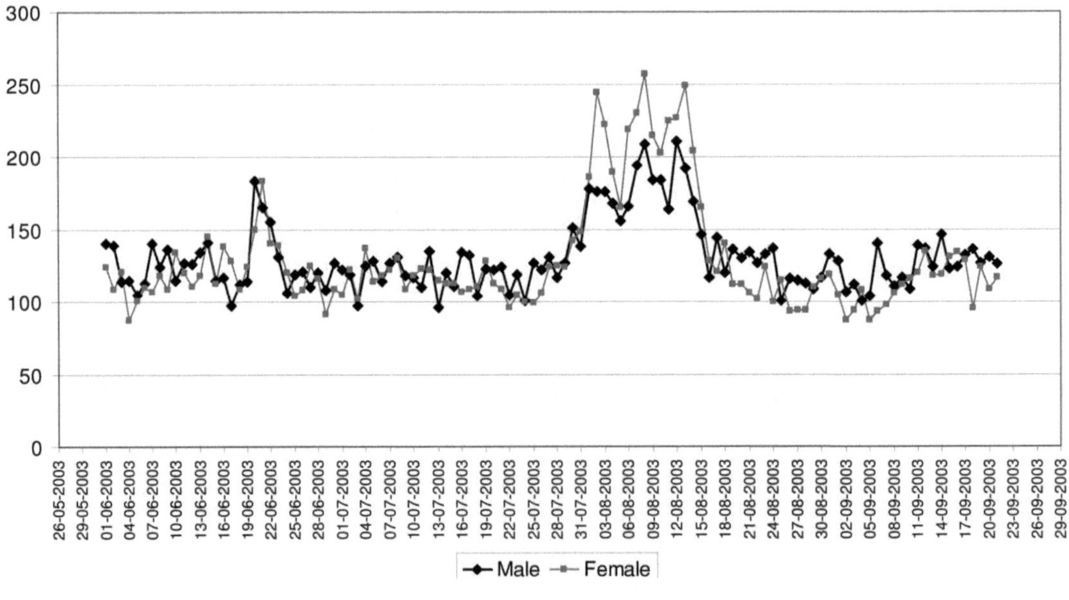

Daily Mortality by sex from 1st June to 21st September 2003 in Portugal mainland

Fig. 5

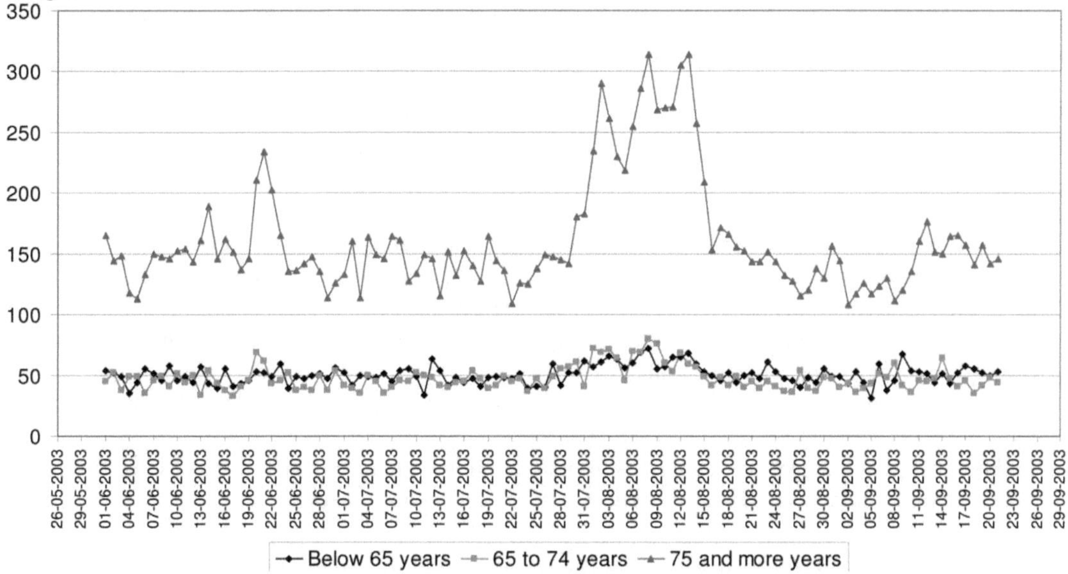

Daily Mortality by Age Group from 1st June to 21st September 2003 in Portugal mainland

Fig. 6

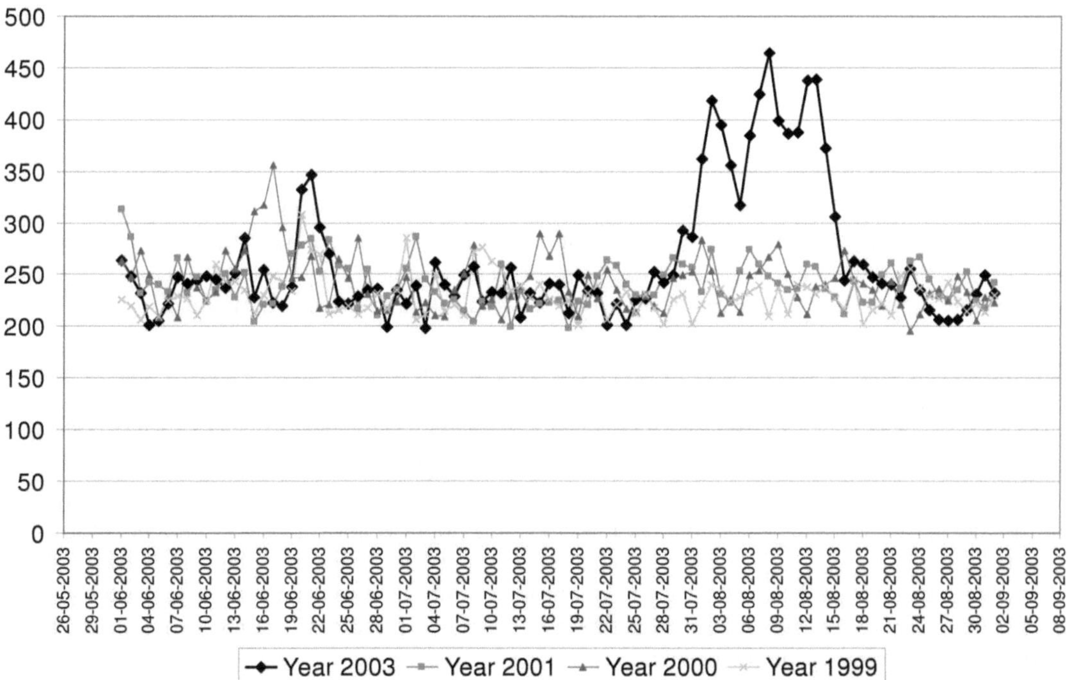

Daily Mortality from 1st June to 1st September in 1999 to 2001 and 2003 in Portugal mainland

Figures 3 to 6 show daily summer mortality series for some population strata. The main finding is how close mortality relates with the temperature, especially on the heat period of July – August, where the ups and downs on the mortality processes follow the ups and downs on observed temperatures.

Another important feature that has been described elsewhere (Braga AL, Zanobetti A, Schwartz J 2002, for example) concerning high temperatures' harvest effects on respiratory and cardiovascular mortality, and this had not been previously observed in Portuguese heat wave mortality data, is the apparent harvest effect at the end of August to the beginning of September.

Figure 4 compares mortality between males and females. Several features are noteworthy:

- Both mortality processes are in synchronization with the temperature evolution, same ups and downs, up to the beginning of the July – August heat wave period;
- Heat impact clearly visible around 19th June equally affects both sexes, previous to that and right afterwards mortality levels are also similar;
- On the longer and hotter extended July – August period, female mortality is higher than male mortality;
- In the subsequent time period to the July – August heat wave, female mortality shows a lower level than male, but the consequence of a harvest effect is still visible in both sexes.

Figure 5 shows evident differences between age groups. Mortality is particularly evident in the higher age group, 75 and more years old. The 65 to 74 year old age group also presents an important effect but still far from the one from that of the higher age group. The other age groups also seem to relate to the heat periods in some way but their global mortality is almost irrelevant to the total heat related mortality.

Figure 6 compares several recent years' summer mortality, delimiting clearly the 2003 heat wave mortality epidemic curve.

6 The ÍCARO Index

Obviously, interest always lies in knowing how well predicted heat related mortality compares with observed mortality data. Would it relate well in the summer of 2003, when so many aspects were new? New was the exceptionally full hot summer, with heat periods in May, June, July – August and September and with an early summer period; the heat wave length; as well as the late summer heat wave.

Figures 7 and 8 show the relation of the ÍCARO index and observed mortality for Portugal mainland and for the district of Lisbon respectively. Results are remarkably good.

It is not surprising that ÍCARO results seem better for Lisbon's district level, since model coefficients used were derived from past experience here. At national level, predictions were somewhat over- or under estimated, but thinking on a national heat health warning surveillance system, main excess mortality moments were particularly flagged in a practical way. On the unprecedented long July – August heat wave, ÍCARO's surveillance system correctly predicted the different upheavals (up and down mortality directions), which is also overwhelming since previous heat wave modelled data did not include such long heat waves with mortality changing in several directions. This might speak for the rationale used in the heat-mortality relation proposed on modelling.

At the Lisbon District level, several aspects are also worth noting: The accuracy of 19th – 20th June and 4th August mortality predictions, particularly the latter, where mortality comes down to the non-daily statistically significant level as opposed to what happens to the national level. In fact, both two downward mortality levels in the middle of the big heat wave are particularly well predicted. Discrepancies to the

Fig. 7

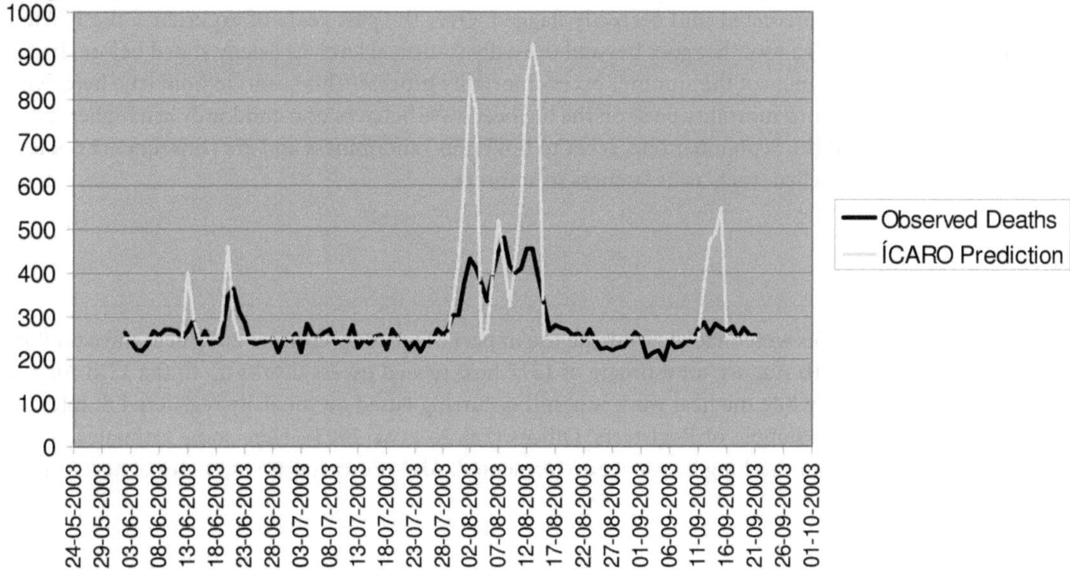

Portugal Continental: Daily ÍCARO Surveillance System predicted mortality and observed daily mortality from 1st June to 21st September 2003 in Portugal mainland

Fig. 8

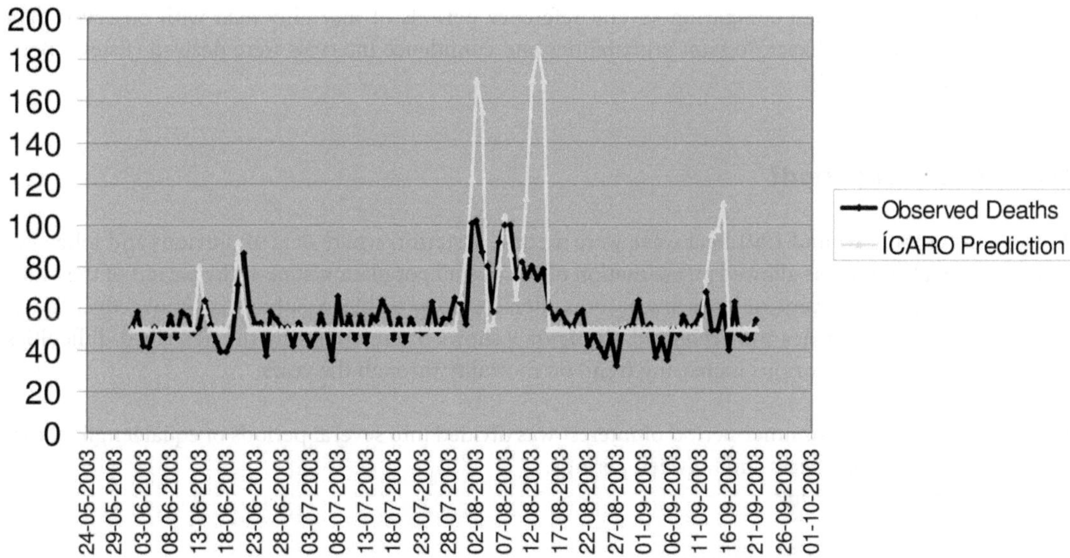

Districts of Lisbon 2003: Daily ÍCARO Surveillance System predicted mortality and observed daily mortality from 1st June to 21st September 2003 in Lisbon district

whole mainland level can be related to the fact that overall districts remained above their threshold risk levels, contrary to what happened in Lisbon.

At both levels, some of the predicted mortality peaks seem excessive. This is not particularly threatening as long as the risk is predicted und correctly flagged. Over the past years of experience the ÍCARO index has revealed itself as a tool that goes beyond the daily statistical level, as briefly stated before. It also flags alteration on randomness of the summer excess mortality process. This seems to hold true here, with some new features. The third mortality peak on the big heat wave behaves non-randomly at a higher mean level. This late peak, and the September one, seem to hold non-randomness and are consequences of features not previously modelled, such as its lateness in summer.

7 Early estimates of excess deaths

Very early, serious attempts were made to evaluate the impact of the long summer 2003 heat wave on Portuguese mortality. On 20th August, an estimate of 1317 heat related excess deaths up to the 12th August was presented, evaluated while the heat wave was still occurring based on the daily registered number of death data collected for a sample of Registrars' Offices (Falcão et al. 2003). Here some estimates of observed excess deaths due to this long heat wave are presented. Also presented are estimates of heat impact period.

Data and statistical Analysis

Mortality data used for 2003 are those discussed above obtained by the General Health Directorate, for all other years National Statistics Institute (INE) mortality data were used. Data on climatic conditions were kindly supplied by the Meteorology Institute.

Data analysis was done comparing several reference periods of mortality data with observed data in the summer of 2003. Exact Poisson probabilities and confidence intervals were derived (Esteve, Benhamou & Raymond 1994).

Which reference period?

Evaluations of the 1981 and 1991 heat wave were made a posteriori where data of previous and following years were available. These allowed an estimation of the overall population state at the period of the event occurrence. During the event, or just a few months after, it is not possible to gather such knowledge. Therefore, some other approaches were required. ➲ *Figure 9* shows the reasons why the described difficulties occur, since there is a notorious increasing trend on mortality through the years.

On a first approach, the summer period of interest was divided into several periods of equal length allowing comparisons among years and between periods.

Period –2: 18 Jun to 7 Jul
Period –1: 8 to 27 Jul
Heat wave period: 28 Jul to 16 Aug
Period +1: 17 Aug to 5 Sep
Period +2: 6 to 25 Sep

The rationale behind these periods was not very theoretical though. The idea was simple: the highest temperature period begun on the 29th July up to the 14th August and the heat impact is known to last for one or two days after. Following this rationale, a rough period from 28th July to 16th August would include the heat stress period and most probably its full impact. The other periods were just equal in length.

Fig. 9

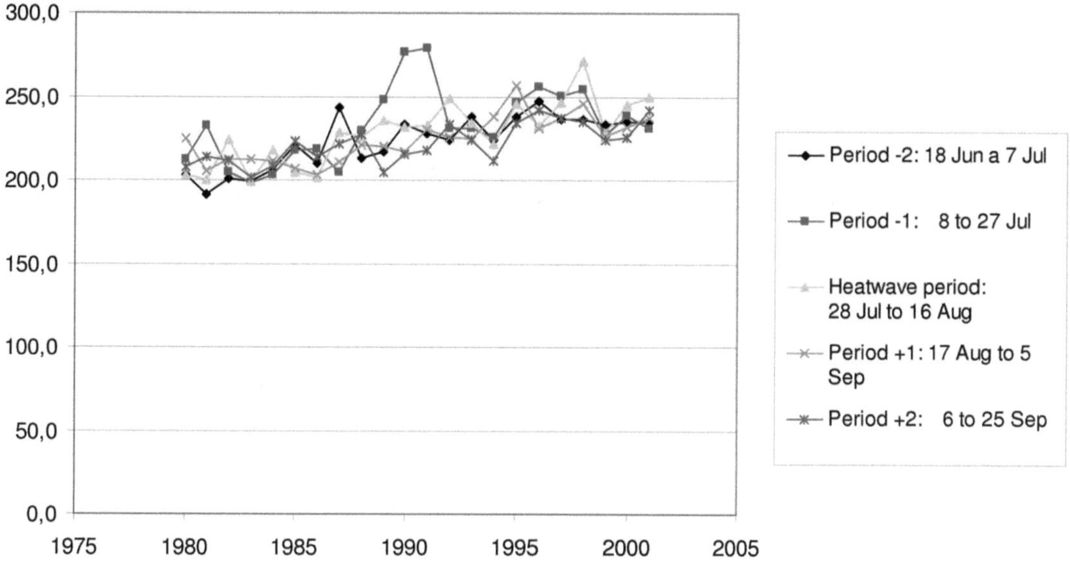

Evolution of daily mean mortality from 1980 to 2001 by different summer periods in Portugal mainland

Tab. 3

Estimates of excess mortality in 2003 by summer periods using homologues time periods in several time intervals

Year/Period		Period –2: 18 Jun a 7 Jul	Period –1: 8 to 27 Jul	Heat wave period: 28 Jul to 16 Aug	Period +1: 17 Aug to 5 Sep	Period +2: 6 to 25 Sep
2001	Daily mean	260,2	240,5	259,5	254,6	248,3
Expected in 19 days		4943	4569	4931	4838	4718
O/E ratio		1,00	1,01	1,46	0,94	1,02
	Excess deaths	-2,0	63,0	2259,0	-282,0	77,0
	IC 95 %	(-136,5 ; 137,8)	(-67,8 ; 198,4)	(2104,5 ; 2427,5)	(-411,8 ; -147,7)	(-55,8 ; 214,7)
p-value (two tailed)		0,987	0,352	0,000	0,000	0,262
1997–2001	Daily mean	253,5	252,9	260,0	246,3	244,6
Expected in 19 days		4816	4804,2	4939,8	4680,6	4647
O/E ratio		1,03	0,96	1,46	0,97	1,03
	Excess deaths	125,0	-172,2	2250,2	-124,6	148,0
	IC 95 %	(-9,5 ; 264,8)	(-303 ; -36,8)	(2095,7 ; 2418,7)	(-254,4 ; 9,7)	(15,2 ; 285,7)
p-value (two tailed)		0,070	0,013	0,000	0,069	0,028

Tab. 3 (Continued)

Year/Period		Period –2: 18 Jun a 7 Jul	Period –1: 8 to 27 Jul	Heat wave period: 28 Jul to 16 Aug	Period +1: 17 Aug to 5 Sep	Period +2: 6 to 25 Sep
1995–2001	daily mean	253,3	256,1	258,1	249,6	244,6
Expected in 19 days		4813,0	4865,3	4904,3	4743,0	4648,0
O/E ratio		1,03	0,95	1,47	0,96	1,03
	Excess deaths	128,0	-233,3	2285,7	-187,0	147,0
	IC95 %	(-6,5 ; 267,8)	(-364,1 ; -97,9)	(2131,2 ; 2454,2)	(-316,8 ; -52,7)	(14,2 ; 284,7)
p-value (two tailed)		0,063	0,001	0,000	0,006	0,029
1980–2001	daily mean	237,4	243,8	252,8	235,1	230,6
Expected in 19 days		4510,7	4631,6	4803,2	4466,0	4381,8
O/E ratio		1,10	1,00	1,50	1,02	1,09
	Excess deaths	430,3	0,4	2386,8	90,0	413,2
	IC95 %	(295,8 ; 570,1)	(-130,4 ; 135,8)	(2232,3 ; 2555,3)	(-39,8 ; 224,3)	(280,4 ; 550,9)
p-value (two tailed)		0,000	0,998	0,000	0,178	0,000

Source: Mortality database - Instituto Nacional de Estatística

Table 3 shows excess mortality within the several defined periods for different reference time periods in the past. The year 2001 could be thought of as a useful reference, since it provides the most recent data and relates better with current population structures. But it had its down side and did not reflect the June heat wave. In fact, recent years show mortality impacts in June (*Fig. 5*).

All other possible reference time periods have similar problems, beyond the fact that awareness exists that higher excess death estimates are obtained from wider reference periods (*Fig. 9*). But on the positive side, within heat wave period all excess deaths estimates are similar.

When did the mortality impact occur?

To assert the period of heat impact on daily mortality, data from 26th July to 21st August for 1999 to 2002 and 2003 was used, and Poisson probabilities were used to compare daily observed mortality in 2003 with period mean daily mortality for each other year and for 1999 to 2001 averages. *Table 4* illustrates that a clear pattern of 17 days, from 30 July to 15th August, arose.

Tab. 4

Estimation of the heat wave mortality impact period and estimates of excess mortality in 2003 using preceding years similar periods

	Year					Comparison with daily mean			
Date	2003	2001	2000	1999	Mean 1999–2001	p 2001	p 2000	p 1999	p mean 1999–2001
27-Jul	253	230	220	218	223	0,62	0,50	0,07	0,32
28-Jul	242	250	213	202	222	0,89	1,02	0,27	0,78
29-Jul	250	266	246	226	246	0,76	0,63	0,10	0,42
30-Jul	292	260	250	231	247	0,00	0,00	0,00	0,00
31-Jul	286	257	253	202	237	0,01	0,01	0,00	0,00
1-Aug	362	233	283	221	246	0,00	0,00	0,00	0,00
2-Aug	418	274	254	240	256	0,00	0,00	0,00	0,00
3-Aug	395	231	213	236	227	0,00	0,00	0,00	0,00
4-Aug	356	223	226	224	224	0,00	0,00	0,00	0,00
5-Aug	317	254	214	228	232	0,00	0,00	0,00	0,00
6-Aug	384	274	250	233	252	0,00	0,00	0,00	0,00
7-Aug	424	260	254	239	251	0,00	0,00	0,00	0,00
8-Aug	464	248	267	210	242	0,00	0,00	0,00	0,00
9-Aug	399	241	279	237	252	0,00	0,00	0,00	0,00
10-Aug	387	235	251	212	233	0,00	0,00	0,00	0,00
11-Aug	388	236	228	240	235	0,00	0,00	0,00	0,00
12-Aug	438	260	212	238	237	0,00	0,00	0,00	0,00
13-Aug	439	258	238	231	242	0,00	0,00	0,00	0,00
14-Aug	372	236	240	238	238	0,00	0,00	0,00	0,00
15-Aug	306	228	246	227	234	0,00	0,00	0,00	0,00
16-Aug	244	212	273	217	234	0,99	0,92	0,21	0,68
17-Aug	263	243	245	244	244	0,26	0,19	0,01	0,11
18-Aug	260	223	241	202	222	0,35	0,26	0,02	0,15
19-Aug	247	223	235	216	225	0,91	0,77	0,15	0,54
20-Aug	241	249	220	222	230	0,84	0,98	0,30	0,83
21-Aug	240	261	243	211	238	0,79	0,93	0,33	0,88
Total observed deaths	8667	6365	6294	5845	6169				
Mean	333,3	244,8	242,1	224,8	237,3				
Excess deaths in 2003		2302	2373	2822	2498				
Excess deaths in 30.7 to 15.8, 2003		2219	2269	2540	2342,7				
95 % Confidence Intervals for Excess Deaths		(2070 ; 2378,3)	(2120 ; 2428,3)	(2391; 2699,3)	(2193 ; 2501,3)				
p for overall excess		0,00	0,00	0,00	0,00				

Source: 1999–2001 Mortality data – Instituto Nacional de Estatística; 2003 mortality data – Direcção Geral da Saúde

How many people died?

Overall excess death numbers for the estimated 17 days of heat impact mortality (table 4) do not substantially differ from previous estimates (table 3). The estimate of 2219 excess deaths obtained for the 17 days of heat impact and for the reference year of 2001 is a reasonable figure for observed excess mortality.

The report Direcção Geral da Saúde – Direcção de Serviços de Informação e Análise & Instituto Nacional de Saúde Dr. Ricardo Jorge – Observatório Nacional de Saúde (2004), correcting for population's structure using mortality rates to calculate expected deaths, shows that the number of excess heat related deaths is 1953.

8 Intervention

The summer of 2003 accounted for an unusual amount of effective intervention and an unusual effort to convey heat wave information to the population. Very early response from intervening partners was very positive. Efforts were undertaken to assess conjoint information from both institutions able to intervene. As early as June there had been warnings on sun exposure at the beach and on how heat could affect younger and elderly people.

Later in the summer it seemed that information on the heat wave and forest fire warnings got mixed by the media. Overwhelming forest fires with very visible effects took over the media, overshadowing information on the heat wave effects on health and mortality.

The pragmatic implementation of the ÍCARO Surveillance System always relied on the institutional missions of each partner, rather than designing an exceptional plan of intervention. Otherwise the system would exceed the natural bounds of individual institutional missions. Nevertheless, this summer experience showed that in the future additional efforts must be made to inform the population of heat risks. The surveillance system must not only rely on passive informational schemes.

Conclusions and discussion

9 Conclusions

The Portuguese Heat Health Warning System – also referred as the ÍCARO Surveillance System – has been in place for five years. In the past it has issued several heat wave warnings. The 2003 summer warning was the most severe.

The 2003 summer was unusually hot, having several periods with excessive temperature. Excessive temperatures were observed in May, June, July–August (continuously for about two weeks) and September.

The ÍCARO Surveillance system (Lisbon Heat Health Warning System) gave a realistic picture about the true effects that occurred. The heat risk predicted through the ÍCARO-index showed a close relation with observed mortality. This fact supports the model rational defined by the ÍCARO methodology.

It has been shown that high heat-related mortality occurred in Portugal in the summer mainly in the July–August heat period. This article shows that this heat wave had a duration of 17 days, generating about 2200 excess deaths that, when accounting for structural population, results in an estimated number of 1953 effective heat related excess deaths. This mortality mainly affected elderly individuals, especially above 75 years old; also women; and in all Portuguese mainland districts.

The surveillance partners had difficulty on conveying messages to the population, using the media, late in the heat stress period. This was mainly due to highly visible and remarkable concomitant forest fires that occurred in most districts of the mainland country.

The ÍCARO Surveillance System experience led to the conclusion and advice that more active ways are required to convey information to the population at risk, so that substantial health and life gains can be achieved.

10 Discussion

The results presented in this article show that the summer of 2003 was an outstanding event, with generally hot temperatures and severe consequences on human health and mortality. Other studies showed that there was also an alteration on hospital and health emergency services (Paixão et al. 2003). These aspects point to the event's complexity, showing that a full report on such an occurrence might not be promptly possible.

This summer heat wave occurrences had dramatic consequences in Portugal, but past experiences and experiences elsewhere (in France for example) allowed for far worse expectations, since this major heat wave was about twice as long as previously known severe heat waves. In fact, in the 2003 summer there was no visible, known or reported health or cemetery service breakdowns.

Intervention in 2003 might not have been as extensive as wished, mainly due to other aspects that are surely also heat related. It was evidenced that reliance on passive systems to reach and inform individuals has its drawbacks; while consequences of forest fires may not have such severe health consequences as the heat wave itself, it has a visual and socially dramatic side that is appealing to the mass media, and also conveying the same information during a long heat period reduces informational interest. So system partners felt a need for and have already started a further investment on intervention planning for the near future.

The main objective of this paper was to present the ÍCARO Surveillance System as an example of a full operational heat health warning system, showing its past and current experience. Obviously, excess death estimation is one important task, since it helps the system's evaluation, but it is not the surveillance system's main concern. The excess deaths herein presented might not reflect the full population's age structure prior to the heat wave; in the winter of 2002/2003 there was not the usually recurrent influenza epidemic (Centro de Virologia – Centro Nacional da Gripe/Observatório Nacional de Saúde – Rede Médicos-Sentinela (2003) and Falcão (2003)) which might have spared some individuals at risk in the winter and may have increased their probability of dying in such a hot summer.

A full report on this 2003 July – August heat wave related mortality by the General Health Directorate and the National Health Observatory (ONSA) is available at URL [www.onsa.pt]. This report presents definitive as well as global mortality ratios and excess death estimates by gender and by age group, standardized by district level.

It is interesting to notice that for all the different time periods used for reference (table 3) the heat wave period is the one with higher expected daily mortality. In fact, if modelling daily air temperature data from 1980 to 2000 with polynomial regression, a summer maximum occurs for all the Portuguese mainland districts between the 4th and 11th August (Nogueira 2004). This suggests that heat might have an endemic nature that has not yet been referred to elsewhere. Nevertheless, comparisons with previous or later periods can lead to higher estimates and past heat wave excess death estimates can be underestimated if the August period is used as reference.

Several times in this article, the mortality process is described using the term "visible non-randomness" when there is an ÍCARO index at a non-statistically significant level. This is just an empirical comment that needs further proofing. This might be accomplished by analysing data comprising two or three days. Such proofing has not been attempted here, but awareness exits of such a necessity, and this will be done in the near future.

This article shows that the ÍCARO Surveillance System, in its current operational version, gives a correct prediction while heat waves are occurring. Such confirmation is in favour of the system existence since 1999. This summer occurrence showed the relevance of the ÍCARO Project research line, which might bring interest to other research areas and related institutions, further research efforts are needed and will allow more knowledge, further developments and more health and life gains in the future.

Acknowledgments

A special word for everyone at ONSA with special thanks to Dr. José Marinho Falcão, Dr. Baltazar Nunes and Dr. Eleonora Paixão. This article would not have been possible without the General Health Directorate's collaboration – a special thanks to Dr. Rui Calado and Dr. Jaime Botelho. And, as always in this adventure of the ÍCARO Surveillance System, a special word for the Portuguese Meteorology Institute is due – and individual thanks to Dr. Teresa Abrantes and Dr. Fátima Espírito Santo.

This work was funded by *Fundação para a Ciência e Tecnologia* (FCT-Portuguese Foundation for Science and Technology) and its *Programa Operacional "Ciência, Tecnologia e Inovação"* (POCTI), and co-funded by European Community Fund FEDER. (POCTI/ESP/39679/2001)

References

Braga AL, Zanobetti A, Schwartz J (2002) The effect of weather on respiratory and cardiovascular deaths in 12 U.S. cities. Environ Health Perspect 110(9):859 – 63

Centro de Virologia – Centro Nacional da Gripe/Observatório Nacional de Saúde – Rede Médicos-Sentinela (2003) Gripe em Portugal - 2002/2003 Relatório Anual: Sistema Nacional de Vigilância da Gripe [Influenza in Portugal – 2002/2003 Annual Report: Influenza National Surveillance System] (http://www.onsa.pt/conteu/publicacoes/pub_relat_gripe-portugal-02_03_onsa.pdf, accessed 8 April 2004)

Direcção Geral da Saúde – Direcção de Serviços de Informação e Análise & Instituto Nacional de Saúde Dr. Ricardo Jorge – Observatório Nacional de Saúde (2004). Onda de calor de Agosto de 2003: os seus efeitos sobre a mortalidade da população portuguesa – Relatório. [August 2003's Heat wave: its effects on the Portuguese population's mortality – Report], Lisboa, Portugal, (http://www.onsa.pt/conteu/fontes/onda_2003_relatorio.pdf, accessed 7 September 2004)

Esteve J, Benhamu E, Raymond L (1994) Epidemiology (Statistical Methods in Cancer Ressearch, Vol. IV). IARC Scientific publications 128, pp 64

Falcão IM (2003) A pouca gripe do Inverno passado. Observações 19. [Last winter's little flu] (http://www.onsa.pt/conteu/publicacoes/pub_observacoes_019.html, accessed 8 April 2004)

Falcão JM, Castro MJ, Falcão ML (1988) Efeitos de uma onda de calor na mortalidade da população do distrito de Lisboa. Saúde em números 3:2:9 – 12

Falcão JM et al. (2003) Projecto ÍCARO. Onda de calor de Agosto de 2003: Repercussões sobre a saúde da população. Estimativas Provisorias (até 12.08.2003) [The ÍCARO Project. Effects of the heat wave in August 2003 on the health of the population. Preliminary estimates as of 12 August 2003]. Lisbon, Onsa. Observatório Nacional de Saúde, Instituto Nacional de Saúde Dr Ricardo Jorge (http://www.onsa.pt/conteu/fontes/proj_ÍCARO.html, accessed 8 April 2004)

Garcia AC, Nogueira PJ, Falcão JM (1999) Onda de calor de 1981 em Portugal: efeitos na mortalidade [Effects of the heat wave in June 1981 in Portugal on mortality]. Revista Nacional de Saúde Pública, volume temático 1:67 – 77

National Institute of Public Health Surveillance (2003) Impact sanitaire de la vague de chaleur en France survenue en août 2003. Bilan et perspectives – Octobre 2003. National Institut de Veille Sanitaire, Saint Maurice, France,

Nogueira PJ (2001) Acreditando no … ÍCARO [Believing … ÍCARO]. Observações, 12. (http://www.onsa.pt/conteu/publicacoes/pub_observacoes_012.html, accessed 8 April 2004)

Nogueira, PJ (2002) 4 x ÍCARO: um balanço [4 x ÍCARO: a balance]. Observações, 17 (http://www.onsa.pt/conteu/publicacoes/pub_observacoes_017.html, accessed 29 October 2003)

Nogueira P, Paixão E (2003) Evaluation of the Lisbon heat health warning system. cCASHh Workshop on Vulnerability to Thermal Stresses, 5 – 7 May, Freiburg, Germany

Nogueira PJ et al. (1999)Um sistema de vigilância e alerta de ondas de calor com efeitos na mortalidade: o índice Ícaro [A heat wave surveillance and warning system based on the effects on mortality: the ÍCARO index]. Revista Nacional de Saúde Pública, volume temático 1:79 – 84

Nogueira PJ (2004) Modelação do limiar superior da temperatura diária máxima durante o verão por distritos de Portugal continental - Documento Interno. [Summer maximum daily temperature upper threshold modelling by district for Portugal mainland - Internal Document]. ONSA-Lisboa.

Nunes B, Castro L (1998) Não morrer de calor! ... Será uma questão de habituação? [Not dying from heat! ... Is it a question of habituation?]. V Congresso Anual da Sociedade Portuguesa de Estatística 1997. in: Miranda SM, Pereira I (eds) Estatística: A diversidade na Unidade. Novas tecnologias, 7. Edições Salamandra, Lisboa, Portugal

Paixão E, Nogueira PJ (2002) Estudo da onda de calor de Julho de 1991 em Portugal: efeitos na mortalidade: relatório científico [A study of the July 1991 heat wave in Portugal: effects on mortality - scientific report]. Lisbon, Observatório Nacional de Saúde (http://www.onsa.pt/conteu/fontes/proj_ÍCARO_relat-cientifico-out02_onsa.zip, accessed 29 October 2003)

Paixão EJ, Nogueira PJ (2003) Efeitos de uma onda de calor na mortalidade. Revista Nacional de Saúde Pública 21;1:41 – 53

Paixão EJ, Nogueira PJ (2003b) Avaliação preliminar da onda de calor de Maio 2001 - Documento Interno. [Preliminary evaluation of May 2001's heat wave - internal Document] Lisboa, ONSA.

Paixão EJ, Nogueira PJ, Contreiras T, Falcão JM (2003) Onda de calor de Agosto de 2003: Estudo da Utilização de Cuidados de Urgência. Lisbon, Observatório Nacional de Saúde - Instituto Nacional de Saúde Dr. Ricardo Jorge. (http://www.onsa.pt/conteu/proj_icaro_ondacalor03_onsa.doc, accessed 15 April 2005)

WHO Regional Office for Europe (2004). Heat waves: risks and responses. WHO Regional Office for Europe (Health and Global Environmental Change Series, No. 2)

Lessons from the Heat-Wave Epidemic in France (Summer 2003)

L Abenhaim

Summary

This article relates the experiences of the exceptional heat wave of the summer 2003 in France to heat related events as a whole. It focuses on five issues: first, whether heat-waves should be considered as epidemics or as endemic events in order to achieve efficient mitigating strategies. It is recommended to consider heat-waves as both, although it is clear that epidemic situations cannot be dealt with properly without a more general program to decrease the effects of endemic heat-related ill health. Second, it is stressed that there is no model available for predicting the occurrence of future epidemics with the required sensitivity and specificity: More detailed analysis has to be carried out. In addition, it is recommended that the threshold for future relief actions be lowered. Third, the issue of what surveillance system is required to detect epidemics of heat-related illnessis addressed. Fourth, the efficacy and effectiveness of prevention plans are briefly evaluated as well as what must be done during epidemics. Finally, the relationship between epidemics and polical crises is outlined.

The heat-wave related epidemics experienced in various countries of Europe during the summer of 2003 has put the issue of extreme heat on the agenda of public health professionals and decision-makers. As with any threat to the health of large populations, the issues to be addressed are those of the prediction and the prevention of the consequences of heat-waves on the one hand and those of the detection of adverse effects and emergency measures to be taken during the episodes on the other. This paper draws conclusions from the lessons learned mainly in France, but also attempts at broadening the scope to heat related events as a whole.

1 Should we concentrate on epidemics or endemics of heat-related events (HRE)?

A key issue is to decide whether we should be concerned only by extreme weather events or with any excess death rates associated with heat. In the first instance, that is if one focuses on episodes such as those experienced in the summer 2003 in several European countries, the model for public health action (including prevention and detection), is that of the management of epidemics. In the second instance, that is if we consider that any elevated temperature, such as those experienced every year in many countries around the world, may represent a threat, the model is that of struggling with endemic situations. At the first glance, these would seem obviously to be quite different approaches. Yet a thorough analysis of previous experiences, and in particular that experienced in France, shows that there are a certain number of common issues between the two approaches. Moreover, it seems unlikely to institute efficient mitigating measures against epidemics without a long-term policy targeted at endemics of heat-related events.

2 Can epidemics of heat-related events be predicted?

A number of investigations have shown that epidemics occur when a certain number of conditions are met: temperature rising above a certain threshold, lack of air conditioning, lack of mitigating measures, and a number of other risk factors. The threshold seems to vary from country to country and, within countries, from region to region or even from town to town. This depends on the usual local temperatures experienced in the summer which, in turn, are correlated to the presence or absence of some of the other factors. Besides possible physiological habituation, the architectural designs of cities and homes, and factors such as the availability of air conditioning in nursing homes, hospitals and private homes, are obviously related to the frequency of occurrence of HREs. Yet epidemics occur from time to time even in countries otherwise accustomed to high temperatures. Even if one were able to accurately describe what factors were usually associated with the occurrence of HRE epidemics in the past, there is no current model able to predict with the required sensitivity and specificity the occurrence of future epidemics. In a sense, this is not very different from what is usually observed in epidemics of other diseases, even some very common ones (such as flu).

If one can quite reliably predict that very high temperatures will produce an increase in the mortality in a given region, available models fall short of predicting with accuracy the exact time of occurrence, the magnitude, and the duration of the epidemics. There are several reasons for this. The first reason is that, although the factors associated with past epidemics are quite well described and reproducible, their predictive value is quite low. Indeed, there have been only a limited number of epidemics studied so far, which doesn't represent a satisfactory statistical basis to produce sensitive and specific parameters for the prediction of the occurrence of epidemics in the future, considering the number of factors to be taken into account. This calls for an in-depth analysis of the summer 2003.

For example, in France the temperature had been way above normal summer levels for at least 6 weeks when an epidemic of 15,000 deaths suddenly occurred in mid August 2003 although the temperature raised by only a few more degrees (❯ *Fig. 1*). ❯ *Figure 2* displays the night temperature in the Paris Region and the number of excess death between August 1 and August 21. It shows that the epidemic started when the night temperature surpassed 23 °C, which is much lower than the "danger level" used in Chicago, for instance.

❯ *Figure 2* also shows that the sharp increase in the number of deaths was highly correlated with the sharp increase in the night temperatures. The epidemic occurred very suddenly, in a few days, only to subside just as abruptly when the temperatures decreased. The weather forecaster (MeteoFrance) issued a warning on August 7, when the epidemic had already caused more than 2000 deaths in the country. Moreover, this warning was only that "heat could represent a risk", without mentioning the possibility of a major disaster.. The French General Directorate for Health issued a somewhat more worrying communiqué in the following two days, and mentioned the possibility of "hundreds of deaths" while still not envisioning the magnitude of the on-going disaster. The French Institute of Public Health (Institut de veille sanitaire) did not issue any warning, nor did the French Agency for Environmental Health and Safety. In other words, everybody was working under the hypothesis that heat would probably produce an increase in mortality, but not a disaster. The "endemic" model and not the "epidemic" one was obviously taken into account.

In 2003, France experienced night temperatures which were way above those of other Northern European countries and, even, in certain regions, those of the northern parts of Mediterranean European countries. When looking carefully at the temperatures and excess deaths, it appears that the northern part of France experienced an increase of deaths, not different from that of countries such as The Netherlands, and that the Southern part of France had the same excess rate of deaths than Northern Italy. On average, the excess rate of death was similar to that of Portugal, which experienced very high temperatures also as compared to the usual level.

In conclusion, it seems necessary to lower our threshold levels for action.

Fig. 1

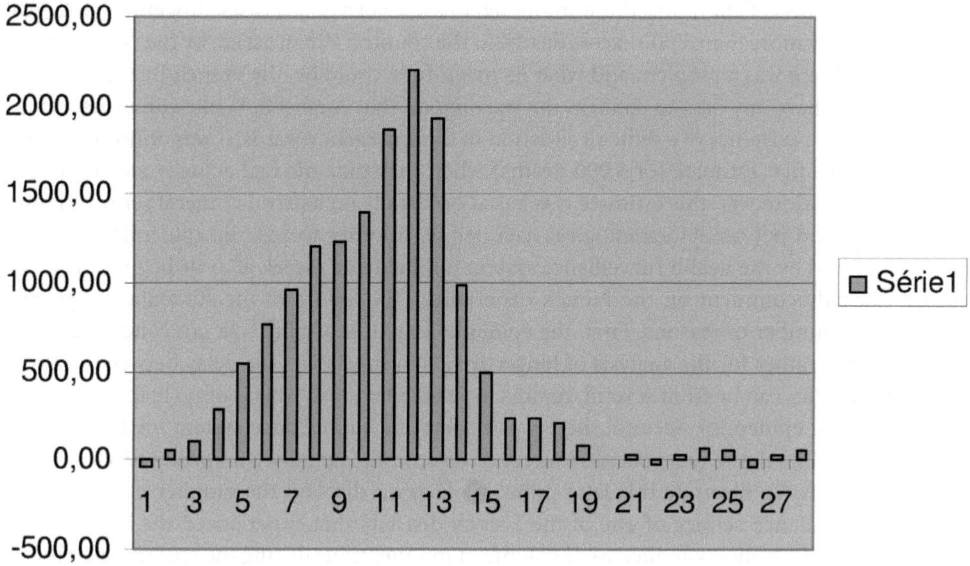

Total number of excess deaths August 1 – August 28, 2003, France (heat wave)

Fig. 2

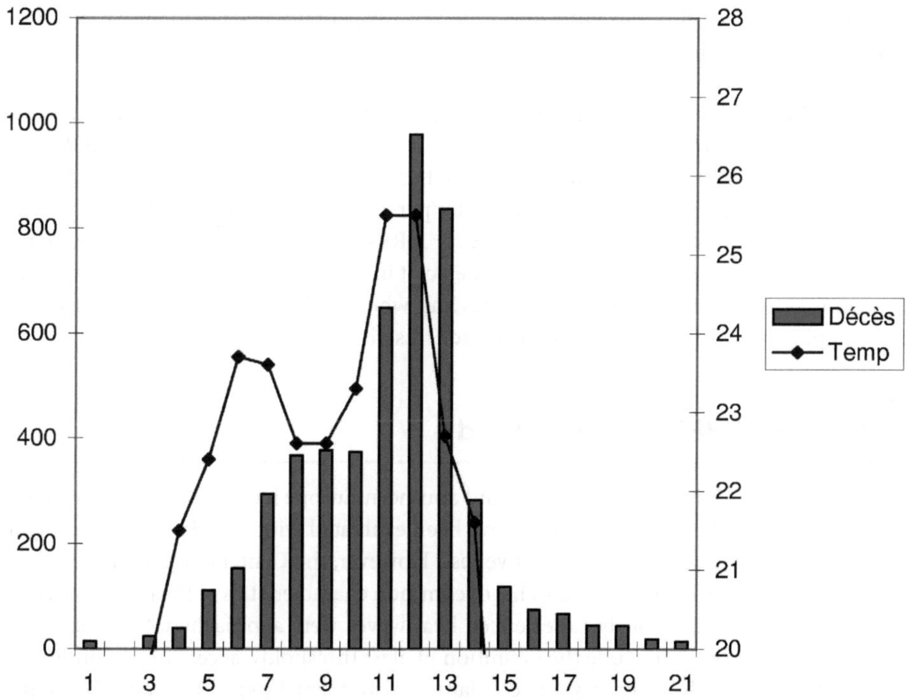

Minimal night temperatures and number of deaths in the Paris Region

3 Can epidemics of HREs be adequately detected?

A couple of French local health departments (DDASS) reported cases of heat related death beginning August 6. Added to the reports of clinicians, these amounted to less than a dozen cases altogether on August 8 when there were already more than 3,000 excess deaths in the country. When asked, by the General Health Directorate whether there was a problem and what its magnitude could be, the French Institute of Public Health (InVS) did not have any data to confirm the warning on that August 8. While clinicians seriously alerted on August 10 the existence of a difficult situation in the emergency wards, it was only on August 13 that the InVS issued its first estimate (of 3,000 deaths) when the epidemic had actually already resulted in 12,000 extra deaths. Moreover, this estimate was based on data from a private funeral service (Pompes Funèbres Générales) and not on epidemiological data per se. In other words, the epidemic was neither detected nor recognized by the health surveillance system until at least a week after its beginning.

This paper can only comment on the French experience. The failure of the surveillance system in France stems from a number of reasons. First, the epidemiology of mortality was not conceived for surveillance purposes but rather for the analysis of longer trends in mortality. The delay between death and the availability of statistics can be from several months to years. Obviously, this is of no help for the detection of rapidly growing epidemics. Second, there was no specific surveillance system for HREs (hyperthermia, for instance). Third, the non specific sources of data on the activity emergency services did not produce relevant signals. To illustrate this later point, ➋ *Figure 3* displays the number of interventions of the emergency ambulance service of one of the French districts that experienced the highest rate of excess deaths (SAMU 93). It shows an increased volume of interventions during the epidemics (day 77 on the graph, corresponding to August 10). But the increase in activity on August 10, for instance, was not significantly higher than that experienced at several occasions in the previous two months. Similar data are found with the emergency activities of General practitioners (SOS-Médecins, not shown). Surprisingly, the total activity of the hospital emergency wards was similar in the first two weeks of 2003 to that of similar periods in the previous years! It is only clinicians, who saw the patients, and could notice that they were actually suffering from heat related diseases, who were in a position to alert. However, emergency physicians spoke of "at least 50 deaths" when there were more than 6,000.

One other reason for this lack of warning signals is the dispersion of deaths over the country. As important they may be in total (60 % increased mortality for the 3 first weeks of August 2003), the epidemics represented in most places (nursing homes, hospitals, smaller cities…) one or two extra deaths which were not quickly detected also because they concerned mainly older persons suffering from other diseases.

In summary, there was no surveillance system in France for HREs available at the time of the epidemics with the required specificity and sensitivity to be able to detect it in a timely way. I do not know whether such surveillance systems exist in other European countries and whether they were more efficient in the early detection of their epidemics. This should certainly be addressed.

4 Can epidemics of HREs be prevented?

A limited number of cities in the United States, Canada and Southern Europe have developed prevention plans against the effects of heat on health. So far, none have been evaluated with sufficient scientific rigor to be able to draw definitive conclusions on their effectiveness. However, the Centers for Disease Prevention and Control of the United States has established a list of common characteristics of the plans reviewed in their country (18 cities who had previously experienced heat waves were approached, 5 had detailed plans). In summary, most of these plans include the definition of alert thresholds, access to air conditioning and retrieval of isolated vulnerable individuals. Such plans did not formally exist in France but some cities like Marseilles which had experienced a lethal heat wave in 1983 issued an early warning in 2003,

Fig. 3

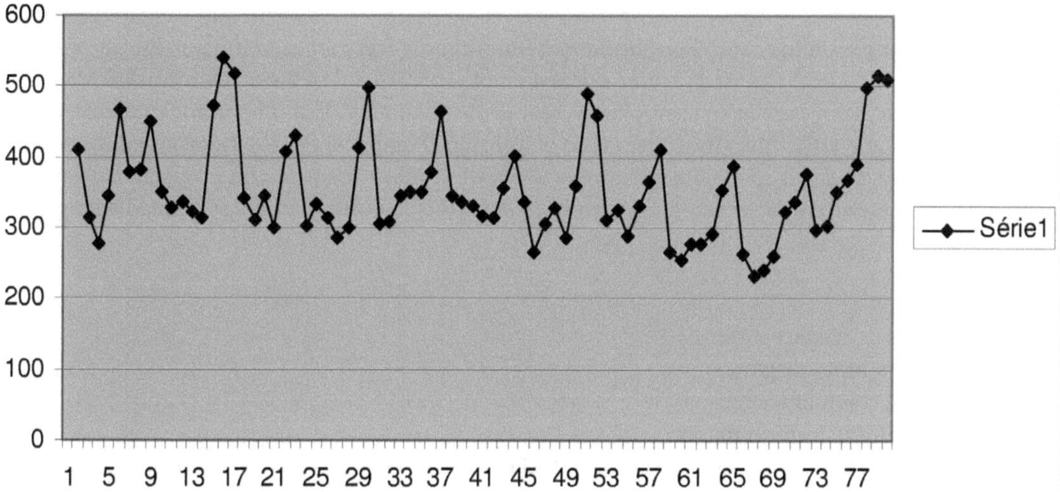

Number of emergency interventions, Department #93, France (1 = May 25, 2003; 77 = August 9, 2003)

with no significant difference in morbidity and mortality observed when compared to cities with similar levels of heat stress.

Access to so-called "cooling centres" has not been proved to be very effective either, because those who use them usually suffer from a lesser degree of disability or handicap. Yet, they may be of use at least to bring some comfort to those who can reach them. The only factor which has the efficacy and effectiveness required to diminish the body temperature in a sustainable way for all populations is air conditioning. The large availability of air conditioning in American cities may explain part of the difference in the mortality observed at certain temperature levels. The large use of air conditioning may be associated with side effects (legionelosis, increase in outside temperature, stress on power supply), some of which are preventable (legionelosis) or manageable (power supply) or not sufficiently large to surpass the benefits in the fragile populations (increase in outside temperatures). Note that individual air conditioning sets are not associated with a risk of legionelosis. This should be carefully assessed. France is a country with a very low access to air conditioning not only in private homes, but also in nursing homes and hospitals.

5 What can be done during epidemics?

In a country like France, 6 to 7 million individuals can be considered to be at high risk of HREs, when adding persons 75 years old and over to persons living with very serious diseases (cardiac, pulmonary, renal, etc). Among them an unknown but probably large share live in isolation. To be able to retrieve such a number of persons in emergency is certainly extremely difficult. A large US city which benefited from such a plan was able to retrieve only 400 persons in a week during a cold episode. Experiences from other countries have not been reported in detail nor scientifically assessed. Although retrieval of isolated vulnerable persons should be organized ahead of time, it is during the heat wave that it has to be applied. Again, the feasibility of retrieving several millions of persons in a timely fashion and offering them access to preventive measures such as air conditioned rooms and overnight accommodation is debatable. Other measures (development of air conditioning in nursing homes and hospitals, installation of individual sets in private homes) have obviously to be planned ahead of time.

6 Epidemics vs crises

Not every epidemics result in crisis. There are flu epidemics almost every other summer in Europe without major political crises. On the contrary, some crises are not proportional to the epidemic itself. Mad cow disease or SARS in Europe are such examples, were it is the possibility of an important epidemic which was taken into account rather than the observation of its reality. France experienced a major political crisis around the heat wave related epidemics of summer 2003, while other countries, hit proportionally just as much, did not. These are some elements frequently associated with political crises associated with public health issues:

- Surprise
- Lack of alert
- Uncertainty as to dangers and risks
- Lack of efficient mitigating measures
- Deficiencies in communication
- Distrust in political decision makers.

All of these were experienced during the summer 2003 in France, but maybe also in other European countries which did not experience the same political crisis. The fact is that the French public health system has been regularly under attack since the so-called "Tainted Blood Affair" and that political crises are more likely to occur whenever trust is at stake.

Conclusion

The prediction, detection and prevention of heat related epidemics suffers from a lack of scientific knowledge and experience. Most mitigating measures, and certainly the most efficient one, air conditioning, are the same for the prevention of epidemics and for the prevention of morbidity and mortality related to more usual heat conditions experienced regularly. It is suggested that the most efficient way to avoid the consequences of HRE epidemics associated with extreme weather events is to address the question of less extreme weather situations in a continuous fashion.

How Toronto and Montreal (Canada) Respond to Heat

T Kosatsky · N King · B Henry

Summary

It is only during the last five years that Canadian cities have begun to develop formal programs to protect the public's health from the effects of summertime heat. Toronto's (Ontario) Hot Weather Response Plan followed recommendations from advisory committees for seniors and for the homeless. The public health department was confronted with a rain-storm on the first day, in 1999, that it issued a heat alert. Toronto has since instituted a two-level alert and emergency response with action levels based on the estimation of mortality impacts through a synoptic model developed at the University of Delaware. Key to the program is media alerts and community partnerships to aid vulnerable people. Montreal's (Quebec) approach has been to issue public advisories based on real and apparent temperature thresholds elaborated in collaboration with the Canadian Meteorological Service. Montreal has instituted a program of research and action designed to inform the population and to identify and mitigate population vulnerabilities in order to make residents more resistant to the effects on health of heat. Priority areas for health protection include hospitals and nursing homes, few of which are now air-conditioned; rather than retrofit air conditioning, relative air-cooling and air dehumidification have been proposed where feasible. In the community, local health centers target their vulnerable elderly clients requiring follow-up during heat waves based on the identification of factors such as dehydrating medications, social isolation, and lack of access to a nearby cooling room. A heat wave emergency response plan, based on the mobilization and updating of existing programs, is coordinated by civil defense authorities, advised by the city's public health department.

Introduction

Toronto and Montreal are Canada's largest urban centers. Various climatologic, environmental, demographic, and socio-economic factors place the populations of Toronto and Montreal at risk of illness and death from summertime heat (Smoyer-Tomic and Rainham 2001). Toronto's (Ontario) Hot Weather Response Plan followed recommendations from advisory committees for seniors and for the homeless. The public health department was confronted with a rain-storm on the first day, in 1999, that it issued a heat alert. Montreal's (Quebec) approach has been to issue public advisories based on real and apparent temperature thresholds elaborated in collaboration with the Canadian Meteorological Service. Montreal has instituted a program of research and action designed to inform the population and to identify and mitigate population vulnerabilities in order to make residents more resistant to the effects on health of heat. This paper briefly describes the Montreal and Toronto responses to heat. Since 1998, the two cities have initiated active heat preparation and response strategies. The contrasting experiences with and responses to summertime heat by authorities in Toronto and Montreal should interest European colleagues.

Toronto

Toronto (2001 population: 2.4 million) has a continental climate with cool winters and warm summers. In 1998, following several summers hotter than those of the 1980s, and alerted by the devastating effects of recent summer heat waves in Philadelphia, St. Louis, and Chicago, US cities whose climate is similar to Toronto's, Toronto Public Health (TPH) drafted a Hot Weather Response Plan. Advisory Committees for Seniors and for the health of Homeless and Socially Isolated Persons promoted the initiative. The plan was based on two parallel activities: the identification of threshold weather conditions for the implementation of population heat alerts, and the development of an emergency response plan in partnership over 800 City of Toronto agencies, health care organizations, and non-governmental groups.

For 1999, TPH set interim action levels of 40° humidex equivalent (humidex is an apparent temperature equivalent calculated from dry temperature and relative humidity and is reported by the Canadian Weather Service) for putting out heat alerts, and 45° humidex equivalent for the implementation of emergency measures.

Under the Hot Weather Response Plan, TPH monitored humidex levels, declared the heat alert or heat emergency, disseminated the notification of heat alerts to the media, and coordinated the response plan. TPH collected information on the health effects of hot weather; strategies to decrease risk of heat related illness; safe fan use in hot weather; and the interaction between some medications and hot weather. This was disseminated through four fact sheets available on the TPH website and sent to community agencies. Community partners make Hot Weather Tip Sheets available to clients and area residents and post the heat "alert" press releases. They contact their vulnerable clients during heat waves to ensure they are in good health and provide advice on how to lessen heat stress. The Parks and Recreation Department and the Library Board identified local community centers and libraries as places for people to cool off. Parks and Recreation increased access to public (swimming) pools and agreed to relax the restrictions on homeless people staying overnight in city parks. The Out of the Cold program, Community Health Centers, Community Care Access Centers, and Ontario Community Support Associations provided outreach to vulnerable members of the community including the homeless, under-housed, and frail, isolated, seniors. For example, Anishinawbe Health Toronto, a Community Health Center focusing primarily on the aboriginal population, stepped up their street patrol in order to identify and help the homeless during a heat alert. Tokens for Toronto Transit were made available to help people on the street get to a cooler place. The Red Cross provided training on heat-related illness and first aid to staff and volunteers of community agencies. (Basrur, 2002).

The plan was put into effect in May 1999 and TPH monitored weather conditions daily. The first heat alert was called in early August 1999. Although the weather forecast met conditions for the calling of a Heat Alert, the day itself was "gray, rainy, and cool."

In 2000, Toronto Public Health, together with the Toronto Atmospheric Fund, an activist research consortium, was granted funds to develop a more predictive heat/health warning (alert) system (HHWS) for Toronto. The University of Delaware, and Dr. Lawrence Kalkstein, were contracted to develop the HHWS on the basis of a 17-year retrospective analysis of daily mortality in relation to a range of local weather factors. The HHWS was designed to replace the use of humidex-based action levels for the declaration of heat alerts and emergencies.

By 2001, the HHWS was in operation. Based on past mortality experience, the HHWS translated weather forecasts into a likelihood that the number of daily deaths would surpass those expected based on a long-term average. When the HHWS expressed a probability of excess deaths at 65 – 89 % a "Heat Alert" was called by TPH and the response plan was activated. If the system predicted 90 % or greater chance that the number of excess deaths would be more than predicted, a "Heat Emergency" was called. Further, it was decided that Toronto Public Health would call a heat emergency only on days preceded by a heat alert, so as to allow enough time for the response plan to be put into effect.

By 2001, new partners had joined the response team, and new functions added. The Red Cross provided a Heat Information (telephone) Line. Toronto's Emergency Medical Services established a program whereby a paramedic and a Red Cross volunteer could visit the home of callers to the Heat Information Line judged to be in distress but at a level not grave enough to warrant emergency transport; where indicated, medical advice and recommendations for personal and residential cooling were provided. The City set up staffing and plans to open four cooling centers during heat emergencies, including one with overnight capacity. At these centers, bottled water, snacks, cots, and air-conditioned space are made available to all who need it. In addition, bottled water is provided to the homeless through "street patrols" run by some community partners.

The summer of 2001 put the Hot Weather Response Plan to the test. Six heat alerts and three heat emergencies were called. 401 persons called the Heat Information Line during the emergency days: of these, 28 were referred directly to emergency responders, and 23 received a home visit. Approximately 1700 people visited the cooling centers and 20 – 36 stayed overnight at the cooling center open 24 hours. Extensive media coverage provided heat safety to the public. In 2002, the weather was even hotter, and TPH called Heat Alerts on 15 days while two days reached Heat Emergency conditions. During the two Heat Emergency days, 1800 people used the cooling centers.

Toronto Public Health has, with its partners, promoted research into such urban heat adaptive measures as the use of cool surfaces and shade vegetation. A joint research project with the Canadian Meteorological Service is designed to target alert and emergency days on the basis of the joint effects of weather and air pollution. In addition, TPH and Toronto Emergency Medical Services are partnering with hospital emergency departments to measure the effects of heat on emergency room use in order to incorporate morbidity concerns into the Heat Watch Warning (Alert) System.

Montreal

Montreal (2001 population: 1.8 million) is 500 km northeast of Toronto with a climate that is slightly cooler and is affected by weather fronts which typically arrive a day later. Montreal occupies a river island; the design of housing reflects insulation from the cold more than accommodation to summertime heat. As with Toronto, both the number of very hot days and average summer temperatures were higher in the 1990s than during previous decades: while average summer mean temperature was 19.4 ° during the period 1971 – 2000, it was 20.5 ° in 2002, and 20.7 ° in 2003 (Jennifer Milton, Canadian Meteorological Service, Quebec Region, personal communication). Also as with Toronto, although there has been a general awareness that summer heat is harmful to health and well being, there is no record that Montreal has suffered a killer "heat wave" such as occurred in Chicago in 1995. In fact, few diagnoses compatible with heat injury were recovered in a retrospective review of coroner's reports, death certificates, hospital discharge summaries, and emergency room files for Montreal residents during the period of the 1990s (Kousavilis AT and Kosatsky T 2003).

During the period 1998 – 2003, the Direction de Santé publique de Montréal (DSP or Montreal Public Health Authority) has become active in planning for summer heat waves. Various factors have motivated this initiative:

- Ongoing studies showing an association between numbers of deaths and maximum daily temperature (➜ *Fig. 1)* (Smoyer-Tomic, 2003).
- Advocacy by the Public Health Authority for reduction of carbon emissions as part of a larger strategy supportive of the Kyoto Accord and for the lowering of airborne particle counts.
- Media pressure during summer 2001: the implementation of emergency measures by Toronto during a heat wave in August of that year (described above) led to questions as to why Montreal, in the grips of a similar heat wave, had not also opened cooling centers.

- The French heat wave of 2003 raised concerns another notch. Montreal, as the largest French-speaking city in North America, is particularly affected by what happens in Paris.

Fig. 1

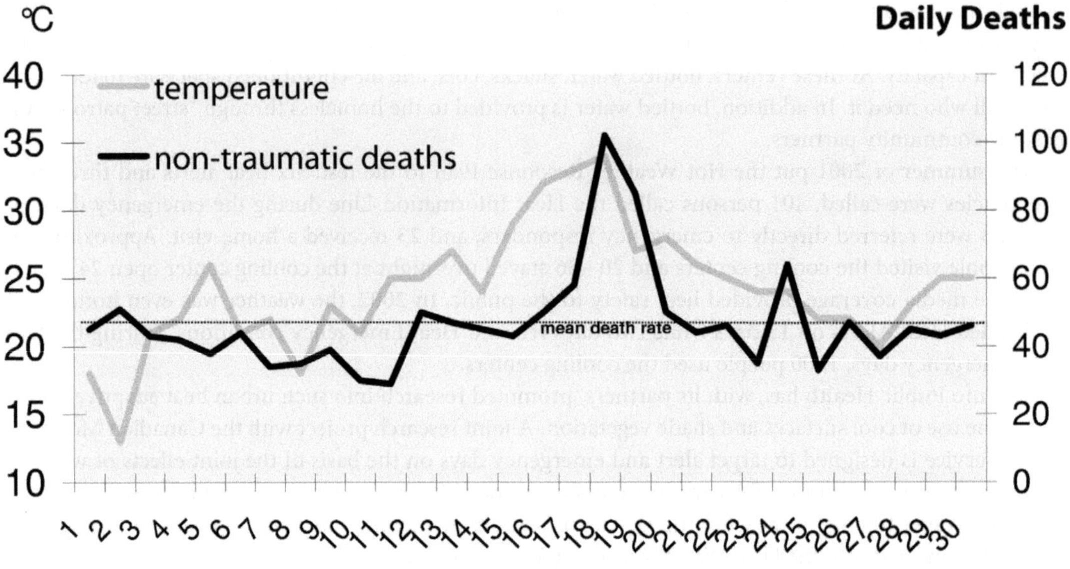

Daily deaths in relation to reported temperature (°C, not humidity adjusted), Montreal (Canada), June 1994

As of 2003, Montreal's heat response plan included public warnings whenever an air temperature of 30 °C or more and an apparent temperature of 40 ° humidex or more were forecast, along with cooling measures such as prolonged opening hours for municipal swimming pools, and encouraging shopping malls to accommodate elderly patrons seeking respite from the heat. An informational pamphlet titled "Heat Waves are Deadly Serious" has been distributed to the public through pharmacies, doctors' consulting rooms, community health centers and hospitals. Authorities, both civic and public health, have been anxious to identify and apply a more "scientific" emergency response threshold and to add to those preventive and response measures already in place.

A joint program of action and research, initiated in 2003, involves the following components:

1. The identification of weather parameters associated with excess mortality employing both time-series and synoptic approaches. Based on this research, the action level for calling alerts will be re-evaluated for 2004 and modified if necessary, and a heat emergency action level will also be defined for the summer of 2005;
2. The expansion of health surveillance activities to include emergency ambulance transports, and calls to (telephone) Health Information and Help Lines, in order to establish early which segments of the population are being adversely affected by heat, and to gage impacts on their health;
3. Development of a geographic information platform to represent jointly the urban heat island differential across the City, differences in housing quality, and the proportion of the population by sector known to suffer from conditions (cardiovascular, respiratory, renal, etc.) which increase vulnerability to the effects of heat. The platform will serve as a means to identify priority sectors for intervention before and during heat emergencies, and to guide research;
4. Evaluation of air conditioner use, medication practices and patient hydration in chronic care centers;

5. Assessment of the knowledge, attitudes and practices of elderly and chronically ill persons on issues related to heat and its effects on health.

Priority actions were suggested on:
1. Integration of the emergency heat health response into Montreal's overall civil protection plan;
2. Advice on patient management for physicians and pharmacists;
3. A campaign designed to inform the general public about the effects of heat on health and appropriate preventive measures;
4. Development of social networks to support and protect isolated elderly and chronically ill persons;
5. Measures to marginally lower temperatures in non-air conditioned hospitals and chronic care centers where feasible;
6. Development of client-specific heat health management plans (optimal hydration and medication use, cool respite, danger signs).

References

Basrur S (2002) Community response to extreme summer heat. Staff report, Toronto Public Health

Koutsavilis AT, Kosatsky T (2003) Environmental-temperature injury in a Canadian metropolis. J Environ Health 66:40–5

Smoyer-Tomic KE, Rainham D (2001) Beating the heat: development and evaluation of a Canadian hot weather health-response plan. Environ Health Perspect 109:1241–1248

Smoyer-Tomic KE, Kuhn R, Hudson (2003) A. Heat wave hazards: an overview of heat wave impacts in Canada. Natural Hazards 28:463–85

Flooding:
The Impacts on Human Health

Lessons to be Learned from the 2002 Floods in Dresden, Germany

D. Meusel · W. Kirch

When I was asked to deliver a speech at the Meeting on Extreme Weather Events and Public Health Responses here in Bratislava, I felt particularly grateful to present you the lessons we learned from the events that the 2002 flood brought to Dresden and its neighbouring areas.

Firstly, let me begin by summarizing the main chronological events that we experienced in the Dresden region in the summer of 2002. Afterwards, I would like to draw your attention to public health actions that had to be solved immediately in the course of these events. Finally, I would like to present you with some arguments and perceptions on how these public health actions appear from the perspective 2 years afterwards.

Chronological events

As most of you know and many of you even experienced, July 2002 started with unusually intense rain and violent thunderstorms, causing high waters and floods in many parts of Europe. In the second week of August, the catastrophic dimensions of those forceful weather events had firstly become evident in Lower and Upper Austria, in Slovakia, in the Czech Republic as well as Bavaria. Cyclone Ilse, which later became recognized as a meteorologically perfect cyclone with plenty of warm humidity in its lower spheres and a cold higher sphere, arrived in the mountains surrounding Dresden on the 10th of August 2002. More than 100 litres per square metre rain at night causing small mountain rivers to collapse and water reservoirs to be overfilled. The weather station of Zinnwald, 50 km south of Dresden, registered a 24 hour measure of 312 mm rain between the 12th and 13th of August, being the highest 24 hour measure in Germany since weather had been recorded on a regular basis. This amount of rain falling within 24 hours equalled a third of the yearly average and experts believe that such a quantity is the physically possible maximum amount of rain for the region.

These huge amounts of water caused destruction all the way between the mountain villages at the summit of the Erzgebirge (Ore Mountain) to the cities located in the valley of the River Elbe. The usually small River Müglitz caused many villages to be isolated for hours and to be destroyed to an extend far beyond imagination.

One example showed a family sitting on a remaining house wall in the village of Weesenstein, after the River Müglitz had destroyed the rest of the household within minutes. Furthermore, you can see the family having rescued themselves on top of the wall. In this hopeless situation, they even had to hold out for several hours because rescue helicopters were in operation in the nearby Dresden.

On the morning of 13th August, Dresden was surprised by the severe flood wave on the River Weißeritz that abandoned its century old river bed and took its way directly through the historical parts of central Dresden.

Fig. 1

Dresden's central train station flooded by the River Weißeritz (Source: Sächsische Zeitung 2002)

Figure 1 shows the River Weißeritz taking its way through the central train station, causing most of Dresden's infrastructure to collapse that morning. In the nearby city of Meißen, which is well-known for its porcelain and wine, the small River Triebisch destroyed huge parts of the town. Especially small businesses lost their stores and facilities. Within hours, many of them faced an uncertain future of their enterprises or even immediate ruin. Even more devastatingly, the River Mulde destroyed most parts of the city Grimma, 80 km west of Dresden, while running through the main road of the town. *Figure 2* shows the city of Grimma.

Many more examples could be mentioned here to illustrate the detrimental effects of this first flood wave throughout Saxony. Small rivers grew to raging torrents, leaving behind devastated cities and villages [1, 2].

All the masses of water that had flown down valleys in Bohemia and Saxony in the preceding two days, summoned up in the River Elbe the next days. Following the 15th of August, a second, more silent but nevertheless detrimental, flood wave passed through the cities of Prague, Pirna, Dresden and Meißen and affected all the regions located at the River Elbe in the following weeks.

On the 17th of August, Dresden's water level indicators at the River Elbe showed 9.40 metres as the highest level ever registered. Normal water levels in Dresden circulate around the 2 metre mark and, ironically, reached in the following summer of 2003 their lowest level ever at 0.80 metres. This corresponds to a difference of 8.60 metres or 1175 % taking the minimum as baseline!

Complete streches of farm land on the banks of the River Elbe between the cities of Dresden and Meißen were flooded. An observer from air would have seen sloped trees standing in the water that marked the normal course of the river. Complete regions were converged to a temporary seaside.

The historical city centre of Dresden was totally flooded for several days. *Figure 3* shows the baroque Dresden Zwinger (centre of picture), the world famous opera house Semperoper, in which the composer Richard Wagner conducted the Dresden Orchestra between 1842 and 1848, as well as the Dresden castle.

Fig. 2

The city of Grimma after having been flooded (Source: Sächsische Zeitung 2002)

Fig. 3

The flooded historical part of Dresden (Source: Deutscher Depechendienst (ddp) 2002)

Many art galleries that are located in these buildings had to evacuate most of their collections in the course of the flood. The opera house was not able to perform for many months, because costumes and scenery were damaged by the water.

Even the excavation site in front of the Frauenkirche, the well-known church that was destroyed in the bombardment night of February 1945 and is now being reconstructed, was filled by the river. Only 5 months after the flooding, the neighbouring Hilton Hotel Congress Centre hosted the 10th Annual Meeting of the European Association of Public Health, whose organization by that time seemed to be seriously jeopardized by the flood.

Not only parts of the historical Dresden were affected by the flood, but also complete residential districts. As ➔ *Figure 4* illustrates, the Laubegast district in the eastern reaches of Dresden had to be evacuated completely. In most of the cases, evacuations were initiated by the residents themselves and had to be forced in exceptional cases only. About 43 school houses with a total of 7,400 beds served as a temporary home for rescue teams as well as evacuated residents. At the River Elbe about 40,000 people had to be evacuated, almost 33,000 of them in Saxony alone.

Fig. 4

Flooded residential district Dresden-Laubegast (Source: Deutsche Presseagentur (dpa) 2002)

Public Health activities

Let me now draw your attention on public health actions that had to be solved immediately in the course of these events. In the case of Dresden, these can be divided into two parts: (1) issues of public hygiene; and (2) problems involved in evacuating complete hospitals.

Hygienic issues emerged in the course of the second flood wave. Whole parts of Dresden were cut off from all supplies. In practical terms this meant the breakdown of electric power, light and even news programs. Refrigerators did not work. Drinking water could not be boiled. Much more problematic was the breakdown of emergency phone numbers. Residents had to be urged to call in severe emergencies only. For nearly two weeks the tide of the River Elbe remained high. ➔ *Figure 5* illustrates the water level of the River Elbe in Dresden with a peak of 940 cm while the normal water level is about 165 cm.

Fig. 5

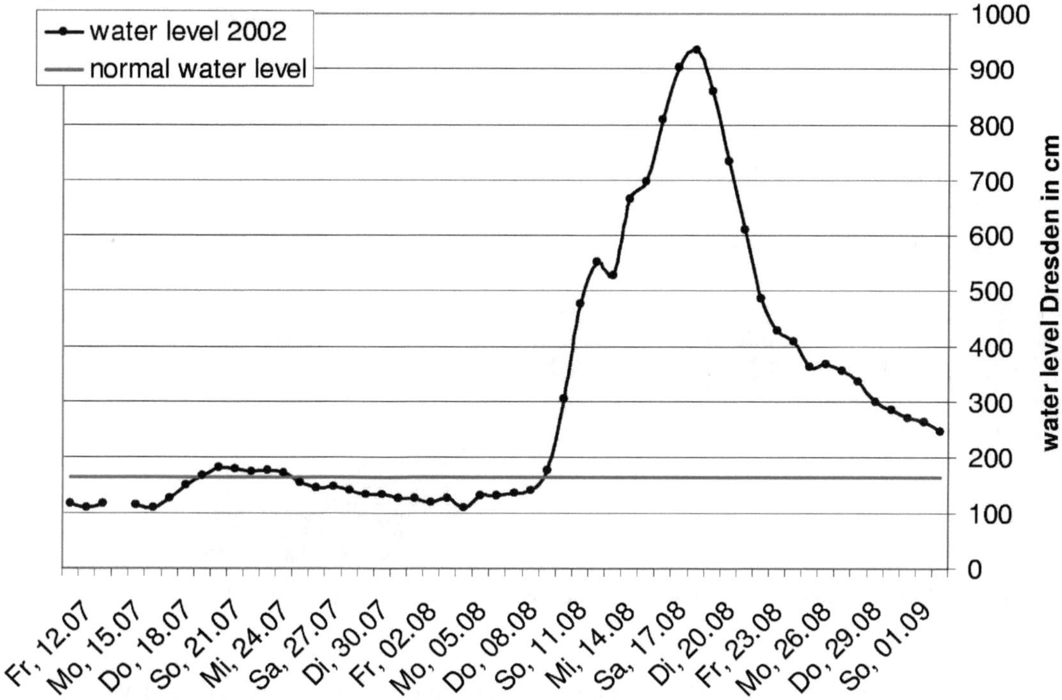

Water level of the River Elbe in the city of Dresden during July/August 2002 (Data source: [http://www.wetteronline.de/pegel/Elbe/Dresden.htm])

After the waters left, a new flood of problems emerged in making evident the real dimensions of the two flood waves. The waters left unhealthy, evil smelling and sometimes toxic layers of mud [3]. In particular, clearance worker had intense contact to polluted mud (➔ *Fig. 6)* that endangered public hygiene by containing, among others, the coli bacillus. Furthermore, furniture damaged by the flood and other garbage piled up to huge dump heaps providing insects and rodents a welcomed home.

Since a flood of such severity presented a completely new situation for both residents and local authorities, information sheets and press releases were issued for the assurance of public hygiene and the protection against epidemic plagues.

The following recommendations were issued by the Ministry of Health and Social Affairs Saxony and Saxony-Anhalt [4]:

- For consumption purposes, only water from the central water supply should be used
- Prior to consumption water should be boiled, especially when preparing food for infants or babies
- Water from private wells should not be used at all
- Non-waterproof wrapped food should be regarded as contaminated
- Especially parents were advised to prohibit children from bathing or playing in flood waters for hygienic and injury reasons
- Direct contact with mud should be avoided by wearing gloves
- Contaminated areas in houses etc. should be cleaned carefully and disinfected; for that reason, chloride and aldehyde preparations were recommended to be obtained from pharmacies

Fig. 6

After the flood left a residential home (Source: Deutscher Depechendienst (ddp) 2002)

- Flooded basement rooms should be dried up on a longer time scale to prevent mould attacks
- Flooded gardens should be dug over to prevent reproduction of insects as well as unpleasant smell.

Fortunately, the broadcast news that a flooded chemical factory in Spolana (Czech Republic) contaminated the waters of the River Elbe with huge amounts of dioxin [5] could not be proved and the residents had been informed that they were facing no threat. Furthermore, it was recommended that persons encountering first symptoms of diarrhoea, vomiting or fever as well as persons having injuries from clearance work should contact their General Practitioner immediately. In contrast, additional vaccinations against hepatitis A and typhus had not been advised because of the good epidemiological situation in Germany. A post exposition prophylaxis against tetanus had been seen as sufficient in case of injuries resulting from clearance work. In summarizing the hygienic component, it can be stated that at all times public health issues could be handled and solved favourably.

A more severe problem was presented by the evacuation of Dresden's hospitals. Four out of six major hospitals in Dresden are located at the close reaches of the River Elbe and were affected by the flooding. On the morning of 13th, a complete electric power and communication failure cut off the hospital complex Dresden-Friedrichstadt from the city. Within a few hours, the evacuation of about 950 patients had to be organised without the help of computers and telephones despite the limited transportation capabilities. Nevertheless, because of its central location within Dresden's flooded areas and the probability of injuries amongst rescue workers, emergency medical treatment had to be maintained. After evacuation was completed on the afternoon of August 13, regular medical treatment was not possible until the 21st of August [6].

Temporary hospitals were set up to compensate for the limited ability of medical care by the major hospitals in Dresden. In particular, basic surgical and traumatologic treatment were carried out here. A nursing home for elderly people served as host for the temporary hospital. An empty complex of buildings was prepared with emergency ambulance, ultrasonic diagnosis and intensive-care units including artificial respiration. After 10 hours work, the first patients could be treated.

Fig. 7

Evacuations of patients with severe diseases by aircraft (Source: Sächsische Zeitung 2002)

A logistic challenge was presented by the evacuation of the university hospital "Carl Gustav Carus" in the night of 14th to 15th August. Located directly at the river meadow lands, city officials decided to order an emergency evacuation. Facing the approaching second flood-wave hundreds of patients were transported to smaller, less endangered medical houses within Dresden. Seriously ill patients were flown out to the city hospitals of Leipzig and Berlin using military aircrafts (*Fig. 7*). At this point, a conflict between the hospital's management and the city's crisis management team was created by a lack of general emergency preparations. City officials handed over the medical responsibility for the evacuation of the remaining 180 seriously ill patients to a first-aid worker after the hospital's management had decided not to evacuate these patients. In the view of the medical professionals, the evacuation of these patients was expected to cause more harm than positive effects [7]. Some deaths are reported resulting from this over-hasty evacuation. However, names were not published in order to protect the relatives of the patients.

Perspectives 2 years after

In summarizing these events two years afterwards, the most important and most general lesson we learnt from the 2002 flood in Dresden in terms of public health can be seen in the insight that severe floods can take place and that we have to be prepared for it. Reviewing the handling of public hygiene during the weeks of flood and tide, we can state, as I outlined earlier, that general preparations were satisfactory.

Regarding the contamination of farm land on the banks of the River Elbe, new scientific findings reveal only minor pollution with harmful chemicals 2 years after the flood [8]. During the flood, the breaking of about 150 dikes caused old industrial complexes, wastewater treatment plants, oil tanks as well as former mining sites to be flooded and washed out. Initial water quality measurements showed elevated concentrations of heavy metals, arsenic, dioxins, pesticides and harmful organic substances in river waters as well as flooded farm lands. However, the enormous masses of water diluted polluting substances in such a way that environmental damage in relation to public health can be judged as limited. An extensive report on the pollution by various chemicals has been published elsewhere [9].

Nevertheless, the hospital evacuations should give reason to think over the general management of such a crisis. Here, I want to stress two central features: (1) the physical arrangement of hospital equipment; and (2) the preparation of crisis management.

To clarify the first: some severe problems resulting in the failure of the hospital's ability to keep up medical treatment during the flood were caused by the fact that the possibility of heavy floods had not been an issue of serious attention. Necessary technical equipment, such as electric power, telephone and computer network distribution devices, were installed mostly under earth basements because of space consideration. With the second flood wave, the rising groundwater-level caused all this technical equipment to malfunction or to break-down completely. Some believe that these factors contributed to the arising necessity of evacuating hospitals in the course of flood events in the first place. In fact, a hospital with autonomous power supplies does not need to be evacuated at all.

The second point needs as much attention as the first: Dresden officials were surprised by both flood waves in terms of organizing smooth cooperation of all parties involved. A crisis management team was set up sporadically by necessity. In particular, dispute over respective areas of authority revealed a lack of planning in advance that could have been easily accomplished. Therefore, prior to any possible crisis caused by extreme weather events we should be prepared to manage it in a coordinated and settled way. After all, threats to personal or public health is what we fear most of all when facing unforeseen situations.

According to the Dresden Flood Research Centre (2004), the following general Public Health recommendations for flood-risk reduction can be given:

1. Improving knowledge on flooding and damage processes in river and coastal zones
2. Further assessing the relationship between floods on the one hand and climate as well as social changes on the other
3. Providing flood warning systems together with improved weather forecasting
4. Availability of risk maps for endangered flood areas
5. Implementation of a multilevel flood disaster management plan coordinating the central and local decision making processes
6. Training of event handling with regard to flash flood risk
7. Political coordination of transboundary adjustments of flood mitigation between European countries
8. Prevention of new housing and potential toxic emissions in flood-prone areas.

Conclusion

Let me finally summarize the most important points from the Dresden flood: when severe floods occur, as in 2002, the following three points should be considered. Firstly, the public health community has to be prepared with regard to public hygiene. Secondly, important hospital equipment, such as electric power supply, has to be assembled in a "waterproof" environment. Finally, for general crisis management, the decision hierarchy between hospitals and administrative authorities should be set up prior to the crisis.

References

1. Van Stipriaan U (2002) Flutkatastrophe August 2002: Chronologie der Ereignisse (Flood disaster August 2002: Chronology of events). [accessed online 16.01.2004]. http://stipriaan.de/flut2002/chronologie.pdf
2. N-tv online: Chronologie der Flut (Chronology of the flood). [accessed online 16.01.2004]. http://www.n-tv.de/3059556.html
3. Grunewald K, Ungerl Ch, Brauch H-J, Schmidt W (2004) Elbehochwasser 2002 – Ein Rückblick: Schadstoffbelastung von Schlamm- und Sedimentproben im Raum Dresden (Elbe flood 2002 – a look back: harmful chemical pollution of mud and sediment samples in the area of Dresden). Z Umweltchem Ökotox 16(1):7–14
4. Ministerium für Gesundheit und Soziales (Ministry of Health and Social Affairs): Hochwasser-Katastrophe: Maßnahmen zur Sicherung der Hygiene und des Seuchenschutzes in den Überschwemmungsgebieten Sachsen-Anhalts (Flood disaster: actions for maintaining hygiene and epidemic protection in flood affected areas of Saxony-Anhalt). [accessed online 16.01.2004]. http://www.asp.sachsen-anhalt.de/presseapp/data/ms/2002/117_2002.htm
5. CNN.com: Dresden evacuated as waters rise. [accessed online 16.01.2004]. http://www.cnn.com/2002/WORLD/europe/08/15/floods.bratislava/index.html?related
6. Hohaus T; Oehmichen F, Bonnaire F (2004) Die Organisation der unfallchirurgischen Versorgung am Krankenhaus Dresden-Friedrichstadt während der Hochwasserkatastrophe im August 2002 – Evakuierung und Patientenversorgung im Behelfskrankenhaus (The organisation of emergency surgery treatment at the hospital Dresden-Friedrichstadt during the flood disaster in August 2002 – evacuation and patient care at a temporary hospital). [accessed online 16.01.2004]. http://www.egms.de/en/meetings/dgu2003/03dgu0375.shtml
7. Mitteldeutscher Rundfunk: Schlichtungsgespräche im Krankenhaus-Evakuierungsstreit (settlement conversations at the hospital-evacuation conflict). [accessed online 16.01.2004]. http://www.mdr.de/nach-der-flut/sachsen/267787.html
8. Neue Züricher Zeitung (2004) Wenige schwere Umweltschäden durch das Elbe-Hochwasser (Less severe environmental damage by the Elbe-flood). 2004 Jan 15, p 4
9. Geller W, Ockenfeld K, Boehme M, Knoechel A (eds) (2005) Schadstoffbelastung nach dem Elbe-Hochwasser 2002 (Harmful chemical pollution after the 2002 Elbe river flood). [online accessed 01 February 2005] URL: [http://www.halle.ufz.de/hochwasser/]

The Human Health Consequences of Flooding in Europe: a Review

S. Hajat · K. L. Ebi · R. S. Kovats · B. Menne · S. Edwards · A. Haines

Acknowledgements

This review is based on a paper originally published in "Applied Environmental Science and Public Health. (2003, vol 1, page 13 – 21). SH, RSK, KLE, and SE were funded by the European Commission as part of the project: Climate Change and Adaptation Strategies for Human Health (cCASHh) (EVK2-2000-00070)

Keywords

Floods, mental health, climate change, Europe

Summary

Floods are the most common natural disaster in Europe. The adverse human health consequences of flooding are complex and far-reaching: these include drowning, injuries, and an increased incidence of common mental disorders. Anxiety and depression may last for months and possibly even years after the flood event and so the true health burden is rarely appreciated. Effects of floods on communicable diseases appear relatively infrequent in Europe. The vulnerability of a person or group is defined in terms of their capacity to anticipate, cope with, resist and recover from the impact of a natural hazard. Determining vulnerability is a major challenge. Vulnerable groups within communities to the health impacts of flooding are the elderly, disabled, children, women, ethnic minorities, and those on low incomes. There is a need for more good-quality epidemiological data before vulnerability indices can be developed. With better information, the emphasis in disaster management could shift from post-disaster improvisation to pre-disaster planning. A comprehensive, risk-based emergency management program of preparedness, response, and recovery has the potential to reduce the adverse health effects of floods, but there is currently inadequate evidence of the effectiveness of public health interventions

Introduction

In Europe, floods are the most common natural disaster. In recent years, floods have received much media attention. European vulnerability to flooding was highlighted by the loss of life and economic damage from flooding events of the Rhine, Meuse, Po and Oder rivers in the 1990s and in the UK floods of 2000. In addition, summer flooding of the Elbe and Danube rivers in 2002 resulted in some of the worst floods seen in Europe for more than a century

Various mechanisms may cause flooding, and different flood characteristics affect the occurrence and severity of the flood event (Malilay 1997). The Third Assessment Report of the Intergovernmental Panel on Climate Change concluded that, by 2100, the general pattern of changes in annual precipitation over Europe is for widespread increases in northern Europe (between +1 and +2 % per decade), smaller decreases across southern Europe (maximum –1 % per decade), and small or ambiguous changes in central Europe (IPCC 2001). It is likely that intense precipitation events will continue to increase in frequency, especially in the winter. Further to this, the frequency of great floods has been demonstrated to have increased substantially during the twentieth century (Milly et al. 2002). This pattern is consistent with current climate models, with the models suggesting that the trend will continue. These projected changes underscore the need to increase the development and implementation of measures to prevent adverse health impacts from flooding (Baxter et al. 2001).

This paper reviews the epidemiological literature to assess the human health consequences of flooding in Europe and other industrialised countries, and investigates the current adaptation strategies available to the health-sector to minimise effects.

Methods

The search for literature on the health effects of floods was restricted to events in the following regions: the whole of Europe, North America and Australasia. Although the focus of this work is to document the effects of flood events from Europe, examples from other industrialised countries are also used to illustrate the potential range of health impacts from floods. Events from low-income countries are excluded as the effects can be very different to those experienced in the majority of Europe. Literature was obtained by consulting experts in the field and by searching the databases using the following search strategy:

- Objectives: To search for literature relating to the human health impacts of flooding
- Database list: BIDS, Embase, Psyclit, Pubmed, Sigle
- Terms searched for: Flood, floods, flooding, disasters, extreme events, health
- Inclusion criteria: all available years
- Exclusion criteria: events not occurring in Europe, North America or Australasia

Health impacts

There are two main types of river floods that affect Europe (Penning-Rowsell & Fordham 1994). First, there is the rapid rise flood. These are most commonly associated with intense thunderstorm activity and so tend to be mainly local or regional events. Secondly, there are the slow rise events characterised by floods on the large rivers of northern Europe, notably the Rhine, the Vistula, the Thames, the Seine, the Loire, and the Rhone (Penning-Rowsell, Handmer, & Tapsell 1996). These large catchments respond to prolonged periods of rainfall or snowmelt. In addition, there is also a risk of coastal flooding due to sea level rise, the effects of which will greatly impact on low-lying coasts and islands. Different types of floods affect human health in different ways.

The adverse human health consequences of flooding are complex and far-reaching (Tapsell 2001) and can therefore be difficult to attribute to the flood event itself. The health impacts are broadly categorised into one of two groups:

1. Physical health effects sustained during the flood event itself or during the clean-up process, or from knock-on effects brought about by damage to major infrastructure including displacement of populations

2. Mental health effects directly occurring due to the experience of being flooded, or indirectly during the restoration process.

Physical health effects

A database of the occurrence and immediate effects of all reported mass disasters, including flood events, in the world from 1900 to present is available from EM-DAT: The OFDA/CRED International Disaster Database [www.cred.be/emdat/]. The database is compiled from various sources including UN agencies, non-governmental organisations, insurance companies, research institutes and press agencies. In order for a disaster to be entered onto the database, at least one of the following 4 criteria has to be fulfilled: 10 or more people reported killed, 100 people reported affected, a call for international assistance, or declaration of a state of emergency. Between 1975 and 2005, 238 flood events in Europe were recorded on EM-DAT, resulting in 2476 reported fatalities. Between 1980 and 1999, an annual rate of 1.3 deaths and 5.7 injuries have occurred per 10,000,000 population due to inland floods and landslides in Western Europe (McMichael et al. 2004). ❯ *Table 1* presents the ten most disastrous floods in Europe from 1990 to 2001, in terms of number of deaths.

Tab. 1

The 10 most disastrous floods in Europe within the last decade as recorded by the EM-DAT International Disaster Database (1990–2001)

Country	Year	Killed	Location
Tajikistan	1992	1346	–
Italy	1998	147	Campania Region
the Russian Federation	1993	125	Yekaterinburg region (Sverdlovsk)
Romania	1991	108	Bouriatie, Ulan-Ude
Uzbekistan	1998	95	Shahimardan, Yerdan
Turkey	1995	78	Izmir, Antalaya, Isparta
Turkey	1995	70	Ankara, Istanbul, Senirkent
Italy	1994	64	Piedmont, Liguria
Turkey	1998	60	Beskoy (Trabzon province)
Tajikistan	1998	57	Ragun, Ainy, Old Mastchoh, Shahrinav, Muminabad, Penjikent, Kuliyab Central, Vose, Dushanbe, Tursenzade, Varzob, Farhor, Baljuvon, Tursunzade, Leninski, Gissar, Kanibadam, Sharristan, Kurgantube, Kaarnikhon, Khovaling

Over the years the physical health effects of flooding in Europe have been documented in the medical literature when major flood events have occurred (Cervenka 1976; Lorraine 1954). In more recent decades, the number of studies looking at the health effects of such events has increased. These are examined separately for different outcomes below.

Mortality

Most flood-related deaths can be attributed to rapid rise floods, due to the increased risk of drowning (French et al. 1983). In October 1988, a flash flood occurred in the Nimes region of France (Duclos et al. 1991). Although the homes of 45,000 people were damaged and more than 1,100 vehicles destroyed, only 9 deaths by drowning (including two people who tried to rescue others) and 3 severe injuries were reported. The limited death toll can be attributed to the fact that the disaster occurred early in the morning when most people were still home, and also because of the mild temperature, government rescue plans, and most of all, because of rescue operations conducted by civilians. Other health problems and injuries during the post-impact phase may have been limited by the response of trained military personnel and by the distribution of boots and gloves to other responders. In 1996, 86 people died from a flood in the town of Biescas in Spain as a consequence of the stream of water and mud that suddenly covered a campsite located near a channelised river (Marcuello & Estrela 1997). During the 1998 flood in Sarno, Italy, there were between 147 and 160 fatalities caused by a river of mud that destroyed an urban area (Thonissen 1998).

Many slow-rise river flood events have also been associated with fatalities. At the end of 1993 and the beginning of 1994, overflow of the River Meuse and on the middle and lower Rhine caused 10 deaths in Germany and other affected countries (Bayrische Ruckversicherung, 1996a). Thirteen months later, intense rainfall produced a new flood of similar characteristics. The effects, however, were not so great as people were aware of the risk and better prepared (Bayrische Ruckversicherung, 1996b).

In 1997, river floods in central Europe left over 200,000 people homeless, and more than 100 people were killed (Kriz et al. 1998; Saunders 1998). The main countries affected were Poland, Germany and the Czech Republic.

In the UK, the effects of floods have been well documented. On Saturday 31 January and Sunday 1 February 1953, a great storm surge, accompanied by gale force winds, swept over the north of the UK, causing widespread flooding of coastal areas (Greave 1956). The worst effects were in Canvey Island where 58 people died, although it was noted that the death rate climbed significantly during the 2 months following the disaster, as compared with the same 2 months the previous year (Summers 1978). In total, 307 people in the UK died due to the event, and 1795 people in the Netherlands. Other examples in the UK include 5 fatalities in the central England and Wales floods of 1998. In addition 3 people drowned in Glasgow in 1994; 4 people drowned in south Wales in 1979; and, in 1975, 2 people drowned and 2 were struck by lightning during a severe thunderstorm in Hampstead, London, where floods caused damage to houses, cars to float along streets, subways to fill and sewers to burst (Faulkner 1999).

Injuries

Comprehensive surveillance of morbidity following floods is limited, but following the 1993 Midwestern United States floods, surveillance of flood-related morbidity was undertaken (CDC 1993). Five hundred and twenty four flood-related conditions were reported in emergency departments. Of these, 250 were injuries, 233 were illnesses, 39 were listed as 'other', and two were 'unknown'. Of the 250 reported injuries, the most common were sprains/strains, lacerations, 'other injuries', and abrasions/contusions. No comparable data for Europe has been identified.

Illnesses from flood-induced contamination of water supplies

There is a small risk of communicable disease following flooding, although severe occurrences are rare in industrialised countries due to the public health infrastructure in place prior to and following a flood event such as water treatment and effective sewage pumping. One example of an outbreak is that of leptospirosis which occurred after the flooding in the Czech Republic in 1997, although the quality of the data appears to be poor (Kriz, Benes, Castkova, & Helcl 1998). Work from Finland highlighted that a total of 14 waterborne epidemics occurred in the country during 1998 – 1999, resulting in about 7,300 registered cases (Miettinen et al.). Thirteen of the epidemics were associated with undisinfected groundwaters from mostly flood-induced outbreaks. Boiling of the drinking water was one of the first actions taken in almost all of the outbreaks.

Infectious disease outbreaks following flooding may be more common in tropical industrialised countries such as Australia, where conditions such as Melioidosis and other vector borne diseases have been linked to local flooding (Munckhof et al. 2001).

No specific increase in infectious disease was observed following the flash flood in Nimes (Duclos, Vidonne, Beuf, Perray, & Stoebner 1991), and no increase in the incidence of acute gastroenteritis or other possibly flood-related communicable diseases was observed following the 1995 river floods in eastern Norway (Aavitsland et al.). This was attributed in part to measures such as maintaining safe water and providing information on safe management of flood water during evacuation and clean-up. Similarly, another report stated that no dramatic increase in water borne diseases were reported or documented in the aftermath of the major floods of the 1970s, although the attitude of the public, the mass media and of the health services sometimes leads to ineffective mass immunisations (de Ville, Lechat, & Boucquey). The authors stated that an epidemiological system and accurate information on the actual situation are essential in cases of major disasters

Other flood-induced illnesses

Chronic health effects (including respiratory problems) secondary to a flood disaster have been documented in the literature. Investigation of leukemias and lymphomas in western New York State found an apparent space-time clustering of cases beginning 2 years after major floods occurred in the Canisteo River valley in 1972 (Janerich et al. 1981). This, in conjunction with a marked increase in rates of spontaneous abortion, suggested an unidentified flood-related environmental exposure. Possible explanations provided by the authors include exposure to human and animal viruses during evacuation, or perhaps because of substantial psychological or physical stress experienced by people at the time of the flood.

Following the flash flood in Nimes in 1988, 12 cases of carbon monoxide poisoning were reported involving firefighters, civilians, and members of the military who were pumping water and effluents from basements (Duclos, Vidonne, Beuf, Perray, & Stoebner 1991). Authors of a case study of heavy metal soil contamination after the flooding of the river Meuse during the winter of 1993 – 1994 concluded there was a potential health risk for river-bank inhabitants as a consequence of lead and cadmium contaminations of the floodplain soils (Albering et al. 1999). There is also the risk of toxic fungal spread as a result of flooding, both in houses following home water damage (Jarvis et al. 1998). and as an agricultural pest (Rosenzweig et al. 2001).

Following the 1968 flooding in Bristol, Bennet reported a 50 % increase in mortality ($P < 0.02$) in the homes of people who were flooded, with many of these deaths being from chronic diseases such as cancer (Bennet 1970). The author observed similarities with the pattern of health and mortality after bereavement, and argued that the effect of flooding upon mortality and ill-health is largely a result of distress and the psychological effects of the event.

Mental health effects

Comparison of rates of mental ill health between studies are hampered by varying definitions of outcome used. Some investigators focus on Post-traumatic stress disorder (PTSD) as a specific entity; some use rates of common mental disorders such as anxiety and depression, whereas others use standard instruments such as the General Health Questionnaire which do not give a specific diagnosis. There is no doubt however that flooding, in common with other traumatic life events, is associated with increased rates of the most common mental disorders: anxiety and depression (Bennet 1970; Sartorius 1990). Aside from the experience of being flooded, many of the mental health problems stem from the troubles brought about by geographic displacement (Fullilove 1996), the damage to the home or loss of familiar possessions (Keene 1998), and also the stress involved in dealing with builders, etc. during the aftermath (Tapsell & Tunstall 2001). In this context, lack of insurance may be an important factor in hindering recovery. Lack of insurance (or under-insurance) was a common factor in exacerbating the impacts of the 1993 floods in Perth, Scotland (Fordham & Ketteridge 1995).

A panel study of over 200 elderly adults interviewed both before and after the floods in southeastern Kentucky, US, in 1981 and 1984 found that flood exposure was related to modest physical health declines such as functional impairment and fatigue (Phifer, Kaniasty, & Norris 1988). The study also demonstrated that persistence of health effects was directly related to flood intensity: whereas the effects of the 1981 flood generally were limited to the first year post-flood, the more severe 1984 flood continued to have a significant impact 18 months after the flood. Men, those with lower occupational status, and persons aged 55 – 64 were at significantly greater risk for increases in psychological symptoms. Sociodemographic status did not moderate the impact of flood exposure on physical health (Phifer 1990).

A case-study of the health effects of flooding in Uphill, UK showed a consistent pattern of increased psychological problems amongst flood victims in the five years following floods (Green et al. 1985). A survey following the Lewes floods in 2000 demonstrated higher levels of depression (as measured by General Health Questionnaire scores) among flooded households compared to controls in the same area (Reacher et al. 2004). Also in the UK, qualitative work at least one year after the Banbury and Kidlington floods revealed continuing psychological health effects among most of those interviewed (Tapsell 2000). There was also the suggestion that the full health impacts of the floods had not been manifested at the time of the study. It is often only after people's homes have been put back in order that the full realisation of what has happened to them is appreciated.

A pilot cross-sectional study based on the 1992 floods in the Vaucluse was conducted using randomly selected households in two affected towns. This study demonstrated that the long-term psychological consequences of an environmental disaster could be carried out several years after the event but that the feasibility of such a study would depend ultimately on its acceptance by the public and the relevant authorities (Verger et al. 1999). Another pilot study, looking at the effects of the Netherlands flood of 1994 – 1995 on the health and wellbeing of exposed subjects, suggested 15 – 20 % of children were having moderate to severe 'stress' symptoms, and 15 % of adults still experienced very severe 'symptoms of stress' 6 months after the event (Becht et al. 1998).

An increase in psychosocial symptoms and PTSD including 50 flood-linked suicides were reported in the two months following the major floods in Poland in 1997 (IFRC 1998). Based on interviews with Polish children aged 11 – 14 years, these floods were also reported to be associated with long-term PTSD, depression and dissatisfaction with life (Bokszczanin 2000).

By contrast, study of a psychiatric out-patient department after the 1997 floods in the Czech Republic suggested fewer hospital admissions during the investigation period than in other years, and no attempted suicides and suicides were recorded (Kucerova 1999). The author suggests that the absence of suicidal activities may have been caused by a development of positive attitudes of individuals towards themselves and others as a result of the extensive disaster, such as is often observed during times of war.

Contact with health services

Contacts with health services may increase following a flood event, although a further effect of floods upon health could be the likely disruption of 'normal' health and social service programmes. These services are likely to be heavily involved in the aftercare of a disaster, thus removing them from their normal caring activities.

The impacts of flood events on the use of primary and secondary health services have not been extensively investigated in the UK or elsewhere. Primary care attendance following the Bristol 1968 flood rose by 53 %, and referrals and admissions to hospitals more than doubled (Bennet 1970). Similarly, the number of visits to general practitioners, hospitals and specialists were all significantly increased for flooded persons in the year following the 1974 Brisbane floods (Abrahams et al.).

Vulnerability

'Vulnerability' is defined in terms of the capacity of a person or group to anticipate, cope with, resist and recover from the impact of a natural hazard (Blaikie et al. 1994). Certain groups within communities (e.g. the elderly, disabled, children, women, ethnic minorities, and those on low incomes) may be more vulnerable than others to the effects of flooding (Curle & Wiliams 1996; Flynn & Nelson 1998; Fordham 1998; Ketteridge & Fordham 1995; Morrow 1999; Tapsell & Tunstall 2001; Thompson 1995; Ticehurst et al. 1996). Consequently, these groups may suffer greater effects from a flood and may need special consideration by the authorities during the response and recovery periods. However, much of the work is qualitative in nature and robust estimates of differences between groups based on epidemiological studies are lacking.

People on lower incomes may be more vulnerable to the effects of a flood in that they may not have adequate insurance or the financial capacity to recover from the experience. Lack of resources can impede resilience to a disaster as savings or other financial resources can be used as a buffer against the worst tangible and intangible impacts of a flood (Ketteridge & Fordham 1995). Health is affected by the places in which people live, work and interact, and yet many epidemiological studies overlook the characteristics of places and instead focus solely on the people who inhabit them (Smoyer 1998).

Environmental disasters can be particularly devastating to already vulnerable populations such as the homeless and migrants, who, because of social, political and economic constraints, experience special health care needs. In the US, a synthesis of past disaster research showed how various racial and ethnic groups may be differentially affected, both physically and psychologically, and how disaster effects vary by race and ethnicity during the periods of emergency response, recovery and reconstruction (Fothergill, Maestas & Darlington 1999). The authors suggest that racial and ethnic communities in the US such as African-Americans and Mexican-Americans are more vulnerable to natural disasters due to factors such as language, housing patterns, building construction, community isolation and cultural insensitivities of the majority population. A variety of other studies have also demonstrated vulnerability of sub-groups (1990; Enarson & Fordham 2001; Green 1988a; Green 1988b; Green et al. 1994; Huerta & Horton 1978; Ohl & Tapsell 2000).

To ascertain vulnerability, the public health consequences of disasters need to be properly evaluated (2001; Blake 1989; French 1989; Kirchsteiger 1999; Lechat 1990b; Lechat 1990a; Logue, Melick, & Hansen 1981). However, an assessment of human health impacts is complicated due to numerous uncertainties, including the lack of good quality epidemiological data, and the unknown economic, political, and technological response of society.

Strategies to reduce flood impacts

Before, during and after a flood event, activities may be undertaken by the population at risk, by policy makers and by emergency responders to reduce health risks. Traditionally, the fields of engineering and urban planning aimed to reduce the harmful effects of flooding by limiting the impact of a flood on human health and economic infrastructure. This is accomplished by building codes, legislation to relocate structures away from flood-prone areas, planning appropriate land use, and managing costs of floodplains. Mitigation measures may reduce but not eliminate major damage.

Early warning of flooding risk, and appropriate citizen response, has been shown to be effective in reducing disaster-related deaths. From a public health point of view, planning for floods during the inter-flood phase aims to enable communities to effectively respond to the health consequences of floods, and to enable the local and central authorities to organise and effectively co-ordinate relief activities, including making the best use of local resources and properly managing national and international relief assistance. In addition, medium to long-term interventions may be needed to support populations who have been flooded. These should include initiatives such as public health authorities being alerted to the possibility of post-flood diseases and injuries, and the identification of and provision of health services for individuals with post-flood mental health problems. Currently, however, it is unclear whether the mental health impacts will respond best to psychological and/or pharmacological interventions delivered through health services, or whether the interventions would best be targeted at providing financial or other assistance with housing etc.

It seems likely that effective vulnerability reduction will necessitate the involvement of a range of sectors at the local, regional and sometimes national level. There is a need for statistical indicators of vulnerability and the harm caused by major emergencies. Data from these indicators will assist in monitoring and evaluating vulnerability reduction, and identification of communities at risk. It is important that an approach to relief, disaster preparedness and mitigation starts with programmes with a developmental focus on increasing the health and safety of the potentially affected populations (McCluskey 2001). An overall objective of disaster epidemiology should be to describe the health effects of disasters and the factors contributing to these effects (Noji 1992; Noji 1995; Noji 1996; Noji 2000). Results of epidemiologic studies of natural disasters provide clues to diagnosis, help medical care providers match resources to needs, and permit better contingency planning (Binder & Sanderson 1987).

Discussion

Despite floods being the most common natural disaster in Europe, the health risks associated with flooding are surprisingly poorly characterised. This review highlights the dearth of good quantitative data available on the health effects of flooding, resulting in uncertainty in the full range of potential health impacts of flood events. The studies that have been conducted mainly concentrate on assessing effects of large events, even though more frequent smaller events may also have an important health impact. Different types of floods may need different types of intervention strategies to minimise impacts. In general, the reviewed reports suggest two main messages arising when considering the health impacts of floods in industrialised countries (WHO, 2002):

1. The biggest impacts occur as a result of the psychological distress experienced during flooding and in its aftermath
2. In direct contrast to low-income countries, the likelihood of infectious disease outbreaks following flooding in temperate industrialised countries are low. Maintenance of existing public health responses to flooding in these countries are important to sustain the low risk.

A better understanding is needed of vulnerability risk factors. There are suggestions that some people may be more susceptible to the effects of floods than others, but the available evidence is insufficient to allow vulnerability indices to be devised and used operationally. Again the need is for more and better quality epidemiological data, including

- Centralized and systematic national reporting for deaths and injuries from floods using standardized methodology
- Development of instruments to assess health risks
- Identification of data needed to prepare for and evaluate the impacts of such strategies on future events.

There are methodological problems associated with mounting retrospective studies. Such studies are likely to be prone to recall bias as affected subjects may be more likely to remember adverse effects. In addition, the selection of a suitable control population is required to assess the true effects attributable to the flood disaster. It has been suggested that surveillance of flood-related morbidity, mortality, vector populations, and environmental health should continue throughout the response and recovery periods (Malilay 1997). Should any unusual conditions be noted, specialised investigations or surveys should be conducted so that appropriate interventions can be made

The frequency of extreme weather events such as floods is likely on the increase due to changes in the world's climate (Fulton 1999). Significant reductions in greenhouse gas emissions need to be made, however the impact of such reductions would only become apparent in 50 – 100 years time. Preparations need to be made for the climate change that is on its way as a result of the greenhouse gases that are already in the atmosphere (Parry et al. 1998). As a consequence, more emphasis needs to be placed on available structural and policy-oriented control measures, and the activities undertaken by the population at risk and by emergency responders before, during and after a flood event in order to prevent or reduce the risk of flood-related injury, illness or death.

References

Aavitsland P, Iversen, BG, Krogh T, Fonahn W, Lystad A. [Infections during the 1995 flood in Ostlandet. Prevention and incidence]. Tidsskr Nor Laegeforen 116:2038 – 2045

Abrahams MJ, Price J, Whitlock FA., Williams G (1974) The Brisbane floods, January 1974: their impact on health. Med J Aust 2:936 – 939

Albering HJ, van Leusen SM, Moonen EJC, Hoogewerff JA, Kleinjans JCS (1999) Human health risk assessment: A case study involving heavy metal soil contamination after the flooding of the river Meuse during the winter of 1993 – 1994. Environmental Health Perspectives 107:1:37 – 43

Baxter PJ, Moller I, Spencer T, Spence RJ, Tapsell S (2001) Flooding and climate change, in Health effects of climate change in the UK. Department of Health, London, pp 152 – 192

Bayrische Ruckversicherung (1996a) ...13 months later. The January 1995 floodsBayrische Ruckversicherung. Special issue 17

Bayrische Ruckversicherung (1996b) The 'Christmas floods' in Germany 1993 – 94. Bayrische Ruckversicherung Special issue 16

Becht MC, van Tilburg M-A L, Vingerhoets A-JJM, Nyklicek I, de Vries J, Kirschbaum C, Antoni MH, van Heck GL (1998) Watersnood. Een verkennend onderzoek naar de gevolgen voor het welbevinden en de gezondheid van volwassenen en kinderen. [Flood: A pilot study on the consequences for well-being and health of adults and children]. Uitgeverij Boom, Netherlands

Bennet G (1970) Bristol floods 1968: controlled survey of effects on health of local community disaster. British Medical Journal 3:454 – 458

Binder S, Sanderson LM (1987) The role of the epidemiologist in natural disasters. Annals of Emergency Medicine 16:1081 – 1084

Blaikie P, Cannon T, Davis I, Wisner B (1994) At risk: natural hazards, people's vulnerability and disasters. Routledge, London

Blake PA (1989) Communicable disease control. In: Gregg MB, French J, Binder S, Sanderson LM (eds) The Public Health Consequences of Disasters. US Dept of Human Health and Services, Atlanta

Bokszczanin A (2000) Psychologiczne konsekwencje powodzi u dzieci i mlodziezy. [Psychological consequences of floods in children and youth] Poland: Psychologia Wychowawcza.

CDC (1993) Morbidity surveillance following the midwest flood – Missouri, 1993. Mortality and Morbidity Weekly Report 42(41):797 – 798

Cervenka J (1976) Health aspects of Danube river floods. AnnSoc Belg Med Trop 56:217 – 222

Curle CE, Wiliams C (1996) Post-traumatic stress reactions in children: gender differences in the incidence of trauma reactions at two years and examination of factors influencing adjustment. British Journal of Clinical Psychology 35:297 – 309

de Ville dG, Lechat MF, Boucquey C (1977) [Attitude toward the risk of epidemics during sudden disasters]. Rev Epidemiol Sante Publique 25:185 – 194

Duclos P, Vidonne O, Beuf P, Perray P, Stoebner A (1991) Flash flood disaster – Nimes, France, 1988. Eur J Epidemiol 7:365 – 371

Enarson E, Fordham M (2001) Lines that divide, ties that bind: race, class and gender in women's flood recovery in the US and UK. Australian Journal of Emergency Management summer:43 – 52

Faulkner DS (1999) Flood estimation handbook. Institute of Hydrology, Wallingford

Flynn BW, Nelson ME (1998) Understanding the needs of children following large-scale disasters and the role of government. Child & Adolescent Psychiatric Clinics of North America, 7:211 – 227

Fordham M, Ketteridge AM (1995) Flood disasters – Dividing the community [presentation] International Emergency Planning Conference, 1995, Lancaster, UK.

Fordham MH (1998) Making women visible in disasters: problematising the private domain, Disasters 22:126 – 143

Fothergill A, Maestas EG, Darlington JD (1999) Race, ethnicity and disasters in the United States: a review of the literature. Disasters 23:156 – 173

French JG (1989) Floods. In: Public Health Consequences of Disasters. Department of Health and Human Services, Atlanta, U.S.

French J, Ing R, von Allmen S, Wood R (1983) Mortality from flash floods: a review of national weather service reports, 1969 – 81. Public Health Rep 98;6:584 – 588

Fullilove MT (1996) Contributions from the psychology of place. Am J Psychiatry 153:1516 – 1523

Fulton J (1999) Using science to advocate action on climate change. Ecosystem Health 5:110 – 117

Greave H (1956) The Great Flood. Report to Essex County Council. Essex County Council

Green C (1988a) The Human Aspects of Flooding. Flood Hazard Research Centre, Middlesex University, Enfield

Green C (1988b) The Relationship between the Magnitude of Flooding, Stress and Health. Flood Hazard Research Centre, Middlesex University, Enfield

Green C, van der Veen A, Wierstra E, Penning-Rowsell E (1994) Vulnerability defined: analysing full flood impacts. In: Penning-Rowsell E, Fordham M (eds) Floods Across Europe: Hazard Assessment, Modelling and Management. Middlesex University Press, London, pp 32 – 68

Green CH, Emery PJ, Penning-Rowsell EC, Parker DJ (1985) The health effects of flooding: a case study of Uphill. Flood Hazard Research Centre, Middlesex Polytechnic, Enfield

Huerta F, Horton R (1978) Coping behavior of elderly flood victims. Gerontologist 18;6:541 – 546

IFRC (International Federation of Red Cross and Red Crescent Societies) (1998) World disaster report 1997. Oxford University Press, New York

IPCC (2001) Climate Change 2001. Impacts, Adaptations and Vulnerability. Contribution of Working Group II to the Second Assessment Report of the Intergovernmental Panel on Climate Change. Cambridge University Press, Cambridge

Janerich DT, Stark AD, Greenwald P et al. (1981) Increased leukemia, lymphoma, and spontaneous abortion in Western New York following a flood disaster. Public Health Rep 96:350 – 356

Jarvis BB, Sorenson WG, Hintikka EL et al. (1998) Study of toxin production by isolates of Stachybotrys chartarum and Memnoniella echinata isolated during a study of pulmonary hemosiderosis in infants. Appl Environ Microbiol 64:3260 – 3265

Keene EP (1998) Phenomenological study of the North Dakota flood experience and its impact on survivors' health. Int J Trauma Nurs 4:79 – 84

Ketteridge AM, Fordham M (1995) Flood warning and the local community context. In: Handmer J (ed) Flood Warning: Issues and Practice in Total System Design. Flood Hazard Research Centre, Enfield, pp 189 – 199

Kirchsteiger C (1999) Trends in accidents, disasters and risk sources in Europe. Journal of loss prevention in the process industries 12:7 – 17

Kriz B, Benes C, Castkova J, Helcl J (1998) Monitorování Epidemiologické Situace V Zaplavených Oblastech V Èeské Republice V Roce 1997. [Monitoring the Epidemiological situation in flooded areas of the Czech Republic in 1997.] In: Konference DDD ,98; Kongresové Centrum Lázeòská

Kolonáda Podìbrady, 11. – 13. Kvìtna 1998 [Proceedings of the Conference DDD'98, 11 – 12th May, 1998, Prodebrady, Czech Republic.] Prodebrady, Czech Republic

Kucerova H (1999) [Reaction of patients in the psychiatric outpatient department to floods in 1997]. Ceska a Slovenska Psychiatrie 95:478 – 482

Lechat MF (1990a) The epidemiology of health effects of disasters. Epidemiologic Reviews 12:192 – 198

Lechat MF (1990b) The public health dimensions of disasters, International Journal of Mental Health 19:70 – 79

Logue JN, Melick ME, Hansen H (1981) Research issues and directions in the epidemiology of health effects of disasters. Epidemiologic Reviews 3:140 – 162

Lorraine NSR (1954) Canvey Island flood disaster, February, 1953. Medical Officer 91:59 – 62

Malilay J (1997) Floods. In: Noji E (ed) Public Health Consequences of Disasters. OUP, New York, pp 287 – 301

Marcuello C, Estrela T. Floods in Spain: Main features and their effects. In: River Flood Disasters, ICSU, SC/IDNDR Workshop. Koblenz, Germany, BFG IHP/ODP-Secretariat, pp 135 – 144

McCluskey J (2001) Water supply, health and vulnerability in floods. Waterlines 19:14 – 17

McMichael AJ, Campbell-Lendrum D, Kovats RS et al. (2004) Climate change. In: Ezzati M et al. (eds) Comparative Quantification of Health Risks: Global and Regional Burden of Disease due to Selected Major Risk Factors. World Health Organization, Geneva

Miettinen IT, Zacheus O, von Bonsdorff CH, Vartiainen T (2001) Waterborne epidemics in Finland in 1998 – 1999. Water Sci Technol 43:67 – 71

Milly PC, Wetherald RT, Dunne KA., Delworth TL (2002) Increasing risk of great floods in a changing climate. Nature 415;6871:514 – 517

Morrow BH (1999) Identifying and mapping community vulnerability. Disasters 23;1:1 – 18

Munckhof WJ, Mayo MJ, Scott I, Currie BJ (2001) Fatal human melioidosis acquired in a subtropical Australian city. Am J Trop Med Hyg 65:325 – 328

Noji EK (1992) Disaster epidemiology: Challenges for public health action. Journal of Public Health Policy 13:332 – 340

Noji EK (1995), Disaster epidemiology and disease monitoring. Journal of Medical Systems 19;2:171 – 174

Noji EK (1996) Disaster epidemiology. Emergency Medicine Clinics of North America 14;2:289 – 300

Noji EK (2000) The public health consequences of disasters. Prehospital Disaster Med 15;4:147 – 157

Ohl CA, Tapsell S (2000) Flooding and Human Health. British Medical Journal 321:1167 – 1168

Parry M, Hulme M, Nicholls R, Livermore M (1998) Adapting to the inevitable. Nature 395:741

Penning-Rowsell EC, Fordham M (1994) Floods across Europe: Hazard assessment, modelling and management. Middlesex University Press, London

Penning-Rowsell E, Handmer J, Tapsell S (1996) Extreme events and climate change; floods. In: Downing TE, Olsthroon AA, Tol RSJ (eds) Climate Change and Extreme Events: Altered Risk, Socio-Economic Impacts and Policy Responses, pp 97 – 127

Phifer JF (1990) Psychological distress and somatic symptoms after natural disaster: differential vulnerability among older adults. Psychol Ageing 5:412 – 420

Phifer JF, Kaniasty KZ, Norris FH (1988) The impact of natural disaster on the health of older adults: A multiwave prospective study. Journal of Health & Social Behavior 29:65 – 78

Reacher M, McKenzie K, Lane C et al. (2004) Health impacts of flooding in Lewes: a comparison of reported gastro-intestinal and other illnesses in flooded and nonflooded households. Communicable Disease and Public Health 7:1 – 8

Rosenzweig C, Iglesias A, Yang XB et al. (2001) Climate change and extreme weather events: implications for food production, plant diseases, and pests. Global Change & Human Health 2:90 – 104

Sartorius N (1990) Coping with disasters: The mental health component. Preface. International Journal of Mental Health 19:3 – 4

Saunders MA (1998) Central and Eastern European Floods of July 1997. Benfield Greig Hazard Research Centre, London

Smoyer KE (1998) Putting risk in its place: methodological considerations for investigating extreme event health risks. Soc Sci Med 47:1809 – 1824

Summers D (1978) The East Coast floods. David & Charles, London

Tapsell S (2001) The health effects from fluvial flooding. Report to the Environment Agency. Flood Hazard Research Centre, Enfield

Tapsell SM, Tunstall SM (2001) The health and social effects of the June 2000 flooding in the North East region. Flood Hazard Research Centre, Enfield

Tapsell SM (2000) Follow-up study of the health effects of the Easter 1998 flooding in Banbury and Kadlington. Report to the Environment Agency. Flood Hazard Research Centre, Enfield

Thompson N (1995) The ontology of disaster. Death studies 19;5:501 – 510

Thonissen C (1998) Water management and flood prevention in the framework of land-use management.[presentation] Workshop held by DGXVI of the European Commission. 1998, July 2–3, Thessaloniki, Greece

Ticehurst S, Webster RA, Carr VJ, Lewin TJ (1996) The psychosocial impact of an earthquake on the elderly. International Journal of Geriatric Psychiatry 11:943–951

Verger P, Rotily M, Baruffol E et al. (1999) Evaluation of the psychological consequences of environmental catastrophes: A feasibility study based on the 1992 floods in the Vaucluse (France). Cahiers Sante 9:313–318

WHO (2002) Floods: Climate Change and Adaptation Strategies for Human Health. Report of meeting, London, 30 June–2 July 2002. http://www.who.dk/document/E77096.pdf

Mortality in Flood Disasters

Z. W. Kundzewicz · W. J. Kundzewicz

Introduction

There has been a considerable increase in the number of large floods and in severity of their impacts in the recent decades, worldwide. Although floods are not commonly perceived as public health events, they do clearly lead to deterioration of human health over vast flood-affected areas. Flood-related morbidity and mortality raise considerable concern. Each year, the global number of flood-related fatalities is in the range of thousands.

The paper looks into the data on flood fatalities. It does not aim to present a detailed overview of the worldwide situation. It focuses on a couple of national data sets, with primary reference to the 1997 flood in Poland.

Mortality estimates

The two most essential socio-economic characteristics of a disastrous flood are: the number of fatalities and the economic damage. Neither of these characteristics is easy to quantify in a reliable way. Estimates of human and economic losses in a particular flood do considerably vary, depending on the information available to the searching body and the timing at the assessment. The assessment of the number of flood victims is, in a number of ways, problematic. Kundzewicz (2003) gave an example of assessment of the number of flood fatalities in the huge floods in July and August 1931 on the River Yangtze in China, according to various sources. The estimates largely differ, ranging from 145,000 through 400,000, 1,400,000, to 3,700,000 fatalities. That is, the lowest and the highest estimates differ as much as 1 to over 25. It would be of much interest to decipher the reason for such significant differences. Possibly the higher estimates are based on some way of double counting (data from different sources?). Identification of "flood-related deaths" (additional deaths) in a region devastated by a flood must have caused huge problems. The higher estimates may include fatalities caused by flood-related starvation. As stated by Smith and Ward (1998), such estimates are notoriously unreliable.

A worldwide statistics on floods since 1985 has been collected by the NASA-supported Dartmouth Flood Observatory of the Dartmouth College in Hanover, New Hampshire, USA. ➔ *Figure 1* produced by the Dartmouth Flood Observatory, and available on the net, (http://www.dartmouth.edu/~floods/archiveatlas/fatalitiesgraph.htm) shows the number of flood fatalities worldwide in individual years since 1985. According to these data, the number of flood fatalities worldwide, during the 18 years, 1985–2003, was about 300,000. It can be stated that, in general, mortality in disastrous floods has not been curbed at the global scale. This pertains in particular to less developed countries, such as Bangladesh, India, and China, where the average number of flood fatalities remains at a high level.

Fig. 1

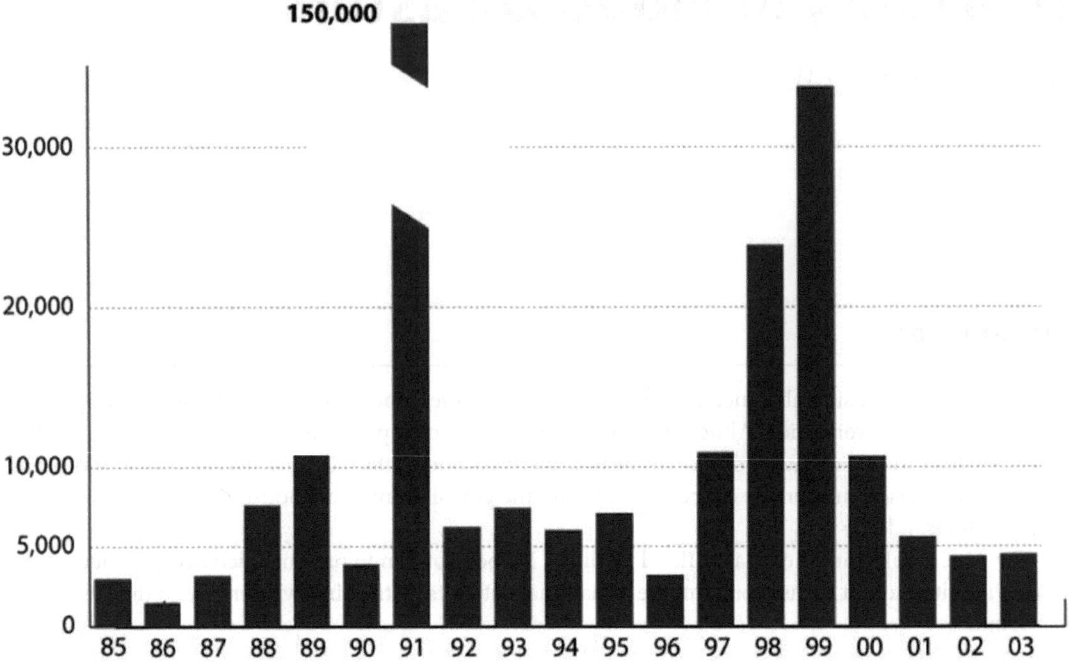

Annual records of flood fatalities worldwide in 1985-2003. Result of compilation by NASA-supported Dartmouth Flood Observatory, Dartmouth College, Hanover, New Hampshire, USA. From: http://www.dartmouth.edu/~floods/archiveatlas/fatalitiesgraph.htm.

The most destructive flood was a coastal surge in 1991 in Bangladesh, when during two days of April, nearly 140,000 people were killed (Munich Re, 1997). In the last two decades there were two further years with the number of flood fatalities in excess of 20,000, namely 1998 and 1999. More than 40,000 people died during three single extreme events. In October – November 1998, Hurricane Mitch caused 10,000 fatalities in Central America, in October – November 1999, a cyclone in Eastern India resulted in 10,000 casualties, while a flood in December 1999 in northern Venezuela and Columbia killed 20,000 people.

Mohapatra & Singh (2003) present data on the number of flood fatalities in river basins of India in the time period 1953 – 2000. *Figure 2,* based on this data, shows that floods continue to be a massive killer in India. In the year 1977, the number of flood fatalities in India was equal to 11,316. In 1953 – 1967, it was rare that the annual number of flood fatalities exceeded 1000 (only once, 1374 in 1961). Now, it is rare that this number goes below 1000. From 1976 to 2000, it happened only once that the annual number of flood fatalities in India dropped below 1000 (576 fatalities in 1999).

Long time series of annual numbers of flood fatalities in the USA have been compiled by the Hydrologic Information Center of the US National Oceanic and Atmospheric Administration (NOAA). The data (time series of values collected over 101 years, 1903 until 2003) can be found under the address http:/www.nws.noaa.gov/oh/hic/flood_state/recent_individual_deaths.html. A disclaimer states that in the earlier years not all isolated fatalities were accounted for due to more limited communication capabilities. That is, even if data are available since 1903, they may underestimate the number of victims in early years, reflecting fatalities associated with major floods and some, but not all, isolated events. In the USA, the number of flood

Fig. 2

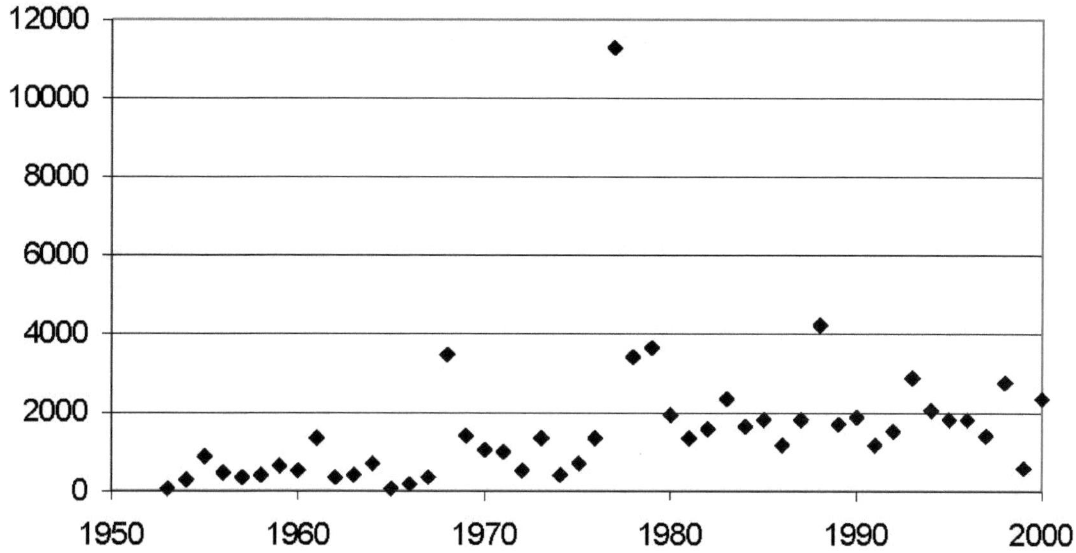

Number of flood fatalities in India, based on data published by Mohapatra & Singh (2003).

fatalities continues to be high. From 1961 to 2000, there were 21 years in each of which at least 100 flood fatalities were recorded in the USA (up to 554 in a single year, 1972).

Figure 3 shows numbers of flood fatalities in the USA as running means over subsequent five-year periods since 1903. A value of the running mean for a year N is equal to one fifth of the sum of values for five adjacent years: N-2, N-1, N, N+1, and N+2. A clear long-term decreasing tendency cannot be deciphered.

In some countries, there has been a tendency of decrease of the number of flood-related fatalities. This is a sign that flood preparedness systems, including forecast-warning-dissemination-response systems may considerably contribute to saving lives. In the first 15 years after the World War Two, destructive floods frequently visited Japan, killing a thousand or more victims almost every year (Kundzewicz & Takeuchi, 1999). In 1953, there were 3000 fatalities, and in 1959 – even 5600. This can be partly explained by the lack of appropriate maintenance of channels and flood defences (dikes), related to the general hardship of a nation, which had lost the war. It also turned out that the existing system of levees led to increase of catastrophic flows, jeopardizing major cities. After 1960, Japan has never got more than 1000 flood fatalities per year, and except for the year 1962, the number of fatalities was always less than 600.

Material damage vs death toll

Annual flood damage statistics in the USA has been also compiled by the Hydrologic Information Center of the NOAA (address: http:/www.nws.noaa.gov/oh/hic/flood_state/flood_loss_time_series.html). *Figure 3* shows the flood-related material damage, in inflation-adjusted dollars, as a running mean over subsequent five-year periods since 1903. Amidst strong oscillations, even in five-year running means, the material losses show an increasing tendency.

Having ignored the earlier data containing several years when the number of flood-related fatalities in the USA was equal to zero, one can observe that the value of the ratio of the running means of the mate-

Fig. 3

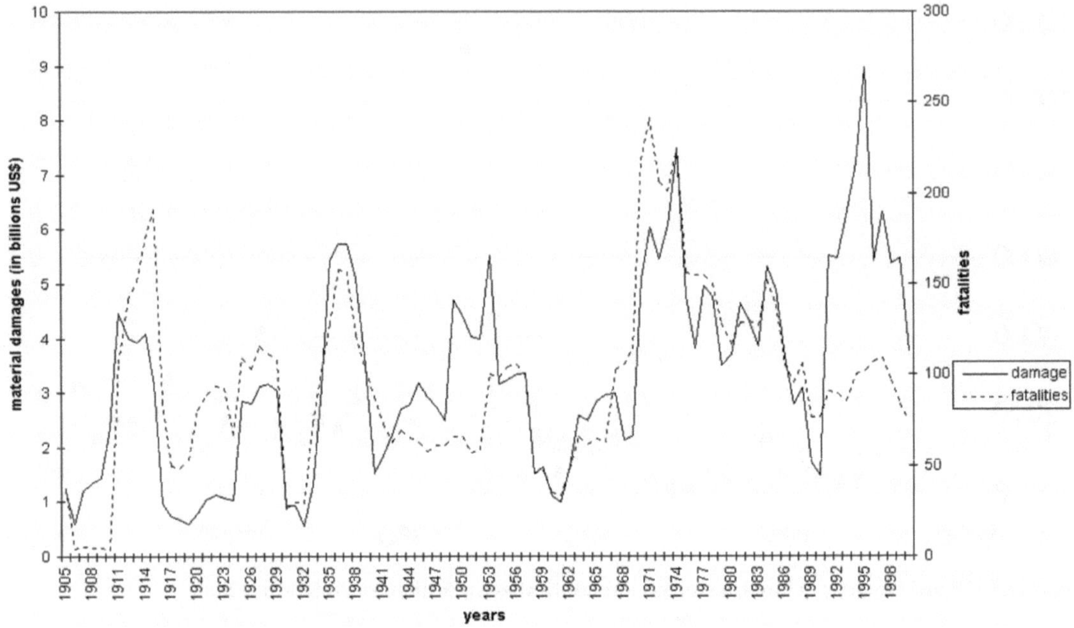

Flood-related fatalities and material damage in the USA (five-year running means). Data source: Hydrologic Information Center of NOAA.

rial damage and the number of fatalities has been somewhat growing with time (*Fig. 4*), though not in a uniform way.

Fig. 4

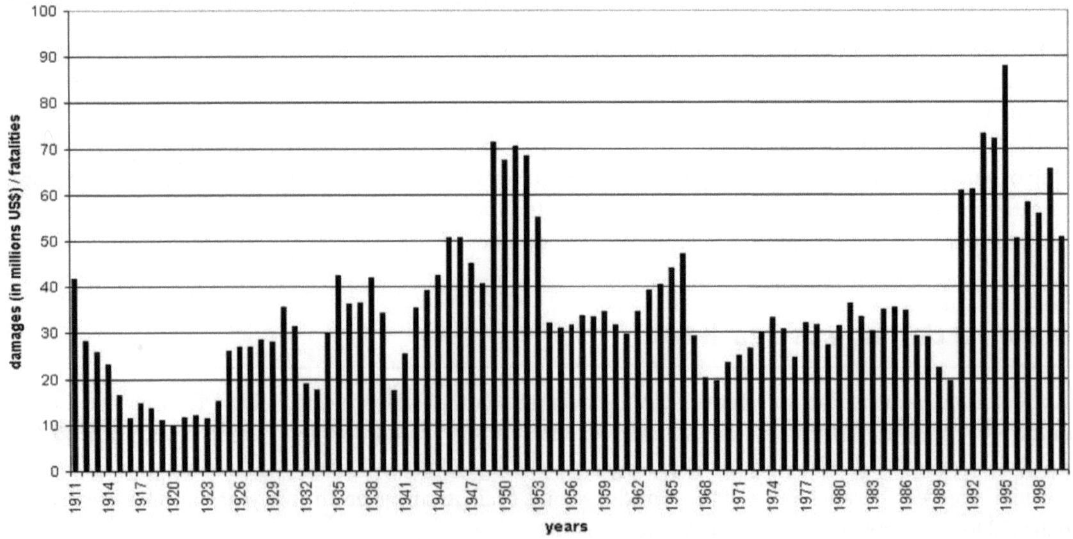

Ratio of flood-related damages (in millions of US$) to number of flood-related fatalities in the USA (five-year running means). Data source: Hydrologic Information Center of NOAA.

Kundzewicz and Takeuchi (1999) studied relationships between the GNP of a country and the value of the ratio of material damage to number of fatalities for individual most dramatic flood events, based on the 1990 – 1996 data from Munich Re (1997). Their observations were as follows:

- in general, the ratio of material losses to number of fatalities grows with the wealth level measured by the GNP per capita of a country
- repeated occurrence of a disastrous flood twice in the same place may visualize existence of a learning effect. The values of the damages-to-fatalities ratio, for the first flood and for the second flood of comparable magnitude (coming not long after the first one) do largely differ. The damage caused by the second flood is usually considerably lower. Occurrence of a flood raises awareness and triggers actions towards improvement of flood preparedness system.

Fig. 5

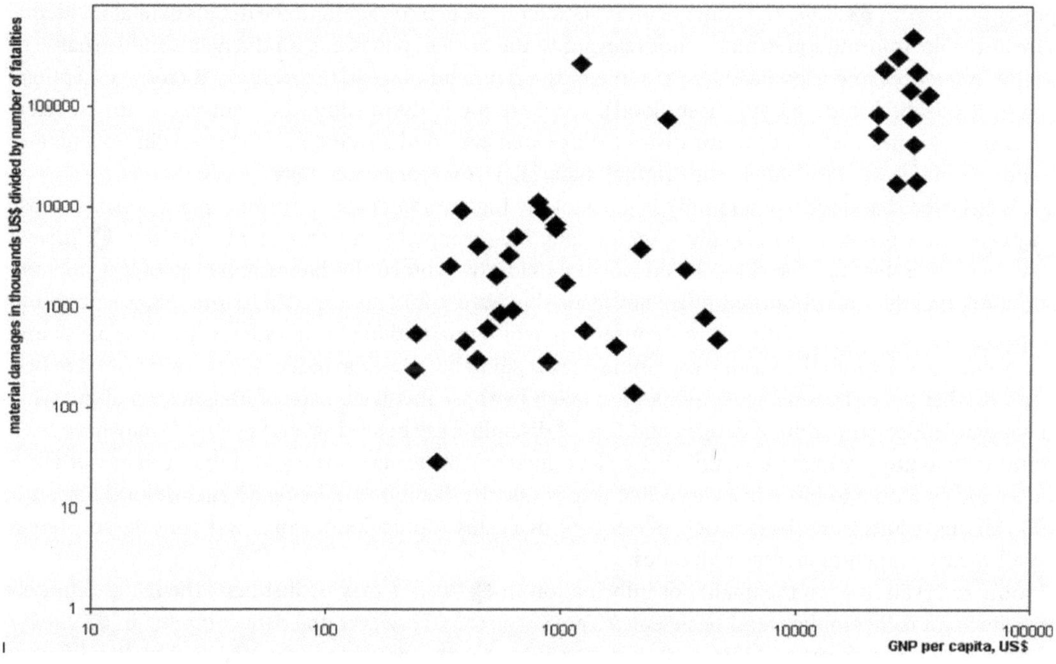

Ratio of economic damages (in thousands of US$) to number of fatalities for national data on most disastrous flood events, as a function of wealth of the country (GNP per capita).

Figure 5 shows an update of the global event-based analysis by Kundzewicz & Takeuchi (1999), of the ratio of flood-related material damage to the number of flood fatalities. The data used in Figure 5 stem from different statistics, collected by Munich Re (e.g., Munich Re, 1997 and a number of more recent annual publications); and from other sources available to the authors, and refer to the time period 1990 – 2002. Only large events were considered, in which 100 (or more) fatalities and/or 500 (or more) million dollars material damage were observed. The GNP data were assembled from various publications of the World Bank. Both flood damage and GNP were adjusted for inflation. Those Munich Re data, which give aggregate estimates of losses for a group of countries (with different GNP per capita), were simply ignored and not disaggregated into individual countries.

Figure 5 demonstrates that the ability of countries to invest in flood defence, being related to the country's wealth, plays a significant role in curbing the death toll. On the top of costly structural measures (levees), such factors as education and awareness, access to information, and early warning help reduce the number of flood fatalities.

How do fatalities occur?

There is a scarcity of information on the direct reasons of flood-related deaths. Interesting material has been assembled by journalists of the Polish weekly Fakty (journal, which is not published anymore) during the devastating summer flood of 1997 in Poland (on the Odra, the Vistula, and their tributaries), which took a high death toll (over 50 fatalities). Actually, the 1997 Odra flood has an international dimension, affecting the Czech Republic (also with more than 50 fatalities and high material losses), Poland, and Germany (material losses only). An edited excerpt from the material published in Fakty (Anonymous, 1997) is presented here, in ❯ *Table 1*. It conveys information on the gender, age, and the details of fatal incidents, where available. Information, which is not relevant to the present paper, e.g. on the name and surname of a victim, references to place of residence, name of the river and place of the fatal event (some exceptions being in Kłodzko, devastated by a flash flood), or where the body was found (including, at times, reference to street names), reference to the district (this extensive flood affected many districts in Poland, and beyond), even if it was available in the original material, is not reproduced here.

It is believed that flood vulnerability is particularly high among elderly citizens, poor persons, those with prior heath problems (especially – motorically handicapped), and with small children. ❯ *Table 1* clearly demonstrates that the flood killed not only elderly, motorically-handicapped people who were alone at night in their bedrooms inundated by fast rising waters, and not exposed to forecast, warning, and evacuation. Some of the fatalities were a consequence of human taking (and underestimating) a risk and literally encroaching into the harm's way. Some deaths could have been avoided it the awareness were better and if other, more fortunate, decisions were taken by those involved. Lack of imagination about what can happen, underestimation of danger, and lack of discipline are behind several deaths. It may have been beyond imagination to many a victim that a river current can be very strong, and that it does not allow even a good swimmer to move in a controlled way (while the likelihood of being hit against hard objects is high). All this results from the lack of experience with a catastrophic flood, which was considerably larger than all events happening in recent decades.

Some reservation as to the quality of information in ❯ *Table 1* may be justified. The list is definitely not uniform, it is incomplete, and some entries are suspicious. However, the data compiled in ❯ *Table 1* is a unique source anyway, and raises interest. Possibly sources related to insurance could tell a different story.

How did flood fatalities happen? In several cases, a silent mystery was taken to grave and reconstruction of circumstances is not possible. However, in some cases journalists could collect relevant information. There were far more male fatalities (43) than female fatalities (6). ❯ *Figure 6* presents distribution of fatalities into age classes (both genders together). The average age of a victim was about 46 years. It is worth noting that, contrary to intuition, a vast majority of victims (over 71 %) were between 20 and 59 years of age.

Tab. 1

Information on the gender, age, and circumstances of flood-related fatalities in the July 1997 flood in Poland (ordered after the age of victims, in ascending order). Data source: Anonymous (1997). Not all flood-related fatalities are included in the source. Notation: NA – information not available

Age	Information about fatalities
Gender male	
Child	Died in result of breast injury, while being transported in a car, which fell into a road breach.
10	On 13 July, fell into water. Helicopter was used. Reanimation action was unsuccessful.
20	On 9 July, evening, drowned falling to a stream.
22	On 11 July, body was found in a river (injury of body and head).
23	Jumped from a bridge to a river in order to win a bet. His body has never been found.
24	Went to watch the flood. Body found on 18 July.
25	On 15 July, left to take flood photographs. Drowned in unknown circumstances.
26	On 8 July, about 18 h, fell from a railway bridge into a river.
27	Went fishing in a stream, in which his body was found.
27	On 8 July about 14.00 he was washed away by a flood wave while walking over a foot-bridge and drowned.
30–40	Unidentified; found in a gravel exploitation area.
32	On 8 July about 19.50, walked along the bank of a river, fell into a river and drowned.
32	NA
32	On 11 July, a mud-covered body was found in a channel of a river.
34	On 8 July about 20.30, having consumed alcoholic beverages, fell from a foot-bridge into a river and drowned.
34	On 7 July, while working on clearance of a bridge, fell into water and drowned.
34	On 9 July, evening, while securing a bridge on a stream, fell into water and was carried away by the stream current.
34	Body found in a stream.
≈ 40	(Unidentified, 175–180 cm), Drowned on 12 July in a river.
43	On 7 July, body found in water in an underground passage (in Kłodzko).
44	On 9 July, jumped into the water of a river in order to save a drowning dog, and was carried away by strong river current and drowned.
45	Drowned in the current of a river.
47	Floating body found.
50	Found in a river.
50	Tried to cross a river, with three children, in a rubber boat in order to collect his property. The boat capsized. Children were saved.
51	NA
52	Body found at a sport stadium.
56	On 10 July at 17 h, was hit by a truck Star A-26 owned by a Fire Brigade, due to lack of adequate attention, while the truck was driven on a rear gear during the flood fighting action.
57	Drowned, carried away by water current after having gone in a canoe in order to collect grain for evacuated farm animals.

Tab. 1 (Continued)

Age	Information about fatalities
58	Body found. Reason of death – drowning.
59	NA
60–70	(Unidentified, height 165-170 cm) Body found in a stream.
61	On 8 July, drowned in his apartment (in Kłodzko).
64	On 7 July, drowned carried away by flood waters.
75	On 8 July, body found under a hedge (drowning).
78	NA
81	On 8 July drowned in his apartment (in Kłodzko).
84	Commited suicide, jumping from a bridge into water. Has left a letter. His body has never been found.
NA	German citizen. Body found on 16 July
NA	Drowned.
NA	On 7 July at 19.50, was captured by the current of a river.
NA	Body found on 15 July evening.
NA	Body found near a river gate.
Gender female	
39	Body found in a river.
51	On 9 July about 23 h died under earth masses after a hillslope slided onto a house.
57	Drowned in a channel. Municipal police reported about it on 13 July at 6.20 a.m.
60	On 8 July, the body was found on the land owned by the victim, in a basement filled with water. No external wounds.
68	On 13 July about 16 h, divers of the police troops found a body in an inundated house by a river.
75	Drowned in inundated apartment. Motorically impaired.

Table 1 shows that some victims died by taking a risk, in a conscious or unconscious way. Some may have walked home, as usual, and crossed a stream using a slippery foot-bridge, perhaps without a handrail, which served adequately in normal times, but the flood time was not normal. Several records (male) contain the classifier: "fell into a river and drowned", possibly, having been carried away and hit against hard objects. Some victims tried to collect their belongings, or to save a life of a dog. Some just wanted to watch an unusual event – the flood (possibly making photographs) and were not careful enough. Fatal accidents during flood-fighting activities are also reported. A death resulting of a highly irresponsible behaviour – a bet between young parties – is mentioned. Flood-related suicide during the flood was also recorded.

Hajat et al. (2003) reviewed psychological distress of the July 1997 flood in Poland and Czech Republic. No suicides or attempted suicides during the flood were recorded in the Czech Republic. In Poland, 50 flood-related suicides were reported in the two months after the flood. Psychological health effects, such as post-traumatic stress disorder (PTSD) – anxiety, depression – can last several years after the event.

It is clear from US data that many flood fatalities were trapped in cars. Hydrological Information Center data show that over half of the number of flood-related fatalities in the USA were vehicle related. Victims literally drove into harm's way. In 1998, 86 out of 136 flood fatalities recorded in the USA were vehicle-related fatalities, in 1999: 39 out of 76, in 2000: 20 out of 41, in 2001: 31 out of 66, in 2002: 31 out

Fig. 6

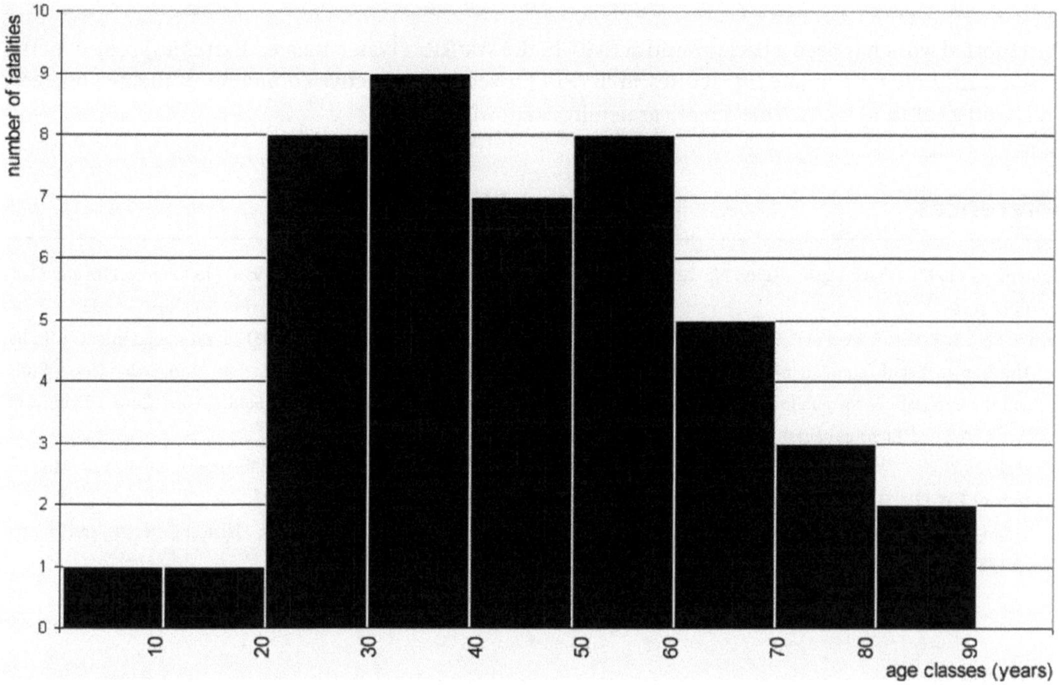

Distribution of flood fatalities (Poland, July 1997) into age classes. Data source: Anonymous (1997). Not all Polish flood-related fatalities are included in the source. In a few cases no information on the age is available

of 50, and in 2003: 45 out of 80. So, over the six-year period, 1998 – 2003: 252 of 449 flood fatalities (over 56 %) were vehicle-related.

In the dry year 2003, a local flash flood was reported in Poland, taking a death toll of five persons, who drowned in a car.

Conclusions

Number of flood-related fatalities is one of the most essential socio-economic characteristics of a flood disaster. A question emerges as to the counting process. The term "flood-related fatalities" is self-explanatory and can be interpreted in a rather broad way. However, there is a substantial difference between a death of an old handicapped woman, who drowned in her bedroom, alone at night, and a young, strong, and self-assured man, who underestimated the danger and encroached into the harm's way. Each death is a human tragedy. Every death has economic consequences (e.g. insurance). Some of the fatalities resulted from humans' taking (and underestimating) a risk. The data compiled in Table 1 demonstrate that many a young life could have been saved, if the awareness were better and if other, more fortunate, decisions were taken by those involved. Lack of imagination about what can happen, underestimation of danger, and lack of discipline are behind several deaths. All this results from the lack of experience with a catastrophic flood of extreme dimension. Obeying simple rules of conduct by car drivers alone could help reduce the number of fatalities in many flood events, even if no vehicle-related drownings were listed in Table 1 in the 1997 flood in Poland (yet, there were two vehicle-related incidents).

Acknowledgements

The reported work has been a background activity in the WatREx (Water-Related Extremes) project of the Potsdam Institute for Climate Impact Research (PIK). Useful constructive comments of the two referees, Dr. I. Gottschalk and Dr. A. Thieken are gratefully acknowledged.

References

Anonymous (1997) Ofiary klęski. Fakty, No 22 dated 24 July 1997, p 16

Hajat S, Ebi KL, Kovats S, Menne B, Edwards S, Haines A (2003) The human health consequences of flooding in Europe and the implications for public health: a review of the evidence. Applied Environmental Science and Public Health 1(1):13 – 21

Kundzewicz ZW (2003) Flood risk growth under global change – Yangtze floods in perpective. Journal of Lake Sciences 15:155 – 165 (China)

Kundzewicz ZW, Takeuchi K (1999) Flood protection and management: Quo vadimus? Hydrol Sci J 44:417 – 432

Mohapatra PK, Singh RD (2003) Flood management in India. In: Mirza MMQ, Dixit A, Nishat A (eds) Flood Problem and Management in South Asia. Kluwer, Dordrecht, pp 131 – 143

Munich Re (1997) Flooding and Insurance. Munich Re, Munich, Germany

Smith K, Ward R (1998) Floods. Physical Processes and Human Impacts. Wiley, Chichester

Key Policy Implications of the Health Effects of Floods

Edmund Penning-Rowsell · Sue Tapsell · Theresa Wilson

Introduction

The frequency and extent of flooding world-wide, and the accompanying losses and related human health impacts, are expected to increase over the next 50 to 100 years owing to the effects of global warming (IPCC 2001) and other factors (Evans et al. 2004). There is cause for concern here, as the impacts of floods are serious and far-reaching.

These impacts include impacts on human health (Tapsell et al. 2002), as elaborated below. In turn these have implications for policies for flood defence, the environment and the provision of local services to assist people in dealing with flood emergencies and their aftermath.

This chapter outlines some of the research undertaken at Middlesex University Flood Hazard Research Centre on the health effects of floods in the UK, and describes how these effects might be mitigated by policies and strategies to identify those who might be worst affected, to warn people of the impacts that they might suffer, and to help them during and after flood emergencies. In addition we show that emergency planning across Europe has not taken on board the impacts that floods can have on health, and this is also cause for concern and needs attention.

Risks to health from floods

Evidence for the health effects of flooding

The human health consequences of floods, particularly relating to flash floods or coastal flooding, can be severe. Health consequences from river or inland flooding would in many respects be the same e.g. those caused by the shock, disruption and inconvenience of the flood, as well as worry about future flooding.

To date there has been no large-scale research in the UK on the health effects from flooding. Bennet's study of the 1968 Bristol floods (Bennet 1970), which demonstrated some significant effects on mortality, was the last population-based epidemiological study on the impacts of flooding, although this only related to one town and one particular flood event. Research carried out in the 1980s further highlighted the seriousness of the so-called 'intangible' impacts of flooding on people's lives and well-being (Parker et al. 1983; Green et al. 1985; Tunstall and Bossman-Aggrey 1988).

A case-control study was carried out by Reacher et al. (2003) on health impacts following the Lewes flooding in autumn 2000. Although the sample size in the study was insufficient in most instances to reach standard levels of significance, results showed that having been flooded was associated with earache, a significant increase in risk of gastroenteritis, asthma getting worse, other respiratory illnesses, skin rashes, and psychological distress. Flooded adults also had a four-fold higher risk of psychological distress compared with the non-flooded.

Several small-scale qualitative studies have also been carried out since the flooding of Easter 1998 in nine communities in England and Wales affected by inland flooding (Tapsell et al. 1999; Tapsell 2000; Tapsell and Tunstall 2001, RPA/FHRC 2003). Two of these were communities flooded at Easter 1998, three were communities flooded in June 2000, and four were communities flooded in autumn 2000. These studies, which covered communities with varying socio-economic backgrounds and who experienced flood events of varying characteristics and impacts, have provided deeper insights into some important consequences on people's health resulting from river flooding. Although the results from these studies cannot be said to be representative of flooded populations generally, due to the small samples involved (a total of 116 people), the same or very similar problems were reported in all communities which indicates a wider applicability of the findings.

What is not clear from the earlier studies is how long the various health effects reported following flooding – both physical and mental health effects – were likely to continue after the flood event. No longitudinal studies on the health effects of natural disasters could be found for the UK except that reported by Tapsell (2000), and more recently by Hepple (2001). Those studies that have been undertaken in the US and Europe have largely focused on the psychological impacts such as post-traumatic stress disorder and associated impairment to physical health (Holen 1991; Hovanitz 1993; Beck and Franke 1996; Bland et al. 1996).

Results from recent research by Risk and Policy Analysts/Middlesex University Flood Hazard Research Centre et al. (RPA/FHRC et al. 2003) in a large-scale study of over 1500 flooded and at risk respondents across England and Wales, demonstrated that flooding causes short-term physical health effects, and reported similar health problems to those reported by Reacher et al. (2003). However, findings also showed short and, more significantly, long-term psychological effects in the flooded population. The long term psychological effects of being flooded were also stressed by Tapsell et al. (2003) in a four year follow-up qualitative study with two communities flooded in 1998. Respondents reported that the psychological health impacts resulting from being flooded were worse than the physical impacts, and many were still demonstrating symptoms such as high anxiety and stress levels, sleeping problems and depression four years after the flood, which they all associated with the flooding.

Factors contributing to serious consequences of flooding on human health

Results from our research reveal that the adverse human health consequences of flooding are complex and may be far-reaching, affecting both physical and mental health, as shown in the following list of health effects of floods in the UK as revealed in qualitative social surveys (from Tapsell et al. 2002):

Physical health effects during, or immediately after, a flood:

- Injuries, e.g. cuts and bruises, due to being knocked over by floodwaters, being thrown against hard objects, or being struck by moving objects
- Injuries from over-exertion during the flood e.g. sprains/strains, heart problems
- Hypothermia
- Electric shocks
- Cold, coughs, flu
- Headaches
- Sore throats or throat infections
- Skin irritations e.g. rashes
- Exposure to chemicals or contaminants in floodwaters
- Shock.

Physical health effects in the weeks or months after a flood:

- Gastrointestinal illnesses/upset stomachs
- Heart problems

- Respiratory/chest illnesses e.g. asthma, pleurisy
- Cuts and bruises
- Sprains and strains
- Skin irritations e.g. rashes, dermatitis etc.
- High blood pressure
- Carbon monoxide poisoning
- Kidney or other infections
- Stiffness in joints
- Muscle cramps
- Insect or animal bites
- Erratic blood sugar levels (diabetics)
- Weight loss
- Weight gain
- Allergies e.g. to mould.

Psychological health effects in the weeks or months after a flood:

- Anxiety e.g. during heavy rainfall or when river levels rise
- Panic attacks
- Increased stress levels
- Mild depression
- Moderate depression
- Severe depression
- Lethargy/lack of energy
- Sleeping problems
- Nightmares
- Flashbacks to flood
- Increased use of alcohol or prescription (or other) drugs
- Anger/tantrums
- Mood swings/bad moods
- Increased tensions in relationships e.g. more arguing
- Difficulty concentrating on everyday tasks
- Thoughts of suicide.

The World Health Organization defines good health as 'a state of complete physical, mental and social well-being, and not merely the absence of disease and infirmity' (World Health Organization 1948). Hazards such as floods can therefore be regarded as potentially multi-strike stressors. The adverse health effects – physical and mental – may result from a combination of some or all of the factors in the list below. Additional components affecting the stress and health impacts of flooding may include socio-economic and cultural factors.

The factors exacerbating the adverse health effects – physical and mental – from flooding, either individually or in combination

- The characteristics of the flood event (depth, velocity, duration, timing etc.),
- The type of property e.g. single storey, two storey etc.
- The amount and type of property damage and losses
- The flood victims' pre-existing health conditions and susceptibility
- Whether flood warnings were received and acted upon
- The victims' previous flood experience and awareness of risk
- Any coping strategies developed following previous flooding
- Flood victims having to leave home and live in temporary accommodation

- The clean-up and recovery process (e.g. exposure to moulds) and associated household disruption
- The frustration and anxiety of victims' dealing with insurance companies, loss adjusters, builders and contractors
- The flood victims' increased anxiety over the possible reoccurrence of the event
- The flood victims' loss in their level of confidence in the authorities perceived to be responsible for providing flood protection and warnings
- The flood victims' financial worries (especially for those not insured)
- The flood victims' loss of the sense of security in the home
- An undermining of people's place identity and their sense of self (e.g. through loss of memorabilia)
- Any disruption of community life.

There appears to be a time dimension to the health impacts resulting from flooding. Health effects can be categorised as those resulting at the time of the flood or immediately after, those which develop in the days or early weeks following the flood, and those longer-term effects which may appear and/or last for months or even years after the flood (see above). A common perception is that once the floodwaters have receded the problem is over. For many flood victims, however, this is when most of their health-related problems begin.

Much more research is needed on trying to understand the complex health consequences which may result following flooding. Disease surveillance needs to be increased during floods, and information disseminated rapidly to dispel false rumours of public health epidemics. The longer-term psychological impacts on people's health and social well-being particularly require more investigation, along with the issue of medical and social support during the recovery period.

Socio-economic deprivation may also be an important determinant of the health effects of flooding. The social and community dimensions of flooding, which can have significant impacts on households and individuals, are other factors often neglected in post-flood studies. Community activity often breaks down following serious flooding, and it can be many months before normal functioning is achieved. Moreover, some flood victims have spoken of a long-lasting deterioration in community life (e.g. Tapsell and Tunstall 2001).

A final issue is that of the impacts on the health service from increased flooding in the future. Evidence from the USA following Hurricane Floyd in 1999 shows that health systems can also be badly affected by flood events, particularly if facilities are themselves located on floodplains, or when affected by disruptions to electricity, water supply, and transportation systems. The UK medical community also needs to be prepared to address these concerns and both the short and the long term health needs of people who have been affected by flooding. There is also the issue that healthcare facilities will be stretched at times of disasters, and this will adversely impact on normal service delivery, not just on the healthcare provision for the disaster victims themselves.

Potential policy responses to floods and their health impacts (I): The role of countries' emergency plans

Health impacts: mitigating policies, strategies and actions

Although the risk to life and health would potentially be greater from a major coastal flood in the UK, inland flood events are already affecting many households and communities every year, and are likely to increase in the future. Already individuals and communities are taking mitigation actions.

There are a number of actions that can be taken in this respect to mitigate the adverse impacts of flooding (*Tab. 1*). For example, the Environment Agency has been improving its flood forecasting and warning systems, and increasing public awareness-raising of flood risks through annual campaigns and 'Flood Awareness' weeks. There is now regular use of flood warnings on television and in radio weather reports and promotion here of the Environment Agency's Floodline information service.

Tab. 1

Health-specific responses (adaptation measures) to mitigate the health impacts of flooding

Health Outcome	Measure
Mental health outcomes (anxiety and depression, etc).	Post-flood counselling Medical assistance (drugs, etc) Visits by health workers or social workers to vulnerable people (elderly, disabled, etc)
Infectious diseases and other physical health effects, etc	Treatment of respiratory problems and skin rashes by normal medical methods Treatment for cold and other exposures Treatment for strains and other effects of physical exertions by the normal medical methods Vaccination (Hepatitis A) of general population Boil water notices Hygiene advice (general, e.g. PHLS leaflet). Outbreak investigation Additional public health actions Enhanced surveillance
Both of the above: Pre-flood activities	Targeted warnings (messages) to different groups Pre-flood awareness raising campaigns Emergency planning Inter-institutional co-ordination activities

Mediating factors between stress and health may include flood warning, education, coping strategies and social support; however, where flooding is unexpected, sudden and without warning, these strategies may be weakly developed or non-existent. Self-help measures to reduce the damage to property and the stress caused by flooding are also being encouraged, thereby alleviating some of the negative consequences on people's health created or worsened in the flood recovery phase.

These measures include flood-proofing of properties, development of a family Flood Plan along the lines of those widely used in the USA, and other community preparedness developments. Where feasible and cost-effective, flood alleviation schemes may also be considered, along with development control legislation to restrict new building in the floodplain. The UK is likely to need to adapt to increased risk of flood events in the future and to develop national coping strategies rather than those purely at the local or individual household levels.

The health-related content in emergency plans: overall summary

If floods cannot be prevented or their probabilities reduced (see above) then their effects need to be mitigated. This is done both during the event and after it has occurred, and both matters need to be catered for in pre-event emergency planning. This is generally the responsibility of local authorities, rather than flood defence agencies, since there are a number of emergencies to be tackled (i.e. not just floods).

We have explored this area in a European context. The detailed results from this work with regard to the health impacts content of these emergency plans, and also of knowledge of health impacts research in the flooding field, are reported in Penning-Rowsell and Wilson (2003). This information has been obtained from our research contacts across Europe.

In summary, these results (➔ *Tab. 2)* are disappointing but very informative. It is clear that many countries, regions and states have emergency plans, but it is also clear that very few refer explicitly to health impacts, rather consigning this to the implicit agenda for emergency planning, namely saving life and reducing injuries.

Tab. 2
The health-related information in some European countries' emergency plans

Country	Emergeny plan
France	A comprehensive system of emergency plans exist for France (Hubert and Reliant, 2003). However, with the exception of war wounded, there apparently is no information on the health impacts of disasters in the French material on emergency planning. Psychologists were sent in to areas affected by flooding in the past, but no systematic reports are known to exists as a result of this effort. Therefore, despite a comprehensive and nationally driven system of risk assessment for hazard areas in France, it would appear that no notice is taken whatsoever of health impacts from natural disasters. Some psychological aspects appear to be touched on embryonically, but no systematic attention is given to this area.
Netherlands	There is a comprehensive plan for emergencies concerning river floods, but not for coastal flooding as the return period of such are so long that such floods "are not supposed to occur at all". There appears to be some research on the health effects of flooding at the IVM in Amsterdam but very little other information on research on the psychological and physical health effects of flooding. There are plans for which the water boards are responsible for flood emergency planning. These emergency plans for all sorts of disasters are under the responsibility of the Ministry of the Interior. They concern both water quantity and quality issues, but there is no evidence that these plans include material on the human or health impacts of flooding. On the other hand there was some research on health effects of floods after the 1993 and 1999 floods on the Meuse.
Germany	There is a systematic procedure for emergency planning in Germany, and a system of city and administrative district emergency plans. The Dresden Flood Research Centre had no information on the health impacts of flooding, but contacts there commented that these plans usually focus on the organisation of civil protection in order to protect human and health impacts.
Italy	There are emergency plans in Italy, operating at a regional and city level. It is unclear whether they have an explicit reference to health impacts. There has been some very limited psychological research on flood impacts.
Norway	Our contacts indicated that they thought there were emergency plans covering Norwegian situations, but that there was no health impact information within these plans. Neither is it known whether there is research on health impacts of flooding in Norway, or of any research results from this field.
Poland	All levels of state and individual administration are obliged to prepare emergency plans. More than 87 % of provinces and 64 % of counties have prepared such plans, but that nevertheless these plans are very simple, and quantitative in nature (i.e. they catalogue the facilities that can be deployed in emergencies) rather than being proactive in terms of policy or operation responses. There are generally no analyses of the impact of natural disasters including flooding. Some research on human impacts is reported at the Polish Academy of Sciences in Poznan, which has undertaken some research on the health problems in natural hazard context.

Indeed this is a recurring theme from many commentators who we contacted across Europe, namely that emergency plans are designed to prevent health impacts rather than record them. Therefore, implicitly, the emergency plans have everything to do with health impacts, but explicitly they make no reference to this at all, posing a problem for this research.

These are the results for major countries across the European Union (Germany, France and Italy, etc.) and it is, of course, possible that other countries do have emergency plans which deal with health impacts more systematically, and that these other countries have much more research effort in this field. However, having investigated systematically across more countries than are covered above (including Greece; Denmark; Portugal; Spain; Austria), no better information came to light than is now contained in Table 2 and given in more detail by Penning-Rowsell and Wilson (2003).

The policy implication is surely that this phenomenon needs more explicit attention at all levels rather than implicit coverage in emergency plans. Emergency planning needs to give much more emphasis to floods as 'people problems', rather than focusing as it does now on damage, the disruption of communications, and post-event economic recovery.

Policy response to floods and their health impacts (II): Targeting flood warnings and assistance on the most vulnerable

Given an increasing appreciation that the full impacts of floods have a social dimension we have sought to develop an index which measures the impact that floods could have upon the communities potentially affected.

We anticipate that this index would be used in addition to knowledge of the potential flood damage impacts and losses that are more readily understood and modelled within benefit-cost analyses within investment appraisals. We outline here, therefore, the definition and source of the variables used in the calculation of the FHRC Social Flood Vulnerability Index (SFVI), which can be used to predict those areas and populations that are likely to be most severely affected in terms of health and other 'intangible' flood impacts.

Our conceptual model is shown in ➲ *Figure 1*, and the choice of data was constrained by the need, first, to use secondary source data that is available for the whole of England and Wales, rather than rely on expensive and time-consuming interview or focus group approaches. We also needed, secondly, to use data that is available for small geographical areas, because flood plains are often narrow and short. We used 1991 census data from the Manchester Information and Associated Services (MIMAS) because this data fits the above criteria, being available for England and Wales at the level of the Enumeration District (ED).

The flood vulnerability index and its constituents

The SFVI is a composite additive index based on three social characteristics and four financial deprivation indicators (Tapsell et al. 2003). The rationale for the selection of these variables is given in Tapsell et al. (2003) and ➲ *Figure 2* shows the links between these and other variables against the six case studies that we have undertaken over several years to quantify the 'intangible' impacts of flood events. The linkages shown in ➲ *Figure 2* are based on results from the hundreds of interviews in those studies of the relative effect of the different variables in determining household vulnerability in the types of floods experienced in those localities.

Fig. 1

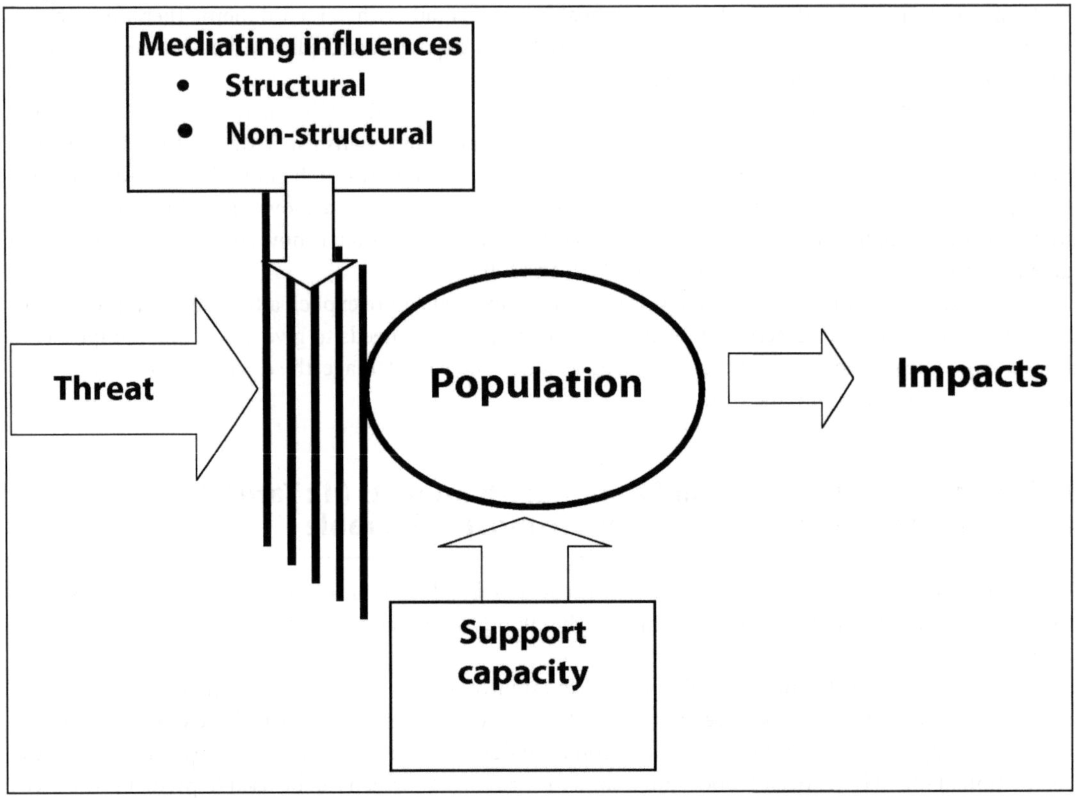

The process model of vulnerability to flooding: threats, impacts and mediating influences (courtesy Colin Green).

1. Green C, Emery P, Penning-Rowsell E, Parker D (1985) Evaluating the 'intangible' effects of flooding on households: a survey at Uphill, Avon. Flood Hazard Research Centre, Middlesex Polytechnic, London
2. Fordham M, Ketteridge A (1995) Flood Disasters – Dividing the Community. Flood Hazard Research Centre, Middlesex University, London
3. Tapsell SM, Tunstall SM (2001) The Health and Social Effects of the June 2000 Flooding in the North East Region. Report to the Environment Agency. Flood Hazard Research Centre, Middlesex University, London
4. Tapsell S, Tunstall S, Penning-Rowsell E (1999) The health effects of the 1998 Easter flooding in Banbury and Kidlington. Report to the Environment Agency, Thames Region. Flood Hazard Research Centre, Middlesex University, London
5. Ketteridge A, Green C (eds) (1994) The Technical Annex for the Full Flood Impacts Module. Flood Hazard Research Centre, Middlesex University, London
6. Hill J, O'Brien P (1999) Disaster in the Community: Emergency Planning for sustainable solutions to long-term problems. Worker and resident perspectives of the North Wales floods 1990 and 1993. Disaster Recovery and Research Team Ltd, Caernarfon, Gwynedd

This result shows that the age and financial status of the affected populations are the most commonly

Fig. 2

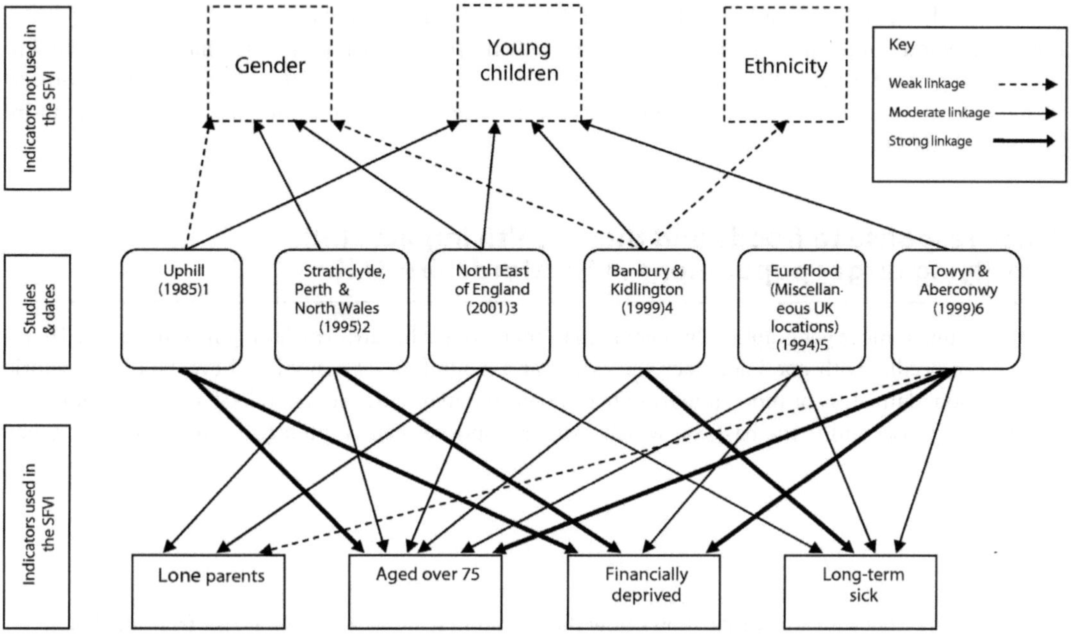

The relationship between results from a range of case studies and the choice of variables for the Social Flood Vulnerability Index (based on qualitative social science methods in focus group research, without control groups).

important variables, followed by the prior health status of members of that population. We have chosen to use the incidence of lone parents as a measure of family structure, because our research points to this causing extremes of vulnerability, although the presence or absence of young children (to which the incidence of lone parents is very closely correlated) had perhaps an equal case for inclusion.

To identify the financially deprived, the Townsend Index (Townsend et al. 1988) was used because, unlike other deprivation indices, it focuses on deprivation *outcomes* (such as unemployment), rather than targeting predefined social groups or using *input* measures such as those in the receipt of social security payments. This enabled us to identify our own social classification. This is important because financial deprivation is only one of several factors that contribute towards vulnerability to flood impacts and it is our intention to target only those social groups which previous research has shown to be particularly badly affected.

The Townsend indicators are all derived from the national census, and are:

1. Unemployment – unemployed residents aged 16 and over;
2. Overcrowding – households with more than one person per room as a percentage of all households.
3. Non-car ownership – households with no car as a percentage of all households.
4. Non-home ownership – households not owning their own home as a percentage of all households.

In order to prevent any undue bias in the SFVI towards financial deprivation, the four Townsend indicators were summed and multiplied by 0.25 before being added to the other variables.

Those other variables and the social groups that they highlight are more relevant to this report. They are:

1. The long-term sick – residents suffering from limiting long-term illness as a percentage of all residents.

2. Single parents – lone parents as a proportion of all residents.
3. The elderly – residents aged 75 and over as a percentage of all residents.

The Social Flood Vulnerability Index is now being used on a regular basis in the compilation of Catchment Flood Management Plans in the UK, and was used in the UK government's Foresight Project on flood and coastal defence, which looked at the flood and coast defence situation some 30-100 years into the future (Evans et al. 2004).

Policy response to floods and their health impacts (III): Identifying target populations in Hertfordshire, U.K.

Floods are much more geographically specific than many other hazards (including heat hazards). Therefore we can predict with good accuracy the extent and magnitude of many flood events, using historic data for flood plain areas or modelling from hydrologic and hydraulic inputs. Indeed, this is a major form of policy response, and is the basis on which further responses can be built (e.g. Environment Agency 2003).

Floodplains and data sources in Hertfordshire

We report here on a case study that shows how those who may be particularly at risk from the health effects of floods can be identified and then be the subject of pre-flood emergency policies and planning in all its forms.

The study was designed to identify, from secondary sources of data, those parts of the population who are most likely to suffer from health impacts in major flood events (e.g. the elderly and those with disabilities). We have studied the incidence of certain at-risk groups, and mapped their distribution in relation to the 100-year floodplain. Clearly these groups need to be targeted for emergency planning effort, and identified after a flood event for the adverse health effects that they are likely to show. What is important about this case study is the potential it shows for identifying those who might suffer health impacts in a major flood, and the possibility therefore of some targeted pre-event preventative actions in response.

The main reason that these groups are not the same as those covered by the SFVI is that these are the groups on which the Hertfordshire local authority has data in their own databases, mainly as clients for their social services. In this respect this case study is perhaps agency led rather than led by the known health impacts of flooding per se, although these criteria overlap in that the clients of social services are by definition some of the most vulnerable people in society.

Mental health patients at risk on the floodplain

➤ *Figure 3* shows the relatively small number of mental health clients of Hertfordshire County Council Social Services who are located in premises on the floodplain. Clearly these clients would be the subject of a particular effort during a flood emergency, and their presence should be planned for in any emergency plan. There is a concentration of these people in the urban areas, notably Hertford and Broxbourne, but the mental health clients in the outlying areas would perhaps represent a greater problem for emergency response staff.

Fig. 3

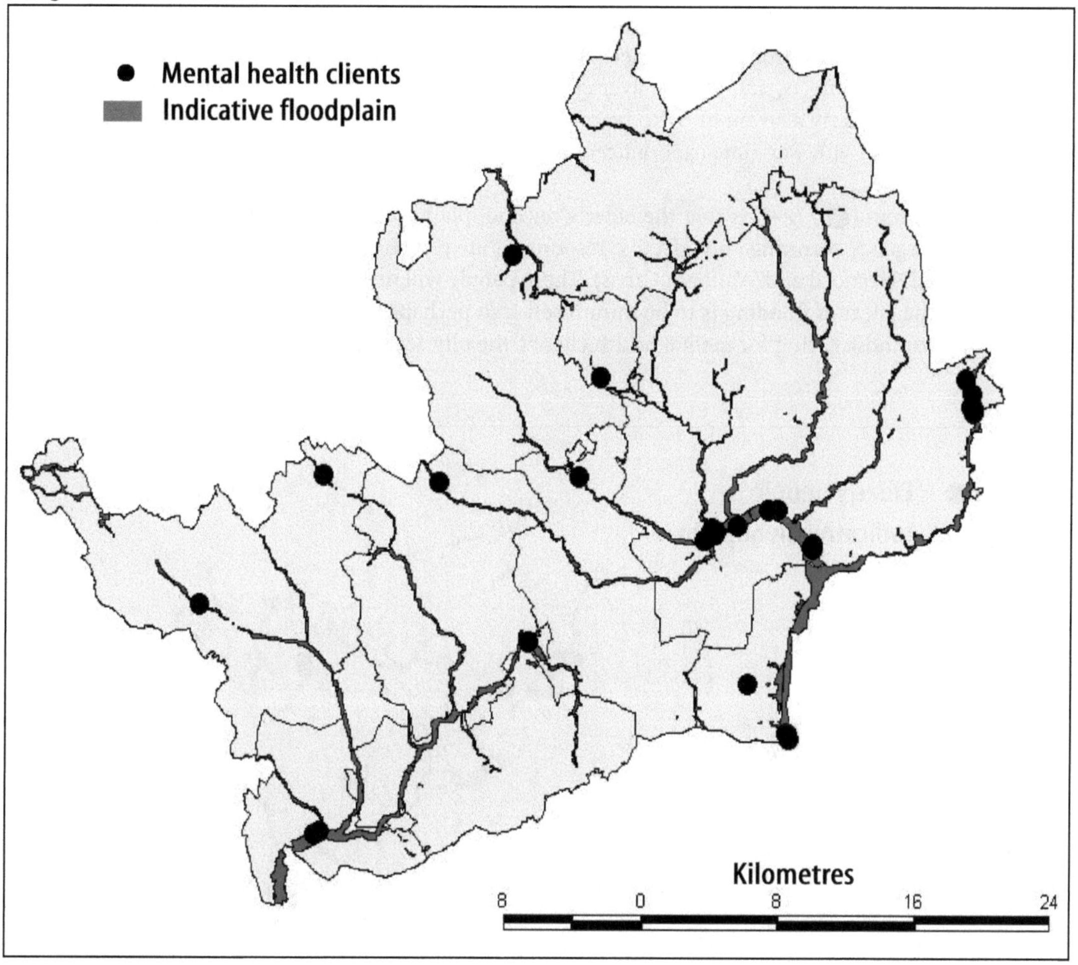

Mental health clients of Hertfordshire (UK) County Council social services in the floodplain (100 year return period flood outline).

The data shows that the County Council needs to consider carefully what response to make to this situation. Numbers are small, but problems could be acute. It may be that the carers of these mental health clients need to be alerted to the fact that they live in a floodplain area, and some sort of emergency evacuation plan discussed with them. Alternatively, it might be necessary to work through the relevant department within the Council, so that they are aware of this potential problem and act and plan accordingly.

This is because this will be a particularly vulnerable group of people, not least because they are unlikely to understand the nature of an emergency when it happens, and to be considerably distressed. Perhaps these matters should be discussed within the Council at the same time as other vulnerable communities are discussed, so that a complete package of emergency planning measures are put in train to tackle this particular issue.

The elderly at risk on the floodplain

We know from previous research that the elderly are particularly badly affected by flooding. They are likely to have adverse health effects, and also on average are likely to be less financially secure than those who are not elderly. The elderly also are likely to be more traumatised during flood events, other things being equal, although those with war-time experience can be more resilient than those without long experience of major catastrophes.

What is apparent (➔ *Fig. 4)* is that the elderly on floodplains are located right across Hertfordshire. This makes emergency planning and effective response quite difficult, although there are concentrations (in Berkhamsted, Hertford and Waltham Cross). These people will need to be the subject of special emergency plans if the effect of flooding is to be minimised, and perhaps these plans need to be worked out in conjunction with those caring for mental health clients, the physically disabled, etc.

Fig. 4

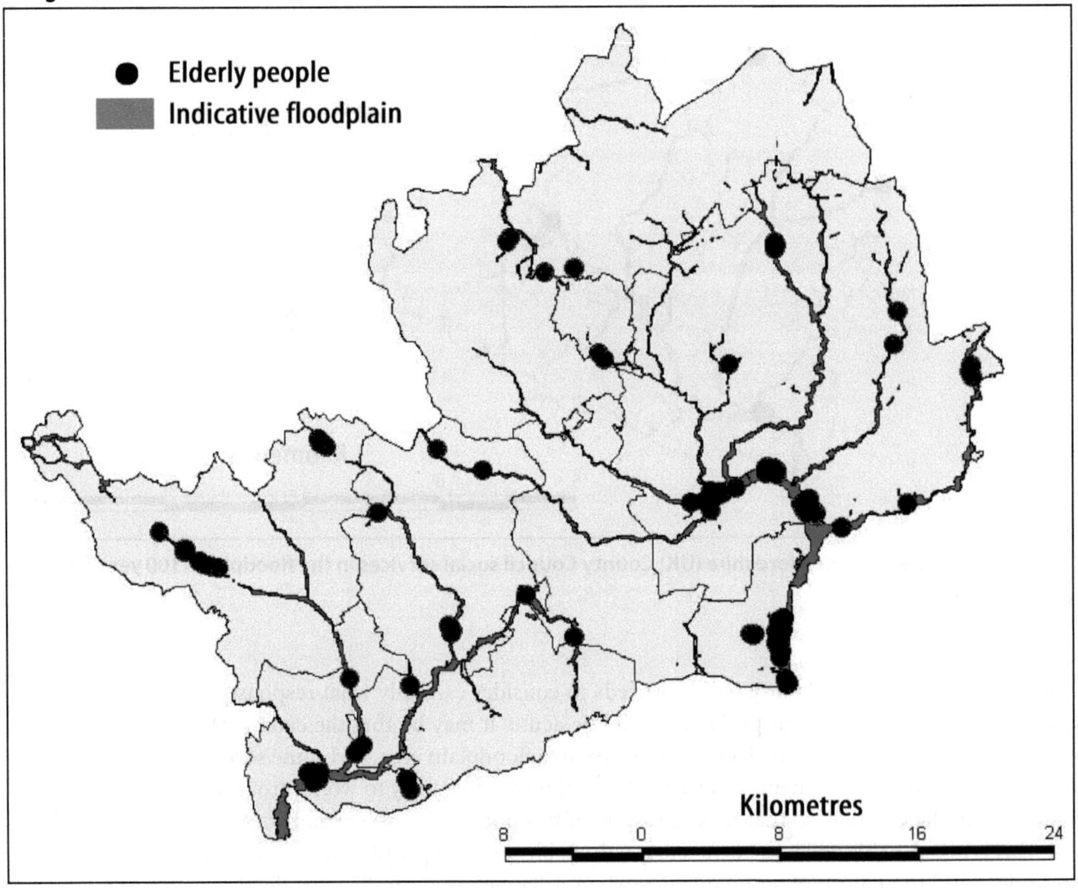

Elderly People in the Hertfordshire Floodplain (100 year return period flood outline).

All in all, the picture as in Figure 4 represents a considerable problem. The elderly are likely to be by far the worst affected by a major flooding instance in Hertfordshire, and in many instances this cannot be mitigated. What will be important is a coherent evacuation plan for these people, carefully managed, so that they escape the worst impacts of any flood of the magnitude shown by the indicative floodplain.

Physically disabled at risk on the floodplain

Figure 5 shows only a small number of physically disabled people live in properties on the 100-year floodplain. Nevertheless an emergency plan should attempt to target these people, because they could be particularly vulnerable to flooding of the magnitude indicated by the indicative floodplain.

Fig. 5

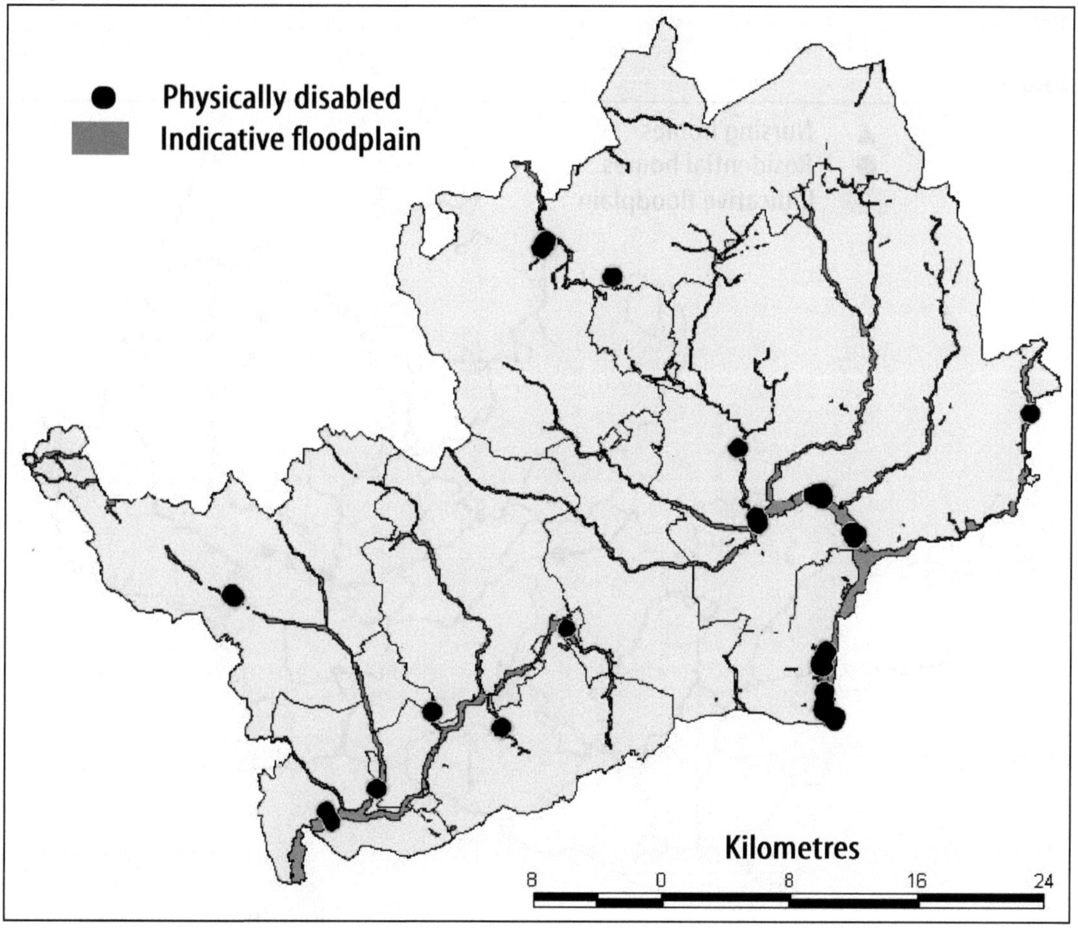

Physically Disabled People in the Hertfordshire Floodplain (100 year return period flood outline).

Not only will the impact on the physically disabled be greater, but the problems of evacuation could be more acute. It is undoubtedly the case that these people will need assistance during a flood, and their local carers may not be able to cope in this respect. It would appear to be necessary to alert the fire brigade and the ambulance services to the presence of these people on the floodplain, so that they can target their energies in this respect at an early stage in the build up of any flood event.

As with the other vulnerable communities discussed here, it could be the case that the emergency plan for these disabled people should be discussed in the round, rather than separately for the different vulnerable groups.

Residential and nursing homes at risk on the floodplain

One of the worrying characteristics of fluvial floodplains is the number of residential homes and nursing homes located there (*Fig. 6*). The reason for this location of nursing and residential homes is that this means they on the flat area of floodplains, adjacent to urban centres. Old people like to walk to the shops along the flat land of the floodplain location, and this is often the reason why nursing homes are located there.

Fig. 6

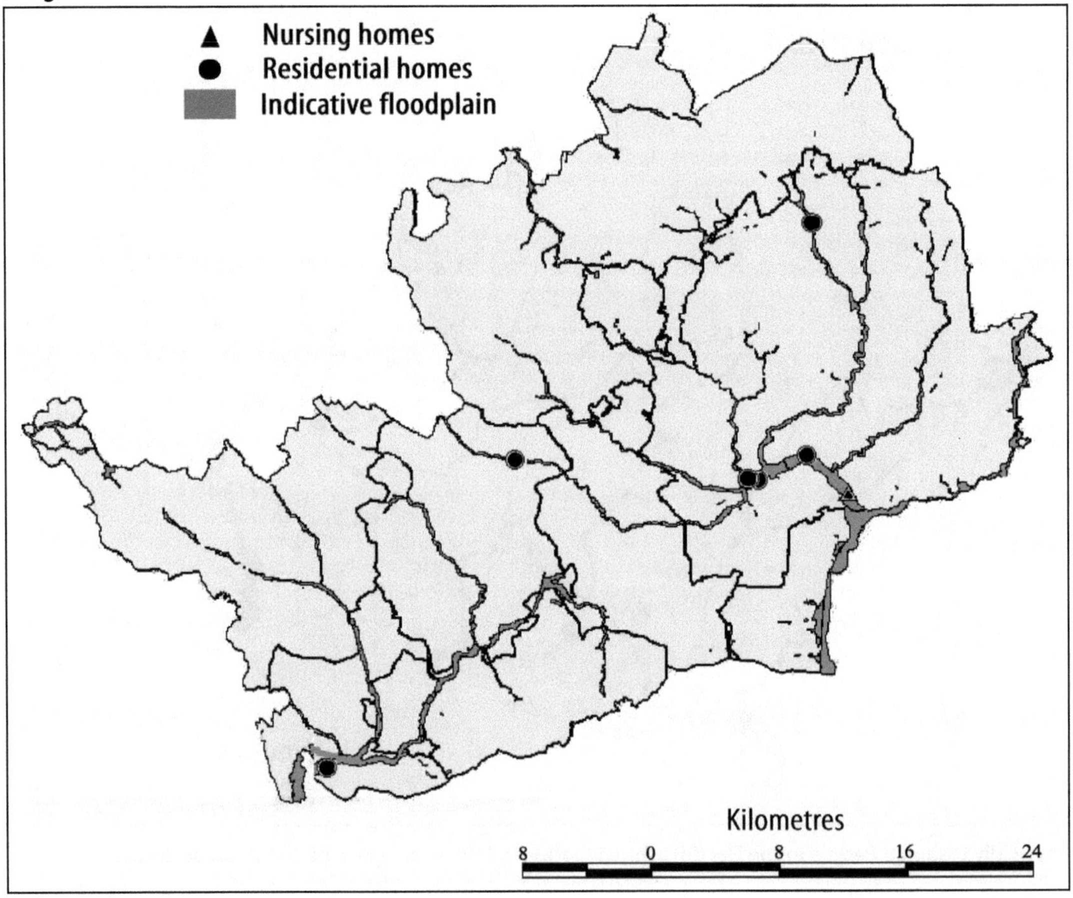

Residential and Nursing Homes in the Hertfordshire Floodplain (100 year return period flood outline).

The map shows a worrying concentration of these facilities in Hertford, but also a residential home in Rickmansworth and a nursing home in the Lower Lee Valley. Particular attention should be given to these facilities in any developing emergency plan for flooding in Hertfordshire. It is likely that an evacuation plan is necessary, and this will have to be carefully crafted to tackle the trauma and distress likely when old people are evacuated from their residential surroundings. Careful links would have to be made with the departments within Hertfordshire County Council which provide assistance to the elderly, and alternative accommodation needs to be identified where these people could be moved at times of major risk.

The effort needed here is despite the fact that the numbers of residential and nursing homes in the floodplain is relatively small. This is a very vulnerable community, generally of elderly and confused people, and they need a level of attention in the emergency plan disproportionate to their numbers.

Summary and conclusions

Policy for flood mitigation is about the prioritisation of responses and actions during and after the flood events, and pre-planning for this activity in multi-dimensional emergency plans. But we have found very little information and guidance, in our Europe-wide survey of emergency plans, that coherent strategies exist for mitigating the health impacts of flooding or other natural disasters, despite the growing evidence of these impacts, and the development of systems and models for targeting those who might be worst affected.

As described above, this is not because these emergency plans do not cover health impacts, and therefore evacuation to hospital and other medical facilities, merely that this is taken as an axiomatic function of emergency plans rather than analysed or catalogued in any detail. Health is something that is *implicitly* the subject of the emergency planning documentation we have analysed but not *explicitly* addressed within that process.

Nevertheless our case studies all show that warnings of floods, to mitigate flood impacts (including health impacts), need to be carefully targeted – and can be targeted – on those who are most vulnerable to flood impacts. They also show that flood vulnerability has a major element of prior health status in its contributory factors, and this can be the basis of a flood impact assessment method (or model) as a guide to policy and prioritisation. In addition our research shows that it is possible to predict flood vulnerability (with its health effects) spatially, from secondary sources, as shown by our analysis of data systems for the County of Hertfordshire, UK.

By way of summary, the list below provides our recommendations that arise from this research, concerning the pre-event planning in terms of warning provision, and post-event care of those who might suffer health deterioration in floods and their aftermath. What is striking is that most of these recommendations are straightforward, but few have yet been built into flood preparedness planning, either in the UK or elsewhere. Perhaps the reason for this is that the responsibilities for these policy areas, and for policy implementation, is inherently split between many organisations (see the list below). What is needed are more coherent and 'joined up' strategies for flood impact minimisation, and for these strategies to take seriously the health effects of flood events for the many people who are unfortunately affected.

Some key policy imperatives arising from implications of the health effects of floods:

Policy recommendation 1: There is a need to provide better information on those especially liable to suffer health impacts in floods, so that they can be (a) located and (b) targeted for assistance:

- The elderly
- Those with prior-event health problems
- The poor
- Those with dependents (especially children).

Likely locus of responsibility for this policy and its delivery: *Local authority social services departments; national census departments.*

Policy recommendation 2: There is a need to provide better warnings of floods before the events occur, and better arrangements for response to these flood warnings:

- Longer warning lead times
- More accurate warnings
- More advisory warning messages: not just facts, but advice as to what to do
- Better warnings for agencies who need to respond.

Likely locus of responsibility for this policy and its delivery: *Meteorological and water/basin agencies; local authorities; emergency services (police; fire; etc)*

Policy recommendation 3: There is a need to provide better post-event care for those who have been affected, even those who appear at first sight not affected:

- Post-event visits to identify health and health-related problems
- Assistance with recovery work phases
- Financial assistance and advice
- Medical/social advice.

Likely locus of responsibility for this policy and its delivery: *Medical authorities; local authority social services departments; insurance and related organisations, etc.*

Acknowledgement

Part of this work has been supported by the EC/WHO study "Climate change and adaptation strategies for human health (cCASHh)".

References

Beck RJ, Franke DI (1996) Rehabilitation of victims of natural disasters. Journal of Rehabilitation 62(4):28 – 32

Bennet G (1970) Bristol floods 1968. Controlled survey of effects on health of local community disaster. British Medical Journal 2:454 – 8

Bland SH, O'Leary ES, Farinaro E, Jossa F, Trevisan M (1996) Long-term psychological effects of natural disasters. Psychosomatic Medicine 58(1):18 – 24

Environment Agency (2003) Thames Gateway and flood risk management – a preliminary assessment. Environment Agency, Bristol

Evans EP, Ashley R, Hall J, Penning-Rowsell EC, Saul A, Sayers P, Thorne C, Watkinson A (2004) Foresight flood and coastal defence (two volumes). Office of Science and Technology, London

Green CH, Emery PJ, Penning-Rowsell EC, Parker DJ (1985) The health effects of flooding: a survey at Uphill, Avon, Enfield. Middlesex University Flood Hazard Research Centre, London

Hepple P (2001) Research into the long-term health effects of the flooding event in Lewes, Sussex of October 12th 2000: a case-control study. Unpublished Masters thesis: London School of Hygiene and Tropical Medicine. London School of Hygiene and Tropic Medicine, London

Holen A (1991) A longitudinal study of the occurrence and persistence of post-traumatic health problems in disaster survivors. Stress Medicine 7:11 – 17

Hovanitz CA (1993) Physical Health Risks associated with Aftermath of Disaster. Journal of Social Behaviour and Personality 8(5):213 – 254

Hubert G, Reliant C (2003) Cartographie réglementaire du risque d'inondation: décision autoritaire ou négociée? Annales des Ponts et Chaussées 105 :24 – 31

Intergovernmental Panel on Climate Change (McCarthy J; Canziani O; Leary N; Dokken D, White K, (eds)) (2001) Climate Change 2001: Impacts, Adaptation, and Vulnerability. Cambridge University Press, Cambridge

Parker DJ, Green CH, Penning-Rowsell EC (1983) Swalecliffe Coast Protection Proposals – Evaluation of Potential Benefits. Enfield. Flood Hazard Research Centre, Middlesex Polytechnic, London

Penning-Rowsell EC, Wilson TL (2003) The health effects of

flooding: A European overview. Report for the cCASHh project. Middlesex University Flood Hazard Research Centre, London

Reacher M, McKenzie K, Lane C, Nichols T, Iversen A, Hepple P, Walter T, Laxton C, Simpson J (2004) Health impacts of flooding in Lewes: a comparison of reported gastrointestinal and other illness and mental health in flooded and non-flooded households. Communicable Disease and Public Health 7(1):1 – 8

RPA/FHRC et al. (2003) The Appraisal of the Human-Related Intangible Impacts of Flooding. R&D Project FD 2005. Defra/Environment Agency

Tapsell SM (2000) Follow-up study of the health effects of the 1998 Easter flooding in Banbury and Kidlington, Final report to the Environment Agency, Thames Region. Middlesex University Flood Hazard Research Centre, Enfield, London

Tapsell SM, Tunstall SM (2001) The Health and Social Effects of the June 2000 Flooding in the North East Region. Report to the Environment Agency. Middlesex University Flood Hazard Research Centre, Enfield, London

Tapsell SM, Tunstall SM, Wilson T (2003) Banbury and Kidlington Four Years After the Flood: An Examination of the Long-Term Health Effects of Flooding. Report to the Environment Agency. Middlesex University Flood Hazard Research Centre, Enfield, London

Tapsell SM, Penning-Rowsell EC, Tunstall SM, Wilson TL (2002) Vulnerability to flooding: health and social dimensions. Phil Trans Roy Soc London 360:1511 – 1525

Tapsell SM, Tunstall SM, Handmer J, Penning-Rowsell EC (1999) The health effects of the 1998 Easter flooding in Banbury and Kidlington. Report to the Environment Agency. Middlesex University Flood Hazard Research Centre, Enfield, London

Townsend P, Phillimore P, Beattie A (1988) Health and deprivation: deprivation and the North. Croom Helm, London

Tunstall SM, Bossman-Aggrey P (1988) Waltham Abbey and Thornwood, Essex: An Assessment of the Effects of the Flood of 29th July, 1987 and the Benefits of Flood Alleviation. Flood Hazard Research Centre, Middlesex Polytechnic, Enfield, London

World Health Organisation (1948) Constitution of the World Health Organisation in Basic Documents. World Health Organisation, Geneva

Learning From Experience: Evolving Responses to Flooding Events in the United Kingdom

Merylyn McKenzie Hedger

Abstract

Flooding has assumed an increasingly high profile in public policy in the United Kingdom. Institutional responsibilities have been changed, investment has been increased in infrastructure and research, more attention is given to preventing development in flood plains and improved warning systems have been introduced. This paper provides a short explanatory overview with reference to experience of catastrophic events in 1953, 1998 and 2000, and increased scientific understanding, particularly, about changing flood risk with climate change.

Introduction

Experience from flooding events in the UK has driven action and investment since the catastrophic East Coast flooding event in 1953. Recent events, combined with increased understanding about climate change, have added momentum to the development of responses. Whilst not driven by concerns for major loss of life, there has been evident major disruption to lives and consequent mental health issues. Property loss has also been a major driver. All institutions concerned in flood defence policy have been involved increasingly in change since a major event in 1998: central Government, the Department for Food and Rural Affairs and the National Assembly for Wales which have overall policy responsibility; the organisation charged with operational responsibilities for flood defence and general supervision in England and Wales- the Environment Agency (EA) and the insurance industry.

This paper provides a short overview of the development of response to flood risk. It does not seek to rest this analysis within the broad body of policy literature [1], but rather provides a simplified narrative constructed from the interaction of the most significant "catalytic" events combined with scientific understanding.

Context

The coastline of England and Wales is approximately 4,500 kilometres long. Many urban and rural areas depend on coastal defences to protect them from flooding or erosion by the sea or tidal waters. The emphasis on the protection of life, and hence urban areas, remains the primary focus of Government flood and coastal defence policy. Currently 10 % of the total housing stock (1.7 million homes) and commercial property worth over £200 billion is located in flood risk areas and 61 % of the best agricultural land is at risk [2]. Defences take many forms. Over one-third of the defences along the coast, mainly in the south and east is man-made but many other areas are safeguarded by natural features such as sand dunes, beaches and salt marshes which may require some intervention to maintain sufficient protection. Flood defence works on rivers and watercourses involving channel modifications, flood embankments, pumping stations, controlled storage and hydraulic control structures are common in man areas [3]. Significant flood damage has been avoided as a result of these works and the low lying districts of many urban areas depend on flood defences for their continued economic viability. Nevertheless, some recent events have caused unexpected damage and been on a scale to cause concern.

East Coast Floods 1953

On January 31 and 1 February 1953, a great storm surge, accompanied by gale force winds, swept over the north of the UK causing widespread flooding of coastal areas. Northerly gales raised tide levels and broke through sea defences at some 1200 sites. The damage extended over 1000 miles of coastline. Over 300 people died; 32,000 had to be evacuated from their homes, and 24,000 houses were flooded [4]. It seems that the surge had an elevation of about 2.6 metres all along the Norfolk coast, although it was only a moderate spring tide. No one was prepared. This event was the most damaging of recent times. By chance the surge did not reach London.

The trauma of loss of life, human misery and property loss alerted the Government to the potential dangers. A radical rethink led to major new flood defence infrastructure being commissioned. Many of the defences were rebuilt using 1953 levels as a maximum. For the Thames estuary, a 1966 report by Government's Chief Scientist recommended that the best solution was a continuation of bank raising in conjunction with a flood barrier incorporating movable gates across the river [5]. Interim bank rising was undertaken in 1971–2. Legislation effected in 1972 gave powers for that solution. Construction of the barrier started late in 1974 and first became operational in October 1982 and was first used in February1983. These defences give protection for up to a 1 in 100 year event until 2030.

During the 1960s and 1970s several major flooding events occurred, notably in 1968 which prompted the engineering industry to improve flood estimation techniques [6] and 1970s manuals were produced [7] which provided guidance for design for over 20 years [8]. However, the issue was largely left to the engineers.

Development of climate change science

For the post 1953 sea defences, the reference context was gradual subsidence of the east and south coasts of the UK due to isostatic readjustment after the last glaciations which means that the southern part of England and Wales is slowly sinking by up to 1 – 2 mm per year. It was the first major Inter Governmental Panel on Climate Change Assessment (IPCC) (1990) that changed approaches to coastal planning throughout the UK with an allowance for climate change being built into all new coastal flooding infrastructure. The "best estimate trend" of the IPPC was used which at that time suggested that a rise in global sea level of about 20 cm by 2030 and 65 cm by 2100. Operating authorities were advised to allow for relative sea level rises which took account of geological crustal movement and climate change Eastern and Southern areas 6 mm/yr, North West 4 mm and elsewhere 5 mm /year [9].

Legislation was passed to set up flood defence and coast protection measures were enabled in 1949. But, it was not until 1991 after publication of the IPCC report this legislation was consolidated and powers were given to provide and operate inland flood warning systems[1]. All the legislation provides the relevant authorities with powers to carry out flood defence and coastal protection activities in a permissive not

[1] Apart from publication of the IPCC report in 1990, there was also a major coastal flooding event in North Wales at Towyn in 1990.

a compulsory way [3]. However the Environment Agency (formerly the National Rivers Authority) was given the duty to exercise supervision over all matters relating to flood defence.

These coastal standards still hold, despite changes in the IPCC assessments – these imply the allowances are on the high side and are thus thought to give better protection. What has changed however has been increased general understanding of what climate change will meant to the UK, largely due to the development of climate change scenarios for the UK Climate Impacts Programme. Scenarios for the programme were developed by the University of East Anglia and the Hadley Centre in 1998 with an improved set being issued in 2002 [10]. Some analyses were undertaken within the Technical Reports about the changes in frequency of extreme events and impacts of increased rainfall on flooding. These scenarios have been used in a variety of regional and sectoral studies within in the UK and have raised awareness amongst decision-makers and national and regional and local scales outside the "traditional" flood defence community. Their publication also combined with experience in 1998 and 2000 of serious fluvial flooding events, so have had an additional impact. The widely read summary of the 2002 scenarios states:

- "For some east coast locations, extreme sea levels could occur between 10 and 20 times more frequently by the 2080s for the medium high scenarios than they do now." "The probability of a present-day extreme sea level occurring in any given increases form 2 % currently to about 33 % by the 2080's and for the highest emissions scenario this probability increases to about 90 %".

- "By the 2080's winter daily precipitation intensities that are experienced once in every two years may become 5 % and 20 % heavier." Winter precipitation may become 30 % wetter." [11].

Easter Floods 1998

In 1998, over Easter, parts of central England experienced levels of flooding not seen since 1947, and it was recognised as an event with a return period in excess of 1 in 100 years. The rainfall in the preceding month had been exceptionally wet with monthly rainfall values of between 1.6 and 3.2 times the monthly average. Prolonged and heavy rain fell across central England flooding many areas in the Midlands and Wales, and resulting in a flood wave resulting in flooding without warning. Five people died.

- "Many thousands of people were severely affected." … "Many were without warning put in fear of death or serious injury. They lost their homes and personal possessions, suffered massive disruption to their lives and livelihoods and some were still without homes six months later." [12].

Following the Easter 1998 floods an independent review was commissioned by the Minister which led to the subsequent implementation of an action plan [12]. The Environment Agency prepared a two year action plan to address the lessons learned, particularly about the lack of warning. Other significant changes were made resulting in changing advice on development in flood risk areas, internal management changes in the Agency, and in the organisation of flood defence nationally.

By far the main outcome was the establishment of a National Flood Warning Centre. The Centre was to lead development of flood forecasting and warning systems, using improved access to meteorological information [13]. A Flood warning system based on four stages of alert was started, linked to the centre to lead development of flood forecasting and warning systems. High profile public awareness campaigns were launched and a major flood defence operational change made.

The Agency has created a national address database of "at-risk" properties based on its Indicative Flood Plain maps for England and Wales in May 2000. The database was used for the first time in the 2000 campaign and enabled the Agency to target people living and working in flood risk areas about the new flood warning codes. This meant that more people knew about the warning system, how to prepare and what to do if they heard a warning [13]. Through key partnerships with the BBC Weather Centre and other TV networks there was prominent broadcast of the new codes and the Floodline telephone number on national and local TV and radio weather bulletins.

All this paid off when tested in the next floods in Autumn 2000 and have provided the basis for better management until now.

Autumn 2000 Floods

Autumn 2000 was the wettest on record for over 270 years. It followed a wet spring and early summer with significant flood events in southern England in May. Significant and extensive flooding over large areas of England and Wales was caused by a series of waves of rainfall which crossed the country over a seven week period in October and November. Catchments soon became water-logged, with the result that until the cold and drier weather followed rivers responded to even modest rainfall. The cumulative effect led to repeated flooding in many places and to prolonged flooding in others [13].

These floods affecting more than 700 locations were multiple events affecting a short period and in at least two cases so severely as to stretch and test the whole system. But the new system worked- flood defences successfully prevented the flooding of 280,000 homes and emergency actions averted catastrophic flooding in South Wales.

However, the costs were huge- in excess of some £1 billion of insured costs. Just under 10,000 homes and businesses were flooded causing damage, trauma and distress to thousands. Many whose homes were flooded are still recovering.

Of the locations flooded 28 % of problems were due to overtopping, outflanking or failure of defences; 40 % occurred where no defences existed and 32 % were due to flooding from ordinary watercourses, inadequate local surface water drainage and third party defences. This has led to changes because it revealed that historic decisions about responsibility for different water-courses had little relevance to the community they drain nor to the householder who didn't know where to turn for advice and support.

It proved necessary to evacuate 11,000 people, including patients from the Worcester Royal Infirmary: a process initiated by police and supervised by local authorities. Many public meetings were held in flooded locations and several Members of Parliament in affected areas used adjournment debates to raise specific problems affecting their constituents. Aftercare groups have been formed in affected areas such as in Lewes where a Flood recovery group was formed to assist traumatised victims (depression was four times above average). Serious financial disruption to business occurred, and significant areas of farmland were flooded in Wales, central and north-east England.

Train services were cancelled with railways covering south west, southern, central and northern England and in south and north Wales with an estimated cost of £320 million net of insurance covering repairs, delays and compensation. Water companies had problems with water sources due to cuts and pollution of groundwater with surface water flooding. "Boilwater" instructions were issued in some places. Major motorways closed, bus services disrupted and schools closed. And sewer flooding proved a particular problem with design standards exceeded both in intensity and duration so sewers could not operate. Electricity substations were flooded and had to be shut down and extensive power cuts were imposed [13].

Current status of responses

After 1998 and 2000 dealing with floods was no longer an issue only for the professional engineer.

"Prior to 1998 we had experienced 30 years without flooding of sufficient experience to capture the attention of the population at large … But it has been the flooding of 1998 and 2000 which has awoken the whole of England and Wales to the realisation that flooding is a natural hazard which on occasions may be of widespread national importance" [6].

Combining with increased awareness of climate change, and actions which were already in place following the 1998 floods, these traumatic floods served to increase momentum on investment, and organisational and institutional change in the following principal ways:

- Strategic policy level.
- Government increased investment on flood defence spending by an extra £150 million each year for flood and coastal defence operating authorities. This has led to some further changes in the methods of allocation.
- For the first time, the Project Appraisal Guidance issued by the Ministry for the assessment of proposed schemes suggested sensitivity allowances for climate change, with a 20 % increase in peak flows over the next 50 years (2001).
- Planning.
- New advice was issued (2001) to planning authorities (in local government) as to how they handle flood risk at all stages of the spatial planning and development process. This replaced guidance which had been issued in 1992.
- Association of British Insurers.
- The insurance industry in its collective form – the Association of British Insurers- the ABI – galvanised its internal appraisal of response and issued an important statement of principles in 2003 about how it would handle flood insurance cover.
- Flood risk warning and self help.
- Further attention was given by the Environment Agency to its handling of flood risk and warning. A new stakeholder organisation has emerged.
- Increased research.
- Existing research programmes were consolidated and extended and novel programmes introduced whose outputs are being fed into policy making.

These developments will be briefly outlined to explain the current status of responses (July, 2004).

Strategic Policy

Flood defence appraisal was changed rapidly post 2000. Whilst an allowance had been made for an increase in sea levels due to climate change in 1991, it was not until 2001 that a sensitivity test was overlaid on flood defence appraisal allowing for an increase in river flows of 20 % over the next 50 years [14]. Research to scrutinise this guidance is underway as it is known to be a crude tool. Increased understanding of climate change is pushing forward the review of the design life of the Thames Barrier: a major re-assessment is underway with the intention of further investment to extend the life of the barrier be extended from 2030 to 2100. Planning started in 2000 for a new study of flood risk management for the next 100 years.

Government commissioned a post incident report on the 2000 floods from the Environment Agency (Lessons Learned) which led to 31 recommendations for further action for the Agency, its partners and Government. By this time, nearly all of the recommendations of the 1998 assessment (Bye report) had been executed. In addition, the National Audit Office investigated inland flooding and reported in March 2001 with four high priorities for action: catchment management plans, completion of floodplain maps and economic evaluation of maintenance.

The main stimulus for further action within the Agency has been the commitment post 2000 floods by central Government to significantly increased spending on flood defence infrastructure by £ 150 million pounds from 2005/6 with the first increase of £55 million in 2004/5. By 2002 good progress had been made [15]. Some institutional and funding changes were introduced by Government to strengthen the Agency's role and to rationalise delivery. Within the Agency a new flood risk management strategy has been developed [16]. Its central premise that whatever the investment it will never be technically, economically or environmentally acceptable to prevent flooding entirely. There has to be a re-assessment of the way that we live and cope with the ever-present threat of floods. And, consequentially we have to change our thinking from defending to managing floods. This will mean reducing the likelihood of flooding and the impact of floods when they occur. New targets are introduced for reducing the risk of flooding, improved warning system, preventing inappropriate development inside floodplains, and to produce Catchment Flood Management Plans for all principal catchments. Flood risk management activities are intended to reduce the probability of flooding through the management of land, river system and flood defences and reduce the impact of floods through effective land use planning, regulation, flood warning and emergency response. A new impetus has also been given to Flood Risk Mapping which is required for planning consultation. So far the maps only show the area affected by a 100-river flood and a 200-year coastal flood, excluding the impacts of existing defences and climate change. Both are to be factored in [17].

Association of British Insurers

Attitudes in the insurance industry have been changing in response to increased understanding of climate change and also actual costly experience. The industry's focus had been on coastal and estuarine areas because of the large value of assets at risk in those areas and because of the historical incidence of the 1953 event and also in Towyn in North Wales in 1990. The industry's attitude changed with several events occurring in the 1990s in Llandudno and Perth 1993, Strathclyde in 1994 [18].

Currently the UK is one of the few countries in the world where insurance companies virtually automatically provide flood cover to all homeowner's insurance policies. This approach has applied over the past 20 years. Flooding had not formed a major component of insurance claim in the country. Weather related claims on average totalled £700 million a year, one-third of the total. However insurers have become increasingly aware of their exposure. Since 1990, weather-related events (flood and storm damage) insurance claims have cost an average of £825 million each year and have risen to levels in excess of £1 billion in four of the past 15 years [18]. Whilst climate change is the foremost reasons, other significant factors are pressures to build houses in the floodplain and the increasing value of goods insured (buildings and their contexts). It was the Easter 1998 floods, which triggered the ABI into action, and it undertook a detailed investigation resolved a number of course of actions, including strengthening its staff resources to handle flood risk [18].

So the insurance industry itself is now a powerful force. It favours a type of consolidation which will support a continuation of the present voluntary system, rather than moving to a European model with state intervention of regulation with state involvement in reinsurance [20]. A formal statement of principles was issued for the first time in September 2002. Where properties in flood risk areas are protected to a minimum standard of 1 in 75, flood cover will be available. Cover will be continued for properties which have less protection than 1 in 75 years, if flood defences are planned in the next 5 years. Where flood risk is unacceptably high and no improvements in flood defences are planned, here insurers will examine risks on a case by case basis. As its price for continued coverage the ABI has demanded an annual review of progress from Government on maintained spending; improved funding arrangements; implementation of procedures on new development appraisal (see below) improved flood risk mapping to be available; implementation of an emergency planning review and the adoption of realistic solutions for sewer flooding [21]. Some homeowners are already faced with very significant premium increases. The industry considers that public funds should be focused on improving protection for existing properties, and it is known that between 15–85 % of existing fluvial flood defences are in fair, poor or very poor condition. It considers that further development in high-risk flood areas should be curtailed and that developers should bear the cost of flood defence measures associated with new building where it is permitted.

Spatial planning and flooding

Local government through its planning departments prepare development plans and control development within their areas through the granting or refusal of planning permission. Comprehensive guidance was first given to planning authorities in 1992 on development and flood risk and local planning authorities that they should consult and take account of surveys of flood risk and advice on sea-level rise and global warming. The guidance also spelled out that local authorities should use their planning powers to guide development away from areas that may be affected by flooding or instability, to restrict development that would itself increase the risk of flooding or interfere with the operations of the flood and coastal defence authorities and to seek to minimise development in areas at high risk. There was however no statutory requirement to do this.

Following the Easter 1998 floods, the "Government looked "for a step change in the responsiveness of the land-use planning system to flood risk management ..." In 1999, high level targets were introduced to ensure delivery and in particular for the Agency to report on coverage of plans with flood risk statements or policies and whether any objections it lodged on planning applications has been accepted. A new detailed Planning Policy Guidance Note 25 was introduced in July 2001, replacing the 1992 advice and giving tougher guidance to local planning authorities as to the importance of taking account of flood risk in their long term spatial planning and control of development [22]. Key points of this guidance are that authorities were advised to apply the precautionary principle to flood-risk, using a risk-based sequence to avoid the risk where possible and manage it elsewhere; and to consider flood risk on a whole-catchment basis. Areas with an annual probability of river flooding of 1 % or above or with an annual probability of 0.5 % of coastal or tidal flooding to be defined at a significant flood risk A sequential test was introduced. Where annual probability of flooding is less than 0.1 % no planning constraints need apply. Where flood risk is 0.1 % to 1.0 % for river flooding and 0.1 % to 0.5 % for tidal or coastal flooding, flood risk assessment should be applied and flood resistant construction and warning procedures required. These areas are not considered suitable for essential civil infrastructure, hospitals, fire stations, or emergency depots needed in times of extreme flooding. High risk areas with an annual probability of flooding with defences from rivers of 1.0 % or greater and tidal and coastal 0.5 % or higher. The guidance states that these areas may be suitable for development if standards of flood defence can be applied unless they are undeveloped areas or functional flood plains.

The guidance on the treatment of high risk areas is still not regarded as adequate by the insurance industry. There is tremendous pressure for development in flood risk areas. In 2001 proposals for developers for development in flood plains comprised £ 51 billion some 20 % of total proposals.

Flood Warning Systems and Self Help

The National Flood Warning Centre was set up by the Environment Agency in April 2000 to provide a central focus for flood forecasting and warning expertise and best practice. Its aim is " to reduce risk to life, distress and damage to property by the provision of an accurate, reliable and timely flood warning service, where the benefits justify the costs and where provision of a service is technically feasible". It promotes grater understanding and awareness of flood risk among at risk public and a focal point for experts, professional partners and stakeholders. It also leads a research programme. Fundamental to its work is the premise that delivery falls within a continuous learning cycle. Priorities are improving the accuracy of forecasting (in conjunction with the Met Office), extending coverage of warnings and more effective response to warnings. A "Multi Media Warning Dissemination System is being developed" [16].

The National Flood Forum was established in 2002 by affected communities. It is a non-profit organisation set up by people with direct experience of flooding, and is dedicated to reducing the suffering of people from flooding throughout the UK and linking together communities at risk from flooding. It receives considerable support form the Environment Agency and has built good links with a variety of organisations including Government and the insurance industry. It shares advice on new products and has a "Golden Sandbag" award system to reward exceptional effort. This organisation's emergence and official support, signals the greater emphasis that is being given to embedding social vulnerability within policy and empowering self-help and awareness. Past policy has been criticised for being overly technocratic [23].

Public awareness- the Agency is now half-way through a ten year public awareness programme. In addition to the launch of flood warning codes and Floodline, the programme has delivered five public awareness campaigns. Its main message is that everyone has to understand that flooding will happen and it cannot be predicted with 100 % certainty [24]. Account is taken to deliver to those for whom English is not the first language and to the elderly. Awareness has been increased in Flood warning Areas: 72 % are now aware that they are at risk compared to 41 % in 2000. Whilst 98 % are aware that they are at risk, but only 16 % have taken action (5 % in 2000). Independent market research organisations are used to probe reasons and motivations and why campaigns have worked or not [25]. The latest 2003 campaign builds on earlier awareness campaigns with the message Act Now: Be Prepared for Flooding. The campaign is being delivered at a very targeted local level though: regional radio and local press advertising, advertising in pubs in at risk areas with beermats and posters, and direct mail. Innovative support materials have been prepared including videos with TV personalities and storylines in TV and radio soaps. Posters have been prepared for doctors' surgeries. A direct mailshot went to 350,000 householders informing targeted householders about steps they can take to protect their property.

The main components [of the day to day warning service are:

- Flood Warning Areas are defined as those where the Agency is able to offer a flood warning service to the public (Flood watch, Flood warning, Severe Flood warning, All Clear. Warnings may be issued directly (AVM, warden, siren, loudhailer) or indirectly (TV/ radio, Floodline). Over one million homes and businesses in England and Wales are covered where local conditions make it possible to forecast flooding (flooding due to groundwater is difficult to predict). It can handle up to one million hits a day.
- Floodline is the umbrella system for all flood-related communications. A 24 hour-telephone helpline service is available and provides the focus for all activity and information about flooding.
- An overall National Summary on flooding is available on the Agency in English and in Welsh and it is updated every 15 minutes. Regional Summaries are available and the site can be searched by local Flood Warning Area, or by town, postcode, or river. Dynamic buttons automatically appear of key pages on the site only when warnings are in force. TV and radio stations are listed which broadcast flooding information, information on the last five flood warning s recorded for each area.

Increased research

Research activity was stepped up after the 1998 events, and given further momentum by the 2000 events. Following an assessment which indicted that there was gap between getting research results into action, an integrated programme between the Government Ministry (DEFRA) and the Environment Agency was initiated based on a themed approach (2001). Gaps were still perceived and other funding players entered the scene. The academic research councils formed a consortium in 2002 which led to a new programme being initiated in April 2004. Most innovatory was the start of a programme in 2002 commissioned by the Government's Chief Scientific Advisor in the Department of Trade and Industry. This aimed to find out how the risks of climate change and flooding and coastal erosion change in the UK over the next 100 years, and to work out the best options for Government and the private sector for responding to future challenges [27]. It reported in April 2004, and is a major driver behind the production of a new Government flood risk strategy, due in Autumn, 2004.

Conclusion

Policy for managing flood risk in the UK is iterative and dynamic. Policy has been responsive to the experience of extreme events and developing science has enabled new standards and regulations to be established. Failure to adequately warn about the 1998 flooding in particular, led to management change at the highest level in the responsible Agency and new flood warning systems being adopted. The new measures proved successful at the time of the 2000 floods, but the potential for future improvement was identified. Government significantly increased funding. A major new approach to planning for flood risk has been established for new investment. And, a coherent new approach to flood warning is in place which is building incrementally on increased public understanding over time, with the aim to ensure people take action to protect themselves. Also, the insurance industry has demanded specific actions by Government, to be reviewed annually, in exchange for its continued provision of cover.

There are still gaps in standards which demand further attention. One is the handling of climate change in infrastructural investment. And, no allowance is yet normally made for storm surge in coastal investments. A major policy issue for clarification is the Thames Estuary where Government is looking to develop significant new areas of housing to relieve housing pressures in the south-east of England- yet the area lies in the flood plain. The shopping list prepared by the insurance industry has not yet been fully delivered. More tools and information must be delivered to local planning authorities to help them with delivery of flood risk assessment. More vulnerable groups need to be given information on flood risk planning. The January 2005 event in Carlisle revealed inadequacies on forecasting ...

All these issues are now receiving attention and this is likely to be an active agenda for some time.

Acknowledgements

I would like to thank the Water Management Communication team, in particular Liz Cook, for their help in supplying the material on the Agency's flood warning campaigns. The information on the experience of the 2000 flood event is drawn from the Agency report "Learning from the Floods", whose chief author was Gary Lane.

References

1. Johnson C, Tunstall S Penning-Rowell E (2000) Crises as Catalysts for Adaptation: Human Responses to Major Floods. Flood Hazard Research Centre Middlesex University
2. Department of Environment Food and Rural Affairs (2001) National Appraisal of Assets at Risk from Flooding and Coastal Erosion including the potential impacts of climate

change, July 2001. Report prepared by Halcrow Group, HR Wallingford and John Chatterton Associates
3. Ministry of Agriculture, Fisheries and Food (1993) Strategy for Flood and Coastal Defence in England and Wales
4. Baxter PJ, Moller I, Spencer T, Spence RJ, Tapsell S (2003) Flooding and climate change. In: Health Effects of Climate Change in the UK. Department of Health, pp 152 – 192
5. Environment Agency (2001) The Thames Barrier flood defence for London
6. Institute of Civil Engineers (2001) Learning to live with rivers. Full report of the Civil Engineer's Presidential Commission to review the technical aspects of flood risk management in England and Wales
7. Flood Studies Handbook 1975
8. Centre for Ecology and Hydrology (1999) Flood estimation handbook, Wallingford
9. Ministry of Agriculture, Fisheries and Food (1991) Advice on allowances for sea-level rise
10. Hulme M, Jenkins G (2002) UK Climate Impacts Programme UKCIP. Climate change scenarios for the United Kingdom. Technical Report, Oxford
11. Hulme M, Turnpenny J, Jenkins G (2002) UK Climate Impacts Programme UKCIP. Climate change scenarios for the United Kingdom. Summary Report, Oxford
12. Bye P, Horner M (1998) Floods report by Independent review team to the Board of the Environment Agency
13. Environment Agency (2001) Lessons learned from the Autumn 2000 floods
14. Ministry of Agriculture, Fisheries and Food (2001) Flood and coastal defence project appraisal guidance: Overview MAFF Flood and coastal Defence with Emergencies Division May 2001
15. Environment Agency (2002) Flood defence strategy issues. Report to Board. October 2002
16. Environment Agency (2003) Strategy for Flood Risk Management. May 2003
17. Environment Agency (2002) Flood Mapping Strategy July 2003
18. Association of British Insurers (ABI) (2000) Inland Flooding Risk- issues facing the Industry. General Insurance Report 10, ABI London September 2000
19. Association of British Insurers (ABI) (2003) Development and Flood Risk
20. PPG 25 (2003) ABI Guidance on Insurance Issues. ABI London July 2003
21. Milne J (2003) Options for Flood Cover: Do European Models Offer Lessons? ABI, London
22. Association of British Insurers (ABI) (2003) Statement of Principles on the Provision of Flooding Insurance. ABI, London
23. Office of the Deputy Prime Minister (2002) Planning Policy Guidance Note 25: Development and Flood Risk
24. Brown J, Damery SL (2002) Managing flood risk in the UK: towards an integration of social and technical perspectives. Trans Inst Br Geog NS(27):412 – 426
25. Environment Agency (2003) Act Now. Be Prepared for Flooding. Information on the 2003 flood warning campaign
26. MMRB Social (2003) Flood Awareness Campaign – Qualitative and Quantitative Work Reports to Environment Agency January 2003
27. www. Environment-agency.gov.uk
28. DTI (2004) Foresight Future Flooding. Executive Summary. Office of Science and Technology, London

National Case-Studies on Health Care System Responses to Extreme Weather Events

Extreme Weather Events in Bulgaria for the Period 2001–03 and Responses to Address Them

R. Chakurova · L. Ivanov

Summary

The authors point to the great weather varieties in Bulgaria, which enable delineation of 5 climatic areas. The climatic specificities of Bulgaria cause the occurrence of extreme weather events such as storms, hurricanes, tornados, extreme cold spells and ice-formation, torrential rains, floods, unusually warm spells, dry spells, etc. The report analyzes the causes and magnitude of disasters occurring for a three-year period (2001–2003) and which posed environmental and population risk. The role and place of the forces and resources of the Civil Protection Service, the ministries and other agencies for addressing the aftermath and curtailing environmental risk and population casualties are highlighted.

Key words

extreme situation, weather disaster, extreme weather events, natural disasters, environmental risk

Introduction

Bulgaria is both a European and Balkan country. It is located in south-east Europe and occupies the eastern part of the Balkan Peninsula, which is a bridge between Europe and Asia. The area of Bulgaria is 111,000 km^2, which is 22 % of the Balkan Peninsula area. Bulgaria occupies 15th place in Europe in terms of territory size.

Although Bulgaria is located in the southern section of the moderate European climatic zones, it demonstrates great weather varieties, which are determined by the geographic location and landscape of the country. There are five climatic areas in Bulgaria: moderate-continental, transitional, continental-Mediterranean, Black Sea, and mountainous. Latitude is the principal driver for the first three climatic areas, the geography for the mountainous climatic area, while the Black Sea area has been shaped by the influence of the Black Sea. The variety of the landscape with its characteristic forms – plains, hilly lands, and mountains, which occupy almost equal areas of the territory of Bulgaria. The northern parts of Bulgaria are under the influence of the continental air masses entering from the east and north, while the southern parts are under the Mediterranean climatic influence (south and south-west). Atlantic ocean air masses carry moisture and enter from the north-west into Bulgaria. The average annual air temperature in Bulgaria is 12 °C. January demonstrates the lowest monthly temperature and July the highest. The lowest temperature measured in Bulgaria was −38.3 °C, while the highest was +45 °C. The annual temperature amplitude is above 20 °C in the majority of the country, which suggests prevailing continental nature of the weather.

The winds are various in terms of direction and force. The north-west winds are the most common. The north-east winds bring dry air masses. In the summer they are hot and in the winter cold. The south winds are typical for the southern parts of Bulgaria, which occur more rarely. Breezes occur along the Black Sea coast. The diverse terrain creates conditions for the occurrence of local winds. The average annual precipitation in Bulgaria is not high at 670 mm. Most of the precipitation occurs as rainfall. The distribution of the precipitation on the territory of Bulgaria is irregular. Highest precipitation occurs in the mountains, and its quantity decreases from west to east. The annual precipitation distribution is of great importance. Abundant precipitation in the spring and beginning of summer facilitates growth of agricultural crops. Frequent dry spells occur at the end of summer and during fall, necessitating artificial irrigation.

The rivers are the biggest water sources in Bulgaria. The river network features high density, however most rivers are short and with inconstant output. The rivers demonstrate highest flow in spring, while water levels decline during the other seasons. Only the Danube river is navigable. The rivers Iskur, Maritza, Struma, Mesta, Yantra, etc. have great economic importance. The waters of the Bulgarian rivers flow down into two water basins – the Black Sea and Aegean.

The climatic specificities of Bulgaria cause the occurrence of extreme weather events such as storms, hurricanes, tornados, extreme cold spells and ice-formation, torrential rains, floods, unusually hot spells, etc. For a 3-year period (2001 – 2003), forest and field fires, storms, strong winds, rainfalls, floods, etc. occurred (➤ *Tab. 1*).

Tab. 1

A comparative table of the extreme weather events, which have occurred on the territory of Bulgaria until 08.10.2003

Event	Year	Number	Deaths	Injured	Collateral damages (in BGN)
Storms, strong winds, and rains	2001	148	3	1	
	2002	388	2	2	3,750,682
	2003	339	2	2	
Hail	2001	32			26,670,929
	2002	39			680,000
	2003	13			
Floods	2001	29			
	2002	136	1		106,457,720
	2003	154			
Fires	2001	30,948	104	260	
	2002	18,404	96	213	5,378,585
	2003	22,109	68	181	14,190
Snow-drifts/ice formation	2001	164	2		
	2002	243	3		
	2003	77	1	20	

Source: Resources of Civil Protection Service's Information and Administration Center, Ministries and Agencies, Sofia. 2001 – 2003, official report.

The greatest number of fires (30,948) was recorded in 2001. The number decreased throughout the following years, and it was lowest in 2002 (18,404). The higher number of forest fires was not always proportionate to the environmental risk and human casualties. Thus, the number of forest fires for 2002 was 18,404, while the affected area was 651,360 quarters of an acre (about 3 times as much as in 2001). The trend in the field fires was retained, about 6,550 quarters of an acre burnt crops of wheat and other agricultural crops. The fires, which occurred during the peak 2001, created emergent situations in the regions of Kurdjaly, Sliven, Stara Zagora and Bourgas, and in 2002 in the regions of Montana, Lovech, Sofia, Pazardjick, and

Kurdjaly. The large number of burnt areas in the affected regions influenced the environmental balance there because unique natural habitats had been destroyed. The principal causes of fire occurrence were: incautious fire handling, high temperatures and dry spells, arson, etc.

Second in terms of significance were storms, snow-drifts, ice formation, and torrential rains. There was an evident trend of increased number of such events. In 2003, their number was 3 times higher (➔ *Tab. 1*). The environmental consequences were: 1 – 2 m high snow-drifts, which blocked and paralyzed road traffic. High-voltage power-lines disruption occurred with impaired power supply of hundreds of settlements; disrupted automatic phone connectivity with isolation of the affected populated localities; disrupted water supply, heating, medical care, and supply of essential products, destruction of buildings, facilities, etc. Snow storms and snow-drifts have been observed at 2 – 3-year intervals, and they have been most pronounced in northern and eastern Bulgaria. 50 – 60 % of Bulgaria is threatened by snow-drifts on an annual basis.

Floods might occur following abundant rains, snow melting, and partial or complete destruction of 30 large and 600 small dams. Following the above, the most complicated situation is anticipated in the lowlands and basins of the Danube, Maritza, Tunja, Mesta rivers, and the Iskur, Batak, Trakietz, Topolnitza dams, and along the Black Sea coast. Huge areas may go underwater in the event of floods (more than 910 square kilometres), and 83 populated localities may be affected. Only the Danube river is a risk factor for 30 populated localities with a population of 515,000 and about 73,000 quarters of an acre of agricultural plots. Local floods occur in the capital city and other large cities in the event of torrential rains. Partial floods occur along the Black Sea coast from tidal waves during stormy and prolonged east winds, and strong earthquakes with sea epicenter.

A significant problem for Bulgaria is the dry spells and the few potable water sources. The Republic of Bulgaria has the lowest water resources per capita among the European countries. Circulation processes during summer and fall cause soil moisture stock decline, which has an indirect impact on the mass field and forest fires, as well. These processes are the cause of occurrence of long-term dry spells in 30 – 40-year intervals, which recently have become one of the main problems of Bulgaria. The increasingly higher drought and waterlessness for the duration of decades in the whole country have caused incalculable losses for the national economy and the population. Water supply restrictions have been imposed on hundreds of populated localities.

The intensity of the reported weather events demonstrated an increasing trend. Fires have become more frequent due to the high temperatures and prolonged dry spells resulting from global warming. From the data for the 3-year observation period (2001 – 2003), the trend was preserved of a greater number of days with abundant snowfalls, snow-drifts and ice formation, which had caused disastrous situations in individual regions on the territory of Bulgaria. The intensity of the abundant precipitation and floods was unstable, and an increasing trend was observed for the period 2001 – 2003 under report. Landslides have been caused due to the abundant and torrential rains, and a sustained trend of increasingly higher landslide number has been observed since 2001 so far (➔ *Tab. 1*).

The weather events occurring on the territory of Bulgaria led to huge material damages, which reflected significantly upon the economy. The greatest material damages were incurred by floods (BGN106,457,720 for 2002), hailstones, landslides, fires, etc.

The management bodies and the units of the Civil Protection Service, and the ministries and agencies have participated with staff and equipment in addressing the aftermath of extreme situations resulting from weather events in compliance with the Regulation for the Organization and Activities for Addressing the Aftermath of Disasters, Accidents and Catastrophes. With the exception of the forces of SA Civil Protection, significant forces and resources of the Ministry of Transport and Communications, Ministry of Health, Ministry of Environment and Waters, Ministry of the Interior, Ministry of Agriculture and Forestry, SA Power Engineering and Resources, BRC, etc. have been involved to combat disasters and accidents (➔ *Tabs. 2 and 3*).

Conclusion

The following inferences may be drawn based on statistical data analysis:

1. Both anthropogenic (caused by human actions) and natural disasters occurred on the territory of Bulgaria during the observation period (2001 – 2003).
2. Highest environmental risk were posed by the following weather events: storms, snow-drifts, ice formation, torrential rains, floods, and dry spells.
3. Casualties among the population were not proportionate on all occasions to the environmental risk.
4. The effective involvement of the special forces of SA Civil Protection, the ministries and agencies, has contributed to limiting the environmental risk and population casualties.
5. The material and human loss during weather events determined the need for the prevention of health hazards of extreme weather events to find place among national priorities.
6. Efforts for prevention of diseases and casualties resulting from weather events should be included in the national and local plans for health care delivery to the population in critical situations.

References

Andreev VM (1994) Applied research achievements for improving the prevention efforts for protection against hydro meteorological disasters. In: Applied research conference on population protection during disasters and accidents. Collected Works XII, Sofia, p 8 – 9

Bulgarian National Committee on the Decade (1994) Natural Disaster Risk Reduction Programme until 2000. Collected Works, Sofia

(2000) CP Report before the Stability Pact Operations Group, Sofia. 19 – 20.10.2000. Official report.

(2003) Resources of Civil Protection Services Information and Administration Center, Ministries and Agencies, Sofia. 2001 – 2003, official report.

Mihaylova I, Chakurova R (2000) The Role of Bulgarian Units for Addressing the Earthquakes in Turkey in 1999, 4th Medical Geography Congress with International Attendance. Collected Works, Sofia

(1998) Regulation for the Organization and Activities for Addressing the Aftermath of Disasters, Accidents, and Catastrophes, DCM No 18/23.01.1998. Official governmental document, Sofia

Chakurova R, Mihaylova I (2001) Disaster Situations in Bulgaria, which Pose Population and Environmental Risk. 7th Medical Geography Symposium with International Attendance. Collected Works, Sofia

Tab. 2

Participation of the management bodies and squads of SA Civil Protection, the ministries and agencies in rescue and emergency recovery operations in Bulgaria from 01.01.2001 to 31.12.2001

No.	Disaster, accident, catastrophe type	Number	Participation of management bodies								Commitment of squads, forces and means					
			SA Civil Protection	Regional Committees	Municipal Committees	Mayor's Committees	Site Committees	Ministerial Committees	OG of Ministries and Agencies	OG of Civil Protection Directorates	SA Civil Protection specialist squads			Squads of ministries and agencies		
											Number of participations	Staff	Equipment	Number of participations	Staff	Equipment
1	Fires (domestic, forest, field, etc.)	30,948	4	123	353	442	18	14	580	189	211	1,152	317	3,142	38,164	5,026
2	Snow-drifts, ice formation	164	0	21	81	207	23	26	201	69	56	244	112	674	10,969	7,612
3	Floods	29	0	1	18	4	0	0	0	10	9	28	11	67	202	59
4	Hail	32	0	3	14	17	0	0	0	5	0	0	0	171	74	18
5	Storms, strong winds and rains	148	3	9	60	33	9	10	220	29	45	210	61	334	3,475	1,343

Source: Resources of Civil Protection service's Information and Administration Center, Ministries and Agencies, Sofia. 2001 – 2003, official report.

Tab. 3

Participation of the management bodies and squads of SA Civil Protection, the ministries and agencies in rescue and emergency recovery operations in Bulgaria from 01.01.2002 to 31.12.2002

			Participation of management bodies								Commitment of squads and means		
								Management bodies of ministries and agencies			Specialist squads of ministries and agencies		
No.	Disaster, accident, catastrophe type	Number	Operations groups as per ROAAADAC	Regional Committees	Municipal Committees	Mayor's Committees	Site Committees	Number of participations	Staff	Equipment	Number of participations	Staff	Equipment
1	2	3	4	5	6	7	8		9			10	
1	Fires	18,451	1	16	108	105	4	129	2,321	148	10,294	85,920	22,302
2	Snow-drifts and ice formation	243		26	57	32	24	83	218	111	263	10,636	4,683
3	Hails	68		3	27	29		17	138	26	79	1,449	153
4	Storms, strong winds and rains	388	3	41	215	67	21	118	354	128	321	2,618	739

Source: Resources of Civil Protection Service's Information and Administration Center, Ministries and Agencies, Sofia. 2001 – 2003, official report.

2002 – A Year of Calamities. The Romanian Experience

A Cristea

Abstract

A short history of the major abnormal weather phenomena that occurred in Romania in 2002 is described in this short article. Although perhaps less harmful then in the rest of the continent and mainly local or in small regions, their unusual frequency and rapacity stimulated the local people to call the year 2002 a year of "calamities". During all the extreme weather events, local Disasters Defense Committees are created to minimize losses, to repair the damages, but mainly to help local people with supplies, medicine, safe drinking water and hospitalization. The general health measures in cases of disasters are presented. They are classified as applied on-place and in-hospitals. The specific emergency measures discussed in this chapter include: drinking water, water and sewerage infrastructure, chemical hazards as well as other implications. Finally, two important education problems are mentioned which were put in evidence by the inappropriate behaviour of the population under calamities and the rather poor efficiency of the medical personnel in case of disasters.

1 Introduction

The beginning of the third millennium finds mankind faced with a considerable number of unsolved problems. The most serious one, by its immediate and long-term effects, is related to the environment. The international scientific community has to give more convincing answers to some questions people ask more and more often: is the climate changing? If so, how quickly and how much will human society be affected? And what can we do in order to minimize the risks?

The paper presents the experience of a year of natural hazards in Romania and the measures taken to counterbalance the risks on population due to disasters.

2 The calamities of the year 2002 in Romania

The year 2002 in Romania was characterized by spectacular changes in weather and by calamities. The events occurred month after month and resulted in important damages (➲ *Tab. 1*).

January started with extremely cold weather in Transylvania and the Republic of Moldova. About 10,000 hectares cultivated with rape and barley were destroyed resulting in a decrease of production of 50 %. Temperatures under 20 ° below zero characterized the month of February. In central and northern Moldova 30 – 35 % of the vineyards were destroyed by frost. Late hoarfrosts at the end of March caused the loss of 40 % in the fruit production of the whole country. The economic losses for the first three months were estimated to be 14 million Euros.

April was a month during which the absence of water in the soil became acute. Thus, for the first time in the past 20 years, irrigation was started in April. A devastating drought and isolated heavy rains and hails in the centre and south of the country characterized the month of May. In June, drought continued to affect the crops. The absence of rain and very high temperatures for this period of the year, 30 – 40 °C, destroyed 50 % of the wheat and barley cultures. The second trimester resulted in losses of about 23 million Euros.

In July, on the background of drought, devastating torrential rains were recorded in some areas. Torrents wiped out 70,000 hectares of agriculture. The vineyards in the Vrancea county, situated in the east Romania, were affected by successive hails, the estimated loss being 30 – 40 % of the vine production. Early August was also characterized by drought and very high temperatures (40.8 °C in the shade in the southern part of the country). In the second part of the year some weather phenomena characteristic of tropical areas occurred. Heavy and extremely violent rains caused catastrophic floods. One hundred and thirty one localities were affected: 3,000 households and 70 houses were destroyed, 70,000 hectares of farmland were damaged, 47 bridges were damaged, of which 4 collapsed, and 10 national roads partially destroyed. Also, 3 persons died (one old person with cardio-vascular disease and two drowned children).

However, the most dreadful phenomenon was recorded in Făcăieni, in southern Romania. It was a tornado, a phenomenon recorded for the first time in Romania. The strong wind totally destroyed 16 houses, left 300 houses without roofs, uprooted tens of trees, jammed railway traffic for 24 hours, injured 22 persons and killed 3 victims (a family was crushed under a collapsed ceiling and a car crashed into a tree).

After the crops had been seriously damaged by drought, heavy rains in September augmented the losses. In all, the economic losses of the three summer months rose to 18 million Euros.

In October the temperatures were much below the normal average. The fact that temperatures were close to zero degrees had a negative impact on the growth of crops seeded during autumn. November and December registered severe cold, with the soil freezing. This had major implications on the cultivated farmlands (20 % reduction of the estimate harvest, 50 % of the barley crops, vineyards and orchards compromised). The economic losses of the fourth trimester went to 3 million Euros.

Tab. 1
Weather phenomena over the year

Months	Events	Economic losses
January – March	cold weather, hoarfrosts	14 mill Euros
April – June	drought, heavy rains, hails, high temperatures (30 – 40 °C)	23 mill Euros
July – September	drought, high temperatures 45°C, torrential rains, floods, tornado	18 mill Euros
October – December	very cold weather, freezing soil	3 mill Euros

3 Responsibilities in case of disasters

Every Romanian county has a Local Disaster Defence Committee (*Tab. 2*). They are activated during any of these unhappy events in order to minimize losses, to rehabilitate water and sewerage pipes, to recover from electricity faults, to repair roads, bridges and railways, but mainly to help people in need with living supplies, medicines, safe water and health care. These committees usually include: local authorities, the Department of Public Health (DPH), Civil and Military Defence Departments, the Police, the Fire

Department, the Environment Protection Agency, Water Supply and Sewerage Companies, Electric Power and Communication Companies, the Department of Roads and Bridges, the Department of Transport, the Food and Agriculture Department and volunteers.

Tab. 2
Responsibilities of the Local Committee in the case of natural disasters

Body	Responsibility
Local authorities	Evaluate the situation, organize the disaster headquarters, supervise the actions of the committee members to reduce the effects of the disaster, discrete local founds distribution
Department of Public Health	Identification of diseases, medical monitoring, special sanitation and hygiene measures
Civil and military defence departments, Police, Fire Department	Imposes the observing of decisions, acts for the diminishing of the disaster effects, helps people in need
Environment Protection Agency	Imposes measures for the protection of the environment (especially in the case of danger of chemical pollution)
Water supply and sewerage companies	Remedy the damaged water supply and sewerage infrastructure
Department of Transport, Roads and Bridges	Remedies the damaged access routes

4 Public Health decisions and actions

The DPH takes action both in the areas affected by calamities and in those with displaced population. These measures include: active identification of diseases and active medical monitoring for categories of people at risk (newborns, children and pregnant women, elderly, people with chronic diseases, disabled and patients who cannot be moved, people with a low income and the homeless).

Special measures apply also in hospitals. Newborns, infants and women who have recently delivered are not discharged until the situation is back to normal. In order to prevent over-crowding in hospitals, a strict selection of patients in need for hospitalization is done, with highest priority given to acute disorders and emergency surgery. Also hospitals and the DPH are responsible for securing medical supplies with drugs, blood and plasma substitutes, diet food, powder milk, disinfecting products, etc.

In the poor urban communities and in most of the rural areas the sanitation and hygiene measures have particular importance. They are intended to prevent water-born and vector-born diseases, gastrointestinal illnesses, dermatitis, conjunctivitis, and wound infections. Cause of these incidents could be: inadequate sanitation and overcrowding among displaced persons, contact with polluted water, rodent infestations, contamination of food, damage of waste deposits and latrines, contact with polluted water. Measures include: sanitary and waste management education, disinfestation and vector control, creation of safe food storage conditions (withdrawal of depreciated products), disinfection of latrines and neighbouring soil, immunization of population (tetanus boosters, case-by-case vaccinations).

In case of natural disaster, special care is given to storehouses of chemical substances. Causes of accidents could be: underground pipe disruption, overflow of toxic-wastes or release of chemicals. Most important for prevention and proper action is the existence of accurate and updated maps of the locations with potential risks (chemical plants and deposits). Special attention is given to stocks of toxic substances. The body involved always in these cases is the Environment Protection Agency.

Equally important is to secure the water supplies. Specific to the Romanian rural areas remain the well water supply. In case of disasters, the water from local sources is first affected. To prevent the incidence of water-born infections during disasters, the measures include the testing of wells, chemically and bacteriologically, and their disinfection. Till the situation is under control, a particular concern of the DPH is to supply the population with drinking water, bottled, boiled or treated and distributed in tanks.

Occasionally, the disasters produced the contamination of the springs from where water is fetched to towns, or the damage of distribution or treatment facilities. In some cases these damages could have been minor if the installations were newer or better kept. Cheap or old water and sewerage utilities increase the risk of accidents. In all cases the lesson learned was that, especially if you are poor, you cannot afford to buy cheap: there is less money spent to build modern water and sewerage supplies than to rehabilitate old installations after disasters.

5 Human behaviour and responsibilities

Highly important in case of disasters is the education of the population, of personnel with responsibilities in such events, and last but not least, of health personnel. The general impression is that people have problems with acting appropriately in case of disaster. They are not used to washing their hands with soap while preparing or eating food, after toilet use, after participating in flood cleaning activities, or after handling items contaminated with floodwater or sewerage. Since the causes are to be blamed on the low educational level of the rural population, remedies could be obtained by a dedicated education acquired in schools and, sometimes, at the work place (key rules to apply in extreme events to prevent falling sick and striking losses of human lives). The second education problem concerns personnel. It includes, of course, all level of personnel, from public administration with responsibilities in cases of disasters to the army and other civil categories. But of a tremendous importance is the proper emergency education of health personnel. It is not at all easy to establish ideal behaviour, and especially to plan and organize a health system adaptation to an emergency.

6 Conclusions

The attitude of contemporary society towards natural hazards is often contradictory. On one hand, huge human and financial efforts are made in order to prevent and reduce their effects. On the other hand, the development of human society sometimes triggers the occurrence of some disasters or amplifies their outcomes (Cheval 2002). Thus, the climate change, a natural process that has occurred many times during the evolution of our planet, is today beyond its natural limits. The increased vulnerability of human society to natural hazards is not so much due to changes in the way the phenomena show, but rather to some anthropogenic causes such as the increasing population, social inequity, military and political nature of economic support, accumulation of economic capital in areas likely to be affected by hazard, the increasing potential of technological disasters (Alexander 1995; May 1997).

The focal point in the present approach of researches on natural hazards is the human dimension. How prepared societies are to cope with extreme natural events differs in many aspects (education, infrastructure, organization, etc.), thus every situation raises specific problems. However, no matter what type of society , these problems have to be solved with best results; a highly current challenge for those involved in the integrated management of natural hazards. Often, the extreme natural hazards only disclose and aggravate a pre-existing latent poverty (Ribot 1966). Unfortunately, any post-calamity evaluation cannot include the real psychological impacts, the pain and the elapse of hope, which natural hazards leave behind when they turn into disasters.

Although touched perhaps less violently than Western Europe, Romania was not bypassed by the vicissitudes of the changing weather. In particular, the year 2002 was felt very severely and had major implications on the economy and on humans. In its way to modernization and integration into united Europe, Romania is also making efforts to minimize the risks encumbered by severe weather phenomena. Learning by action during the ravaging year 2002, the Romanian responsible institutions for the prevention and minimization of risks in case of calamities are now better prepared to face similar events of the future.

References

Alexander DE (1995) A survey of the field of natural hazards and disaster studies. In: Carra A, Guzzetti F (eds) Geographical Information Systems in Assessing Natural Hazards. Kluwer Academic Publishers, Dordrecht, p 1 – 20

Cheval S (2002) Risk in contemporary economy (in Romanian). The Academica Publishing House, Bucharest, p 118 – 121

May PJ (1997) Addressing Natural Hazards: Challenges and lessons for Public Policy. The Australian Journal of Emergency Management 11;4:30 – 37

Ribot JC (1996) Introduction. Climate variability, climate change and vulnerability: moving forward by looking back. In: Ribot JC, Magalhaes AR, Panagides SS (eds) Climate Variability, Climate Change and Social Vulnerability in the Semi-arid Tropics. Cambridge University Press, Cambridge, p 1 – 10

A System of Medical Service to Assist the Population of Uzbekistan in the Case of Natural Catastrophes

Abdukhakim A. Khadjibayev · Elena M. Borisova

The Republic of Uzbekistan is located in central Asia between the Amudarya and Sirdarya rivers, covering an area of 447.4 thousand square kilometers. The population is 25.4 million. It consists of 12 regions and 1 autonomous republic. The capital is Tashkent (population 2.5 million). The Republic borders Kazakhstan on the north-east, Kyrgyzstan and Tajikistan on the east and south-east, and Turkmenistan and Afghanistan on the west and south-west. The total length of the border is 6221 km.

The territory of Uzbekistan is mostly plain (around four fifths of the area). One of the main plains is the Turan plain. The foothills of Tyan-Shan and Pamir are on the east and north-east of the country. The highest peak (4643 m) of the country is also in the same region. In the north of the central part of Uzbekistan there is one of the world's largest deserts – the Kizilkum.

The climate of Uzbekistan is sharply continental with long hot summers which have the potential to adversely affect human health. The average annual precipitation is 120 – 200 mm on the plain area and 1000 mm in mountainous area. As precipitation is very low, agriculture relies mostly on irrigation.

Ecological problems of Uzbekistan and extreme weather phenomena

One of the current ecological problems of Uzbekistan is the shrinking of the Aral Sea. Once, the Aral Sea provided prosperous life to the population of the region, but today it has damaged the overall surrounding eco-system and directly influenced the health of the people of the region.

According to the UN, around 700,000 tonnes of harmful salts originating from the bottom of the Aral Sea are carried over a 1000 km radius, out of which 500 kg is precipitated per each hectare in the Amudarya's delta. The Aral crisis has caused medical, social, economical and domestic problems, which require huge financial expenses. Nevertheless the government is trying to solve these problems. Medical assistance is being sent to the area, and hospitals, schools and houses are being built.

The consequences of industrialisation are manifesting themselves at an increasing frequency. These include: air, water and soil pollution by manufacturing enterprises; shrinkage and swallowing of the water bodies and deforestation. Progressive anthropogenic pressure on the forest eco-systems of Uzbekistan becomes more evident each year, and has resulted in rare and unique floral and fauna spices becoming extinct. The Tugai forests, for example, are degenerating, their area is becoming smaller and geo-systems are changing.

The increase in greenhouse gases in the atmosphere due to human activities is causing global climate change. The main sources of greenhouse gases in Uzbekistan are fuel and energy manufacturers, the building industry, the metallurgy and chemical industries, automobile and railroad traffic, agricultural activities, mining and transportation of fossil fuel, and waste storage and processing.

Unfortunately, climate change, considerable forests reduction and soil erosion have made landslides, floods and mudflows more common in Uzbekistan. Everyone in Uzbekistan remembers the night of August 8, 1998 when two rivers, the Aksu and the Shahimardan in the Fergana valley, flooded. The flood was unexpected and the majority of the population was sleeping. 600 people went missing, thousands remained without shelter and 109 died. Many houses, bridges and power supply lines were destroyed. The first medical aid was provided by more than 20 mobile teams (each team consisted of five people – a doctor, two nurses and two aidmen), who worked closely with 60 rescuers searching for the victims in the mountains. Difficulties experienced during the rescue, especially regarding medical evacuation, showed the necessity for a revision of the existing Emergency Medicine System (EMS) of Uzbekistan.

Nowadays, the Sarezsk Lake is a concern. The right bank of the lake is not stable, and in the event of an earthquake it may collapse. It is predicted that if a landslide occurred, a huge wave (probably 200 – 250 m high) would flow through the lower part of the blocking dam. Enormous mud and stone flows would sweep over (with a speed exceeding 80 km per hour) the the Bartang and Pyandj Rivers (where the Tajikistan-Afghanistan border is located), entering the Amudarya River, and after several days it would reach the Aral Sea. Settlements, villages and towns would be destroyed and it is predicted that around 5 million people would be affected.

Taking into account all of the above, the government of the Republic of Uzbekistan has undertaken large-scale arrangements to reduce the consequences of ecological and climate change on the health of the people and their security. Active debates are ongoing to discuss the reduction of extreme weather events. In particular, the National Strategy on reduction of greenhouses gases between 2000 – 2010 has been initiated. In cooperation with international organizations a number of projects are running to restore biodiversity and to improve water supply in the Aral region. Interstate normative documents on the establishment of regional cooperation of rescue services of CIS countries have been signed. Families living in the flood- and landslide-prone areas have been resettled to secure areas, and housing areas have been distributed for them. For mudflow prevention, local funds have been provided for cleaning lowland and river banks, and barrage construction. However, current knowledge and capability is not able to completely prevent and remove the tragic consequences of natural disasters. This is why it is important to establish and maintain different structural and effective rescue services. In Uzbekistan the Emergency Medicine Service (EMS) is responsible for medical aid to the population suffering from natural disasters, as well as in everyday life. This service differs from similar institutions of other countries.

Emergency Medicine Service

One of the main distinctive features of the Uzbek model of EMS is governmental guarantee of free and accessible high quality medical services in life-threatening cases.

Before the start of the fundamental reformation of the healthcare system EMS was not a separate division of the healthcare system of the country. Different medical institutions and their subdivisions were working separately and independently, providing emergency and systematic medical assistance to the population. There was no unique mission, unique philosophy, and unique methodology for emergency medical service in these medical institutions. There was also no consistency regarding the qualifications of staff, equipment, and medical supplies. The ambulance service was also independent. In other words, there was no unique structure of EMS organization; there was no state policy regarding support and development of EMS. The technical basis was out of date and there was lack of skilled personnel.

Considering the existing system of healthcare and distinctive features of economical transition period the model of the EMS was created. To achieve accessibility, economical and medical efficiency the EMS is organized on a multilevel basis, with all organizational-structural levels united into a single service, with stable organizationally-methodological, vertical and horizontal links. The structure and functioning of the EMS is given in *Table 1*.

Tab. 1

Structure and Functioning of Emergency Medicine Service of the Republic of Uzbekistan

Level	Institution	Type (workload) of provided assistance
Highest level	RRCEM Regional branches of RCEM	Specialized, qualified medical aid **
Medium level	EMS sections in Central Regional Hospitals	Qualified medical aid ***
Low level	Primary unit of healthcare (family polyclinic) Ambulance service	First aid
Non-hospital emergency medical assistance	Community-healthcare system in case of emergency situation (mobile groups of constant preparedness of EMS)	Have an equipment for specialized, qualified medical aid
	Sub units of the Ministry of Emergency Situations Paramedics (policemen, drivers, teachers, firemen and etc.), trained at RRCEM	First aid

* Rural (family) aid posts

** Specialized, qualified medical aid is provided in special divisions by special practitioners (cardio surgeon, neurosurgeon, urologist, etc)

*** Qualified medical aid is provided by general practitioners (surgeon, therapist, pediatrician).

The Head Centre in Tashkent – Republican Research Centre of Emergency Medicine (RRCEM) – provides emergency basic surgery and reanimation-intensive directions of emergency medicine for those who live in the capital and around; Regional Centres provide all types of medical assistance to the population of the regions (see list below). In addition, 171 settlements of the country have sub-divisions of RRCEM in their Central Regional Hospitals (CRH) and Central Town Hospitals (CTH). In addition to the mentioned sub-divisions, the EMS includes an ambulance service with 194 stations and 1485 ambulance teams, medical sub-divisions of the Ministry of Emergency Situations. Altogether around 44,000 staff are working in the EMS, among them more then 7500 qualified doctors and 20,000 nurses.

The standard beds' structuring in RRCEM and its regional branches.

A. Intensive care and surgery profile

1. Intensive care
2. Abdominal surgery
3. Thoracic and vascular surgery
4. Trauma and neurosurgery
5. Urology
6. Gynecology
7. Burn unit
8. Toxicology

B. Therapeutical profile

1. Emergency therapy
2. Cardiology
3. Neurology

C. Pediatric profile
1. Pediatric surgery
2. Pediatry

The integration of institutions providing emergency medical aid into single structure in Uzbekistan is logical considering the financial resources deficit. The advantages of a unique EMS are:
- The opportunity to realize a unique mission and policy on development of EMS
- Concentration of financial, technical, personnel and scientific potential of the healthcare system
- Efficiency and effectiveness of management
- The opportunity to address the financing of EMS

In the framework of the single system, the organizational and methodological work of the Centre is simplified and directed towards development, realization and perfection of structure and methods of EMS organization, based on efficiency, high technology and effectiveness of all levels of the service. Treatment-diagnostics standards are developed taking three levels of emergency aid into account: for general practitioners – primary unit of healthcare, i.e. first aid, primary diagnostic and primary treatment; for surgeons, therapists and pediatricians of the regional unit, qualified medical aid; and for the special practitioners of the regional unit and RRCEM specialized and qualified medical aid.

To ensure consistent functioning and availability of free bed stock, patients in a stable condition are transferred to other medical institutions (including polyclinics) for further treatment and rehabilitation. Optimal timings of hospitalization and signs for the next stage of treatment and rehabilitation are given in written instructions entitled "Treating-diagnostic standards for the doctors of EMS", as well as being regulated by corresponding Acts of the Ministry of Healthcare and regional healthcare administrations.

By the Act of the Ministry of the Republic of Uzbekistan EMS ascribed to the functional subsystem of State System of Prevention and Actions in Emergency Situation. Service of emergency aid at emergency situations became a component of the overall EMS. In 1999 the Act "About State Service of Emergency Medicine Aid at Emergency Situations" was developed. It regulates basic aims, organizational structure, and management, organization of medical aid to the population in emergency situations, technical and financial supply, as well as social and law security of medical staff of State EMS at emergency situations.

For provision of aid in the cases of emergency situations 39 specialized, ever-ready groups have been formed (three in RRCEM, and three groups in each of the 12 regional centers), 128 groups to provide emergency medical assistance on the basis of CRH and CTH exist.

Basic tasks of the ever-ready medical groups are as follows:
- Medical (on-post) sorting, emergency qualified aid and specialized qualified aid on the basis of local sub-division of RRCEM
- Coordination of efforts of institutions and services at the site of the catastrophe
- Organization and coordination of evacuation of victims out of the site of catastrophe.

The EMS with its sub-divisions is integrated into a general system of rescue works in cooperation with the Ministry of Extraordinary Situations, Ministry of Internal Affairs, Ministry of Defense and other departments. This system has not only increased the efficiency and effectiveness of the EMS and "disaster medicine" but has also reduced general expenses.

The EMS treats 500,000 people in hospital and around 500,000 out of hospital, annually. Around 5,000,000 calls for ambulance service are placed every year. It is important to admit that around 17.5 % of patients are treated in the EMS; however the number of beds is only 6.3 % of the overall number of beds in the country. These numbers not only show the intensity of the work of the EMS but also proves the accessibility of the EMS.

In summary, the EMS in the republic of Uzbekistan is single service, gathering technical and scientific personnel and equipment into a single structure, funded by the state, and is intended for the provision of free, accessible medical assistance to the population in emergency situations. By creating a single service it has become possible to create an effective mechanism of continuous supervision, methodological upgrade of quality of medical assistance and monitoring of diseases status in emergency situations. The integration of all institutions providing emergency aid to the population into a single system is economically advantageous as well as providing state guarantees of free and accessible emergency medicine for all groups of society.

Conclusions

Long dry and hot summers are traditional in Uzbekistan, thus the population has developed effective measures against the effects of heat.

Floods and mudflows are difficult to forecast and very dangerous.

Alongside organisational measures such as resettlement of populations to safe areas from the dam construction sites and strengthening of the river banks, it became necessary to form an effective medical emergency service, which is a multi-branch system and is able to provide medical aid to the population in cases of natural disasters.

In the case of limited resources it is expedient to form the medical divisions of rescue of the population on the basis of existing medical establishments.

Uniting medical institutions into a single medical structure providing an Emergency Medical Assistance increases management efficiency and continuity of evacuation thus saving lives of the victims.

Moscow Smog of Summer 2002. Evaluation of Adverse Health Effects

Victor Kislitsin · Sergey Novikov · Natalia Skvortsova

Abstract

The Moscow extreme event in Summer 2002 was not marked by a strong heat-wave, as seen in Europe during 2000 – 2003, although three-to-four heat-waves (temperature 28 – 32 °C) lasting for four to six days each were observed. The main cause of concern was the presence of a haze and a smell of burning for many days together with high concentrations of pollutants produced by the forest and peat bog fires and industrial and vehicle emission. Due to the lack of appropriate and timely data on health outcomes of the smog at that time, the Ministry of Health and the Moscow Authority asked the Institute of Human Ecology and Environmental Hygiene to evaluate quickly possible adverse health effects caused by the smog. The information was needed for further decision-making in an effort to overcome the consequences of the smog. The present article describes the adverse health effects that the pollutants of concern may cause, and presents measured pollutant concentrations and evaluated adverse health outcomes of the smog air pollution.

Key words

smog, air pollution, air pollutant concentration, health risk assessment

Introduction

The impact of chemical compounds produced by forest and peat bog fires together with excessive industrial and vehicle emissions on the health of the urban population has been observed for many times in different regions of the Russian Federation and all over the World (for example, Canada, USA, India, Indonesia and others).

Two kinds of smog have been distinguished. According to Berlyand (1975), the London smog is usually observed in the early hours in wintertime with calm weather where the temperature is between –1 °C and –4 °C and relative humidity is more than 85 %. The distinguishing feature of the London smog is very restricted visibility (less than 30 m). Photochemical smog (frequently observed in Los Angeles, USA) usually takes place in the middle of the day during the months of August through to September. It is characterised by a wind speed of less than 3 mps, ambient temperature between 24 °C and 32 °C, relative humidity less than 70 % and visibility range of between 1.5 and 8 km.

Moscow smog

In July through to the middle of September 2002, there was an extreme weather situation in Moscow similar to the smogs seen in Los Angeles, USA. It was caused by a long-lasting anticyclone that brought high temperatures (up to 32 °C) and small amounts of precipitation (for the period May to September 2002 there were 150 mm, compared with the normal level of 350 mm), many calm days and low humidity (40 – 65 %). The weather conditions caused forest and peat bog fires over an area about 350 hectares in the Moscow region. During that time, and especially at the end of July and beginning of September, the visibility was restricted to less than 2 km.

The protracted fires, industrial emissions and vehicle exhausts resulted in high concentration of toxic air pollutants in the ambient air of Moscow. For many days a haze and a smell of burning were present. During the smog Moscow medical authorities exerted efforts to help the people overcome by the environmental conditions. The number of serviced medical emergency calls during August and September increased by 10 %. Moscow TV and radio reported instructions on how people should behave to avoid damage to their health.

During the smog period many efforts were made to reduce the amount of pollution in the ambient air of Moscow. The main task, of course, was first to prevent the peat bog and forest fires from spreading, and then to extinguish them. It was very difficult task to perform mainly because of the water shortage in the peat bog areas. The spokespersons of the Anti-fire State Services explained at that time they extinguished up to 40 fires each day. However, at the same time many new fires sprang up. The struggle with the fires was stopped only at the end of September. These efforts played an important role in the reduction of air pollution during Summer 2002.

In addition, the temporal reduction of emissions from the most hazardous industrial enterprises in Moscow upon the request of Moscow authorities, helped to improve the ambient air condition during the most crucial period of the event.

Due to the lack of appropriate and timely data on health outcomes of the smog at that time, the Ministry of Health and the Moscow Government asked the Research Institute of Human Ecology and Environmental Hygiene together with the Moscow Sanitary-and-Epidemiological Centre to quickly evaluate possible adverse health effects caused by the smog. The information was needed for further decision-making in an effort to overcome the consequences of the smog.

Health Hazards

Air pollutants together with the increased ambient temperature are the basic factors determining health hazards of such smog. In general, a list of chemical substances that are emitted into the air during this kind of weather event include total suspended particulates (TSP), specifically fine fractions of TSP – particulate matter with the effective diameter of particles up to 10 microns (PM10), and particulate matter with the effective diameter of particles up to 2.5 microns (PM2.5), ozone, sulphur dioxide, nitrogen dioxide, carbon dioxide, benzene, formaldehyde, polychlorinated dioxins and benzofurans and some other dangerous organic chemical substances including carcinogens (Hurley and Donnan 1999).

In this study PM10, PM2.5, ozone, carbon monoxide, sulphur dioxide and nitrogen dioxide were chosen as pollutants of concern because of their high concentration and, consequently, great affect to human health. The main health hazards caused by the studied pollutants are described below (Hurley and Donnan 1999, Vedal 1996).

Particulate matter (PM10 and PM2.5)

Population groups that are most sensitive to particulate matter exposure include asthmatics, the elderly (over 65 years old), children, people with cardiovascular (CVD) and respiratory (RD) diseases, and expectant mothers. Air pollution by PM2.5 accounts for an excessive increase in total, CVD and RD mortality especially in the elderly. Children and adults are subjected to excessive chronic bronchitises and asthmas. Usually at the same time an increase in the number of doctor visits, hospitalisation and emergency calls is observed. Exacerbation of chronic conditions (asthma), respiratory obstruction, and myocardial ischemia are among the causes of visits and emergency calls. The causes for the increased hospitalisation are myocardial infarctions (MI) and cardiac rate disturbances.

People with chronic asthma and angina need to increase the dose of medicine to withstand the disease. Very often the adverse effects of PM pollution reveal themselves by an increase in the respiratory symptom frequency rate (cough, angina pain), and in the number of days of partial or total disability due to morbidity and the days of restricted activities (decrease in capacity for work and indispositions).

Ozone

Usually high ozone concentrations are registered in summer time, especially in hot sunny afternoons. Air pollutants from industrial enterprises, vehicle exhausts and forest and peat bog fires increase ozone formation. The most vulnerable population groups to ozone are asthmatics, children and the elderly, people suffering from respiratory diseases, people performing physical work outdoors and smokers. Typical symptoms of ozone poisoning are irritation of the eyes and respiratory tract, bronchospasm, general intoxication (headache, undue fatigue, chest pain and heavy breathing). If caused by relatively low ozone concentrations the symptoms disappear between 5 and 7 days after the exposure termination.

Air pollution by ozone accounts for excessive increase in total, CVD and RD mortality.

Carbon monoxide

Carbon monoxide (CO) is the result of incomplete combustion of flammable substances. The gas from the ambient air is capable of reacting with the haemoglobin in blood producing carboxyhemoglobin (CoHb). A concentration of 3 % or less of CO in human blood produces no adverse effects, while smokers usually have a concentration of between 3 % and 8 % CoHb. A concentration of between 2 % and 10 % CO in human foetus blood causes low birth weight, whereas a concentration of between 10 % and 20 % leads to the development of the clinical signs of poisoning – headache, eyesight degradation and heavy breathing. Death occurs where a concentration reaches between 50 % and 60 % CoHb. The usual CoHb concentration level for urban population is between 0.8 % and 1.9 % for non-smokers and 3.6 % for smokers. Air pollution by CO accounts for excessive increase in total mortality and acute MI.

Sulphur dioxide

Increase of sulphur dioxide concentration during the smog period causes great concern. This gas is the product of the wood and peat burning and at the same time is emitted in large quantities by industrial enterprises and vehicles. Adverse effects of sulphur dioxide is similar to that of TSP and nitrogen dioxide.

Nitrogen dioxide

Nitrogen dioxide (NO_2) is produced in any combustion process. The main sources of NO_2 in big cities are usually industrial enterprises and powerplants. Young children (5 years old and younger), asthmatics and elderly people, with RD and CVD are the most sensitive to NO_2. NO_2 affects the immune system and increases the sensitivity of humans to pathogens and viruses.

Air pollution data

In order to evaluate the possible adverse health effects of the selected air pollutants during the smog period their concentrations were obtained and analysed.

The concentrations of TSP, PM2.5, NO_2, sulphur dioxide, CO, and ozone were received from the Moscow Centre for Hydrometeorology and Environment Monitoring and the Public Environmental Enterprise "Mosecomonitoring".

In contrast to other studied smogs, the ambient air concentrations of sulphur dioxide in Moscow during the smog period were relatively low (the daily mean for the smog period was at the usual level for that time of year). For this reason sulphur dioxide was excluded from further analysis.

For all pollutants the mean daily concentrations for each day of the monitoring during the smog period were calculated (➊ *Fig. 1–4).* Unfortunately, the obtained data refers to different time-periods during the smog for different pollutants. This fact must be considered while studying the presented data. Each figure shows also the pollutant's upper limit of a harmless concentration, so-called reference concentration (RFC)[1]. This helps to better understand the real magnitude of the concentration values. As a rule the Russian maximum allowable concentration were taken as RFC values. RFC for PM2.5 was taken from the US EPA National Ambient Air Quality Standards (NAAQS).

The figures show a dramatic rise in PM2.5 mean daily concentration (the mean value for the analysed smog period was 0.20 mg/m^3 compared to the mean yearly of 0.07 mg/m^3). Concentration of NO_2 increased from 0.07 to 0.11 mg/m^3, and those of ozone increased from 0.034 to 0.042 mg/m^3. The mean concentration of CO (➊ *Fig. 2)* remained at the level of the yearly mean (2.8 mg/m^3), but at the end of July and the beginning of September increased concentrations were observed, the local maximum of 5.8 mg/m^3 being reached on the 5th of September.

Evaluation of the Adverse Health Effects

The health risk assessment methodology was used to evaluate the main adverse health effects. Specifically, concentration-response functions for selected air pollutants were used to calculate the number of cases for each effect.

A computer program "EpidCalc", developed in the Institute of Human Ecology and Environmental Hygiene, was used for the purpose of evaluation. The program's database contained complete reference information on the 25 most hazardous air pollutants that could be found in the ambient air of cities and

[1] Reference doses (RFD) and concentrations (RFC) are estimates of the pollutant daily exposure to the human population (including sensitive subgroups) that is likely to be without an appreciable risk of adverse effects over a lifetime. Regulatory agencies in many countries calculate (and publish) reference doses or concentrations for non-carcinogens. In the Russian Federation the Ministry of Health has established the national RFCs – MACs (maximum allowable concentrations) for a large list of adverse air pollutants, but not for all pollutants. They have not defined PM2.5 national RFC yet, that is why the NAAQS (USA EPA) RFC is used in this study. The measurements were made by the Moscow Centre for Hydrometeorology and Environment Monitoring and by the Public Environmental Enterprise "Mosecomonitoring".

Fig. 1

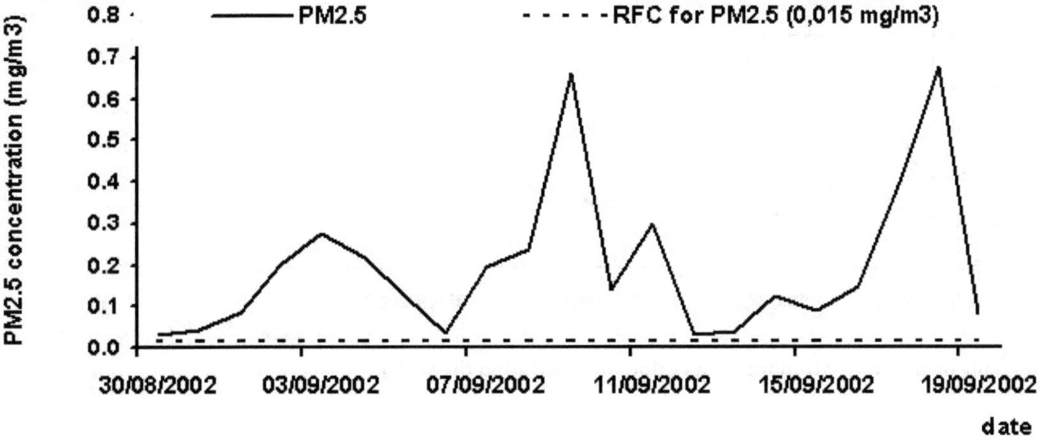

Mean daily PM2.5 concentrations for a period of time during the smog.

Fig. 2

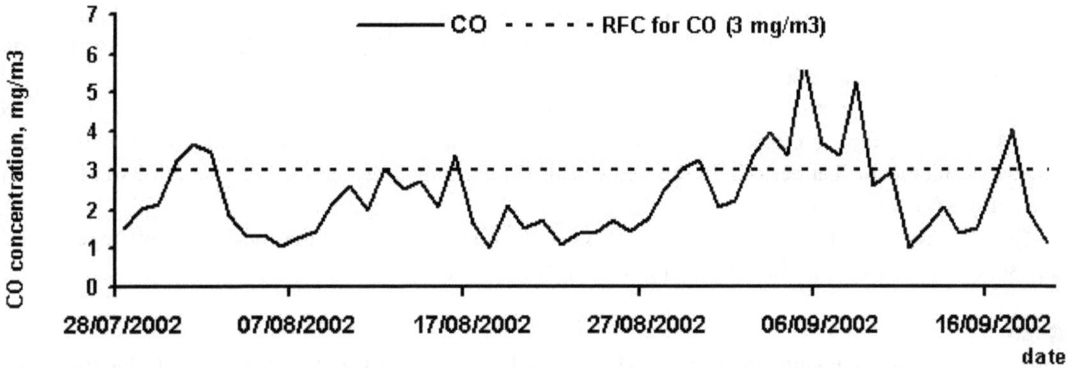

Mean daily carbon monoxide (CO) concentrations for a period of time during the smog.

Fig. 3

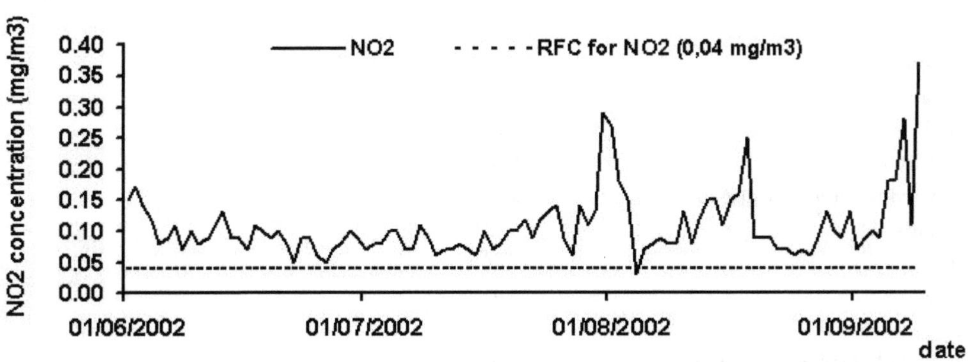

Mean daily nitrogen dioxide (NO_2) concentrations for a period of time during the smog.

Fig. 4

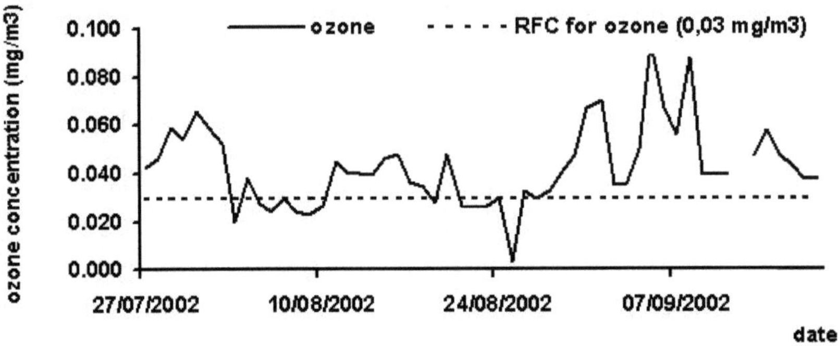

Mean daily ozone concentrations for a period of time during the smog.

towns and enabled calculation of 70 different exposure outcomes. The program used a mean daily concentration value as the input for a concentration-effect function to calculate the daily number of cases associated with an outcome. Then the cumulative effect was calculated by adding up all the daily counts for the time-period under consideration. The calculations were based upon concentration-response functions referenced in WHO (1999, 2000), U.S. Environmental Protection Agency (1999). The same approach is employed, for example, in the software tool AirQ developed by WHO and ECEH in order to facilitate European-wide health impact assessments.

In order to compare the impact of each pollutant on people's health, a ten-day period during the smog from August 30th through to September 8th was chosen. The pollutant concentration data for all pollutants was available for that period. The Moscow health and population statistics for year 2000 were used in calculations if it was required.

Table 1 presents the number of cases connected with the effects that may be caused by the residents' exposure to the selected pollutants during the ten-day period.

Tab. 1

The evaluated adverse health effects, caused by selected air pollutants for a ten-day period during the smog of Summer 2002 in Moscow.

Adverse effect	Number of cases
Carbon monoxide	
Total mortality	11,5
Acute MI	2,1
Ozone	
Total mortality	11,2
CVD mortality	5,0
Respiratory mortality	0,8
Hospitalisation with respiratory disease	2,8
Nitrogen dioxide	
Total mortality	12,6
CVD mortality	5,4
Acute MI	2,7
Hospitalisation with respiratory disease	2,0

Tab. 1 (Continued)

Adverse effect	Number of cases
PM2.5	
Total mortality	98,7
PM10 (recalculated from PM2.5 concentrations*)	
CVD mortality	56,4
Respiratory mortality	4,6

MI, myocardial infarctions; CVD, Cardiovascular diseases; PM, particulate matter with indicated diameter
* It was assumed that PM_{10} = 0.55*TSP and $PM_{2.5}$ = 0.33*TSP.

Table 1 evidently shows that exposure to PM2.5 and PM10 probably was the most hazardous compared with exposures to the other studied pollutants. 98.7 predicted cases of total mortality caused by PM2.5 compared to 11.5 cases by CO; 11.2 cases caused by ozone and 12.6 cases caused by NO_2. PM10 is also one of the main factors of the health damage (56.4 cases of CVD mortality compared to 5.0 caused by ozone and 5.4 by NO_2).

Additional analysis showed that the maximum number of daily total mortality cases from the exposure to PM2.5 and PM10 may have been 32 and 25.7 (on 18/09/2002) respectively, 3.5 cases from exposure to NO_2 (on 04/09/2002) and 1.8 cases from CO exposure (on 05/09/2002). This confirms the significance of PM2.5 and PM10 among studied pollutants.

Conclusions

The preliminary results obtained from the adverse health effects evaluation were used for a number of activities:

- In preparation of the informative letters to the Sanitary-and-Epidemiological Department of the Ministry of Health and Moscow Government.
- In working out the list of studies for deeper analysis of the adverse health effects of the Summer 2002 smog.
- In the development of a proposal to the Moscow Government for improvement of the air monitoring, access to the medical statistics and implementation of warning conditions for a weather extreme event and appropriate alert procedures in order to provide effective interaction among the different services of the authority in case of extreme effects in the future.

Studies are now being carried out on more precise characterisation of the ambient air pollution for the smog period and on analysis of real statistics of the adverse health effects. Unfortunately, due to the absence of daily statistics in computer form (Moscow hospitals and policlinics are obliged to give monthly summary reports only) and the lack of funds to pay for the extraction of the data from the hospitals records, these studies are being delayed. These problems (as well as others) forced the authorities to ask health risk methodology experts to evaluate the smog adverse health effects.

The studies will enable more reliable conclusions to be made about the magnitude of the health damage caused by the forest and peat bog fires in the summer period of 2002 and to improve the adverse health effect evaluation methodology.

References

Allred et al. (1991) Effects of Carbon Monoxide on Myocardial Ischemia. Environmental Health Perspectives 34:804–809

Aunan K (1996) Exposure-response functions for health effects of air pollutants based on epidemiological findings. Risk Analysis 16(5):693–709

Berlyand M (1985) Modern problems of atmospheric diffusion and air pollution. Gidromet, Leningrad, pp 58 (in Russian)

Hurley JF, Donnan PT (1999) Effects of Air Pollution on Health. Report for the ExternE Project, EC DGXII (JOULE Programme). In European Commission

U.S. Environmental Protection Agency (EPA) (1999). The Benefits and Costs of the Clean Air Act 1990 to 2010. Office of Air and Radiation. EPA 410-R-99-001. November

Vedal S (1996)·Evaluation of health impacts due to fine inhalable particles (PM2.5). Contract report prepared from Health Canada, final report November 1996

WHO (1998) Air Quality Monitoring for Evaluation of the Human Health Impact. WHO Regional Publications, European Series, No. 85.,Copenhagen

WHO (1999) Guidelines for Air Quality. Geneva, Switzerland

WHO (2000) World Health Organization Regional Office for Europe. Air Quality Guidelines for Europe, second edn. WHO Regional Publications. European Series, No 91. Copenhagen

Recommendations

Extreme Weather Events: What can We do to Prevent Health Impacts?

B. Menne [1]

Introduction

Extreme weather events are a problem worldwide. As has become clear during the events in 2002 and 2003, European populations are not immune. By definition, the term "extreme" refers to a departure from what is considered the norm. The extreme weather events considered in this book are meteorologic events that have a significant impact upon a local community, region and nation. Included in this definition are temperature and precipitation extremes, whose impact might be enhanced due to local conditions. How much climate extremes might become more frequent or intense over the next decades is highly uncertain. The heat wave that affected many parts of Europe has been seen by many as a "shape of things to come"(Beniston 2003, Schar & Jendritzky 2004). Floods vary in frequency, location and intensity as a result of seasonal and regional variations in precipitation and other weather conditions, and more long-term changes in the climate. Human activity also plays a role. Deforestation in mountainous regions accelerates runoff, thereby increasing the likelihood of flooding. Urban development on former food plains is likely to increase the magnitude of negative impacts of flooding events in the area itself, and to increase the likelihood of floods downstream due to 'canalisation' of rivers (EEA 2005).

Every year these events cause hundreds of deaths in Europe, along with disruption of local environments and communities. The authors of the several chapters in this book highlight that the occurence and severity of health impacts depend on the type of event, the speed of onset, the geographic location, the different population sensitivities, the destructive potential of the event, the baseline hygienic conditions and the existing burden of disease, the effectiveness of measures in place, the preparedness of health systems and its professionals, the rapidity of interventions and the information available and the ability to recover. Many of the health impacts of extreme weather events can be prevented and this is what this chapter is discussing.

From disaster response to risk management

Historically, disaster prevention strategies are separated into the following categories: structural and non-structural mitigation (e. g., building code regulations, warning response systems, disaster policies, retrofitted buildings); preparedness (e. g., individual evacuation); response (e. g., quick and adequate relief efforts); and recovery (e. g., both short- and long-term efforts such as rebuilding correctly or helping individuals and businesses survive) (McGeehin & Mirabelli 2001). A survey of Ministries of Health, carried out by the WHO found that there is a wide variation in preparedness and response activities among the 52 European Region member states. While ministerial activities in this area are mandated by current legis-

[1] with contributions from Paul Becker, German Weather Service, Giovanni Leonard, Health Protection Agency,United Kingdom and Cristiana Salvi, World Health Organization Regional Office for Europe

lation in almost every country, the level of support and of policy development varies greatly. There appears to be more activity in the area of earthquakes than for extreme weather events: among extreme weather events, floods are most likely to be subject of a prevention plan. National activities include programmes to monitor health effects, mobilization and response plans, national prevention plans and cooperative agreement with other agencies, departments, ministries and meteorological services. It is to be noted that several ministries have recently put heat wave prevention and response plans into place. Particularly impressive is the comprehensive nature and preventive orientation of many of these plans, which include strategies for the diagnosis of vulnerabilities, for health monitoring, for population advice, and for financial incentives to encourage vulnerability reduction. Also impressive is the high degree of integration of activities among health and related ministries existing in many countries. An important gap is the lack of formal direct cooperation with the national weather services. This survey showed that integrative actions such as prediction (information gathering on potential risks) and vigilance (information gathering to signal early evidence of stress and/or impact on health) as well as resource coordination are still lacking in some of the countries.

However, even when these systems are in place their success will be fundamentally dependent on a functioning and robust health system. In fact, historically health ministries have directly or indirectly responded to extremes by invoking the three classic public health strategies: disease prevention; the promotion of well-being; and health protection. While each type of extreme weather event and each locale requires a tailored response, public health is most effective where specific programmes form part of a well developed network and approach. Thus, where a strong system of disease surveillance is already in place, public health authorities can more easily incorporate special measures to detect emerging health impacts related to new environmental stressors. Likewise, where the home care network is already active, it can more easily incorporate extraordinary measures to protect the most vulnerable during heat waves and floods. Likewise health promition activities, provision of clean water, and shelter for the homeless, might be important components on which to build on. The participants to the meeting on "Extreme weather events and public health responses" identified several short and long term strategies that need to be strengthened when dealing with public health prevention and preparedness of extreme weather events:

Short term strategies are those actions important before, during and after an event, such as risk assessment, weather warning, rapid detection of impacts, international cooperation, risk communication and research. Long term strategies are those actions that contribute to prolonging our life on earth, supporting systems, and reducing those risk factors that contribute to the aggravation of extreme weather impacts, like urbanization and deforestation.

Risk assessment

It is useful to distinguish between the risk of an event like floods or heat-waves, which describes the probability of occurrence per magnitude, from vulnerability, which describes the inherent characteristics of a system that create the potential for harm [Sarewitz and Pielke Jr. 2002]. Development of effective and efficient responses requires that both vulnerability and risk be understood because the impact of an extreme event is determined by both the physical phenomena and the interaction of those characteristics with social and other systems. The assessment of vulnerabilities includes both environmental factors (examples: residential zones most liable to be flooded, urban areas where heat is trapped) and social and behavioural factors (examples: marginalized or non-autonomous populations, persons with low levels of immunity to infectious disease). For example, although risk of flooding is clearly confined to coastal and mountain and riverbed areas, very little systematic knowledge is available on certain groups within communities at risk (e. g., the elderly, the disabled, children, women, ethnic minorities, those with low incomes and those living alone) or on public and health care infrastructure at risk. This is aggravated by the increasing populations in flood-prone areas and major economic and industrial activities carried out there. A better understanding is needed of vulnerability risk factors, including delayed effect. In order to know more, risk areas need to be characterized for their potential sources of adverse health impacts (the estimated frequency; the location of chemical and nuclear plants and other hazardous sources, drinking water catchment areas, sewage plants, the location of dwellings, the location of public buildings (such as hospitals, kindergartens etc) and transport systems (UN 2001)). Further, focus on only extreme events may miss significant health impacts. Frequent versus extreme events may need different intervention strategies. The level of desired protection might need to be re-understood and be adjusted to frequent events if we consider climate change.

While public health authorities have begun to respond to heat-waves, cold-waves and floods, and in some cases have initiated programmes to prepare their populations and increase their capacity to tolerate extreme weather events, more work needs to be done to describe these responses and to evaluate their effectiveness. This is even more important in those areas were decades of activities are carried out by using structural and non-structural measures. These are rarely accompanied by health impact assessment.

Given the complexity of responses, resources must be mobilized, both from within and outside the health sector, to mitigate these vulnerabilities. In order that these resources truly strengthen population resilience in the domain of health (as opposed to other worthwhile objectives), public health authorities must both assume a coordinating role and act as primary advocates for health.

Weather warning systems for early public health action

Early warning is widely accepted as a crucial component of disaster risk reduction. Awareness of the importance of early warning systems is growing, owing to the recognition that significantly greater populations and assets are exposed to hazards and to concerns that the characteristics of extreme weather may be changing in the future (Yokohama …). Almost all countries maintain services to monitor weather hazards and provide public warnings of adverse conditions. In most of the cases this is done by the national weather services. Classically, weather warnings focus on the delivery of accurate information on specific weather hazards (windstorms, hurricanes, cyclones, etc) in a timely manner. Often more than one warning level is defined depending on fixed thresholds for the parameters of interest. There is no doubt that a lot more could be achieved by deploying resources to strengthen further early warning systems. However, the technological development of the warning system has been much faster than actually the social and policy components of early warning systems. Often technically well developed systems might be weak in knowledge of human vulnerability, inadequate communication of warnings, and lack of preparedness and capacity to act on warnings. Often little is actually known on which measures to use to protect human health. For example, heat health warning systems have now been developed in numerous European cities and vary widely in the methods used. The WHO has developed some indications on how best to incorporate evaluation criteria when the systems are developed, and the evaluation of these systems will show how effective they are. (Koppe, Jendritzky, Kovats et al. 2004) However, as long as we have no robust understanding of the cause-and-effect relationships between the thermal environment and health (epidemiological, statistical and bio-meteorological component); effective response measures to implement within the window of lead-time provided by the warning (public health component); and a community that is able to provide the needed infrastructure (public health component) (Auger & Kosatsky 2002), we might have a fantastic systems in place, but the reduction of health effects might be limited. Historically, there have also been problems with false alarms that might facilitate the loss of trust in the system and lead to inaction.

Flood forecasting does not normally fall within the range of the national weather authorities. However, at European level there are several European Flood Alert Systems (EFASs). These forecasting systems should be used by the major health care providers and emergency managers. So far little information has been provided to the general public on how best to protect themselves from extreme weather and climate events. For this to be done, consensus needs to be gained throughout Europe on the most effective measures and educational tools. Public participation in decision-making is a cornerstone of the successful implementation of integrated intervention plans. Important public information includes guidance on public health measures before, during and after floods, and on behavioural measures before, during and after heat-waves and cold-waves.

Rapid detection of impacts

Many extreme events cause health impacts which are only belatedly recognized. For example, Abendheim describes in this book that the increasing high levels of mortality during the heatwave were neither detected nor recognized by the health surveillance system until at least a week after its beginning. As causes he describes that the "epidemiology of mortality" is not conceived for surveillance purposes but for the analysis of trends in mortality several months to years later. Furthermore, there was no specific surveillance system for HREs (hyperthermia, for instance) and what is most interesting, the non-specific sources of data on the activities of emergency services did not produce relevant signals. It is interesting also to note that deaths were dispersed over the country, with most places only showing one or two extra deaths at the beginning (nursing homes, hospitals, smaller cities ...), and these were also not quickly detected because they concerned mainly older persons suffering from other diseases. Thus only clinicians who saw the patients recognized that the patients were suffering from heat related diseases and it was only after they communicated with each other that this became clearer. Therefore, the question here is what systems are needed to be able rapidly to detect signals of impacts early enough to intervene?

Collaborative work is currently being jointly undertaken by the health protection agency and National Health Service Direct in the United Kingdom, to see whether syndrome surveillance could be useful in early detection of heat related pathologies or mortality. Syndrome surveillance (Henning 2004), covers systems variously described as prodrome surveillance; outbreak detection systems; information system-based sentinel surveillance; bio-surveillance systems; health indicator surveillance; and symptom-based surveillance (CDC 2003). The fundamental objective of syndromic surveillance is to identify illness clusters early, before diagnoses are confirmed and reported to public health agencies, and to mobilize a rapid response, thereby reducing morbidity and mortality. The ability of syndrome surveillance to detect outbreaks/increased numbers of mortality earlier than conventional surveillance methods depends on such factors as the size of the outbreak, the population dispersion of those affected, the data sources and syndrome definitions used, the criteria for investigating threshold alerts, and the health-care provider‘s ability to detect and report unusual cases. In the United Kingdom, telephone calls to primary care advice service NHS Direct can be analyzed rapidly to inform intelligence of hazards potentially affecting public health, such as influenza pandemics or extreme weather events. Results of such analyses can then inform health protection interventions in the health systems and beyond (Nicoll et al. 2004) NHS Direct surveillance can be adapted to the hazards of interest in a given time period and help guide the response. For example, it seems that in the United Kingdom calls characterized as "heatstroke" were the earliest indication of the 2003 heat wave, it has now been agreed to monitor systematically those calls in the summer period, so that the interventions included in the Heat Wave Plan can be triggered timely.

Portugal has put efforts into the development of a computerized system of rapid mortality reporting, so that in the case of abnormal mortality increases these can be rapidly detected. However, both systems have still to be tested and their effectiveness evaluated. Further European collaborative efforts are useful in this regard.

International coordination

Weather and climate do not have borders, and so the entire European Region is at risk from the health impacts of extreme weather and climate events. Preventive measures, policies and strategies must therefore be part of a coordinated international effort to enhance and protect human well-being today and tomorrow. There are many ongoing international activities, some of them have been described in chapters earlier. These international efforts should set goals in their cooperation, for example develop indicators for intra- and intercountry comparison and monitoring of progress, coordinate the processes in developing implementing and evaluation early warning systems, assist and provide guidance in capacity building, as well as inform their communities about progress.

Risk Communication

Extreme weather events and the impact they have on health, highlight the need to keep the community and the general population informed, on the associated risks for individuals and protection measures.

In the heat of a crisis (flooding or heatwaves), effective media communication will directly affect the health outcome of an event. If crisis are difficult to predict, a media communication strategy can and should be planned beforehand. Then it should be adapted to the current emergency. Preparedness gives greater confidence and control in times of emergency that is perceived by the public as an important factor in effective crisis management. "Have a plan ready" is the rule, then be prepared to change it.

As part of risk management, the over-arching communication goal during a crisis is to foster public resilience and guide appropriate public participation to support the rapid containment of the crisis, thus limiting morbidity and mortality, and minimizing the damage to a nation's international standing, its economy and its public health infrastructure. This includes integrating risk communication with risk analysis and risk management from the start; informing and involving various publics (policy makers, media, general public and other relevant stakeholders) early in a crisis, being transparent about what is known and unknown. This helps build trust and credibility, which are strongly associated with public acceptance of official guidance; responding to and validating publics' concerns, empathize with their fears, and show realistic "human" coping behaviour. This helps being perceived as more credible and trustworthy; considering risk perception as a key factor that can lead to poor communication outcomes. People's risk judgments are influenced by many aspects other than just statistical data, such as their values, emotions, group affiliations, socio-economic status, trust in institutions, and sense of control.

Effective media communication includes a media communication plan, trained key staff, and practice of prepared response. The communication plan includes identification of the magnitude of the health problem possibly including a future scenario; goal(s): to inform, persuade, or motivate; lead spokespersons and public information officers respectively trained in risk communication and public health; partners in the communication; key audiences prioritized in order of relevance; appropriate communication channels; a budget for implementation; means for measuring achievement of results.

Monitoring and evaluating the outcomes is a critical activity to measure the risk communication effectiveness. Outcome evaluation provides information about the value of communication activities, e. g., whether the target audience learned, acted, or made a change because of communications and it typically involves before/after comparisons.

The goal of risk communication is not measurable in "quantity" terms: the numbers of articles, radio and TV interviews do not convey the success of the communication. The objective to achieve is to contribute to the control of the crisis, which means that the correct messages at the right time have been delivered.

Research

Assessment of the environmental and health consequences of heat-waves has highlighted a number of knowledge gaps and problems in public health responses. In the past, heat-waves were not considered a serious risk to human health with "epidemic" potential in the European Region. In order to reduce the health impacts of future heat-waves, fundamental questions need to be addressed, such as whether a heat-wave can be predicted, detected or prevented, and how this may be done. Knowledge gaps exist: in characterizing the relationship between heat exposure and a range of health outcomes; in understanding interactions between harmful air pollutants and extreme weather and climate events; in harmonizing episode analyses; and in evaluating the effectiveness of heat-related public health interventions. There is ongoing debate on whether and how to develop heat health warning systems, provide space cooling in particular locations and develop public advice and community-based activities that support the social and medical welfare of the elderly and other high-risk groups in order to reduce their vulnerability to temperature extremes. Cost-effectiveness analyses will be needed. With regard to floods very little is known on flood morbidity and mortality using routine data sources or pre-existing cohorts, impacts of floods on health care systems and flood early warning systems effectiveness.

Conclusions

Ministries of Health and the Environment recognized at the fourth Ministerial Conference for Environment and Health in Budapest the increasing evidence regarding the role of human activities in contributing to climate change and recognized the increasing short-term and long-term hazards posed to human health. They further recognized that climate is already changing and that the intensity and frequency of extreme weather events, such as floods, heat-waves and cold spells, may change in the future. This demands a proactive and multidisciplinary approach by governments, agencies and international organizations and improved interaction on all levels from local to international.

Ministries decided to take action to reduce the current burden of disease due to extreme weather and climate events and invited the WHO, in collaboration with the World Meteorological Organization, the European Environment Agency (EEA) and other relevant organizations, to support these commitments and to coordinate international activities to this end. These activities include the development of guidelines for estimating the burden of disease due to weather and climate extremes; to develop indicators for intercountry and intracountry comparison and monitoring of progress; to coordinate the development of new methods, including sentinel monitoring and surveillance systems, to provide timely information on the health impacts of weather and climate extremes at the European level; to develop and evaluate more effective and efficient interventions, such as early warning systems, to reduce negative impacts; and to harmonize interventions across regions and countries to facilitate the sharing of data and lessons learnt.

At the intergovernmental meeting in 2007, progress within countries and the coordination activities will be discussed (➔ *Annex*).

References

Auger N, Kosatsky T (2002) Hot weather watch/warning system: a proposal for Montreal-Centre. World Health Organization

Bravata DM et al. (2004). Systematic review: surveillance systems for early detection of bioterrorism-related diseases. Ann Intern Med 140:910–922

CDC MMWR (2003) Syndromic Surveillance: Reports from a National Conference

EEA (2005) Climate change and river flooding in Europe, EEA Briefing 1.2005, Copenhagen, Denmark

Henning KJ (2004) What is syndromic surveillance? MMWR Morb Mortal Wkly Rep 53 (Suppl) 5 – 11

ISDR (2005) Hyogo Framework for Action 2005 – 2015: Building the Resilience of Nations and Communities to Disaster, World Conference on Disaster Reduction, Kobe, Hyogo, Japan

Koppe C, Jendritzky G, Kovats RS, Menne B (2004). Heat-waves: impacts and responses.: World Health Organization, Copenhagen

McGeehin MA, Mirabelli M (2001) The potential impacts of climate variability and change on temperature-related morbidity and mortality in the United States. Environ Health Perspect 109 (Suppl 2):185 – 189

Nicoll A, Smith G, Cooper D, Chinemana F, Gerard E (2004) The public health value of syndromic surveillance data: calls to a national health help-line (NHS Direct). Eur J Public Health 14(4)(Suppl):69

Annex

Public Health Response to Extreme Weather and Climate Events

Working Paper of the 4th Ministerial Conference for Environment an Health, Budapest, June 2004

Summary

Severe floods, windstorms, heat-waves and cold-waves have caused dramatic political, social, environmental and health consequences in Europe over the past few years. In response to these events, ministries of health and other public health authorities, along with national and international meteorological services and organizations, are focusing increased attention on developing appropriate strategies and measures to prevent health effects from extreme weather and climate events in the future. Efforts are being made to understand the lessons learnt from recent events, to evaluate the effectiveness of the measures taken and early warning systems in place, and to use the knowledge gained to target future activities. The recent events have also increased interest in whether the intensity and frequency of future extreme weather and climate events could be expected to change as one result of a changing climate.

With this in mind, a working group organized by the World Health Organization (WHO) and the European Environment Agency (EEA) has made the following recommendations.

1. The political, social, environmental and health consequences of extreme weather events have increased in Europe in recent years. We recognize that the climate is already changing, and that the intensity and frequency of extreme weather events, such as floods, heat-waves and cold-waves, may change in the future. These events will continue to pose additional challenges to current and future populations, in terms of health risk management and the reliability of infrastructure, including health services, power supply and others.
2. There is a need for ministries of health and other ministries to recognize that actions must be taken to reduce the current and future burden of disease due to extreme weather and climate events and to include the prevention of health effects due to weather and climate extremes among national health priorities.
3. We urge ministries of health and other ministries, as well as research institutions, to improve our understanding of the regional and national burden of disease due to weather and climate extremes and to identify effective and efficient interventions, such as early warning systems, surveillance mechanisms and crisis management.
4. We urge effective and timely coordination and collaboration among public health authorities, meteorological services and agencies (national and international), emergency response agencies and civil society in developing local, regional and national monitoring and surveillance systems for the rapid detection of extreme weather events and their effects on public health; developing civil emergency and intervention plans, including activities to prevent morbidity and mortality due to weather and climate extremes; and improving public awareness of extreme weather events, including actions that can be taken at individual, local, national and international levels to reduce impacts.
5. We call on WHO, through its European Centre for Environment and Health, in collaboration with the World Meteorological Organization, the European Commission, EEA and other relevant organi-

zations, to support these commitments and to coordinate international activities to this end. In particular, there is a need to develop guidelines for estimating the burden of disease due to weather and climate extremes; to develop indicators for intercountry and intracountry comparison and monitoring of progress; to coordinate the development of new methods, including sentinel monitoring and surveillance systems, to provide timely information on the health impacts of weather and climate extremes at the European level; to develop and evaluate more effective and efficient interventions, such as early warning systems, to reduce negative impacts; and to harmonize interventions across regions and countries to facilitate the sharing of data and lessons learnt.

Background

1. Severe floods, windstorms, heat-waves and cold-waves have affected the World Health Organization (WHO) European Region during the last few years. Their political, social, environmental and health consequences have stimulated debate on whether appropriate action might, at least in part, prevent the health effects of such extreme weather and climate events. In response, the European Environment and Health Committee and the WHO Regional Committee for Europe requested that the WHO European Centre for Environment and Health (ECEH) organize a meeting to exchange information and to discuss and develop recommendations on public health and environmental responses to weather and climatic extremes, to be submitted to the European Member States at the Fourth Intergovernmental Preparatory Meeting in Malta. The meeting on "Extreme weather events and public health responses", held in Bratislava on 9 and 10 February 2004, was convened by WHO, in collaboration with the European Environment Agency (EEA), and hosted by the Ministry of Health of Slovakia. This document presents its recommendations concerning actions to identify, prevent and reduce the health impacts of extreme weather and climate events, in particular floods, heat-waves and cold-waves. In developing these recommendations, the organizers and participants incorporated the available knowledge and experience of the many international, European, national and local initiatives on the subject, in order to avoid overlaps and duplication.

Synopsis of the issue

Potential changes in the intensity and frequency of extreme weather and climate events

2. Europe has experienced an unprecedented rate of warming in recent decades. Between 1976 and 1999, the average number of periods of extreme warmth each year increased twice as fast as the corresponding reduction in the number of periods of extreme cold. Over the same period, in most of Europe, the increase in the mean daily maximum air temperature during the summer months was greater than 0.3 °C per decade. For example, the frequency of very hot days in central England has increased since the 1960s, with extremely hot summers in 1976, 1983, 1990 and 1995. Sustained hot periods have become more frequent, particularly in May and July.
3. It is predicted that the current increasing instability of the climate system may lead to increased climate variability and, with it, a change in the frequency and intensity of extreme temperatures. An unprecedented heat-wave affected the European Region in the summer of 2003. The intensity and duration (two weeks) of the heat-wave in France was exceptional in the country's meteorological history.

4. Climate variability is expected to increase with rising ambient air temperatures. Coldwaves are expected to continue to affect areas already vulnerable to cold temperature extremes. The wind and ice storms that may be associated with cold-waves could cause interruptions in electricity supplies in several countries.
5. Flood hazards and associated health risks have increased in many areas because of a number of climatic and non-climatic factors. The latter include the impact of changes in terrestrial systems (hydrological systems and ecosystems) and economic and social systems. Land use changes, which induce land cover changes, affect the rainfall-runoff relationship.

Deforestation, urbanization and reduction of wetlands decrease the available water storage capacity and increase the runoff coefficient, leading to growth in flood amplitude and reduction of the time-to-peak. Urbanization has adversely influenced flood hazards by increasing the extent of impervious areas. In addition, more industrial and human activities are carried out in floodprone areas. Flood losses in 2002 were higher than in any single year in the past. The floods in central Europe in August 2002 (on the rivers Danube and Elbe and their tributaries) caused damage exceeding € 15 billion. Damage was also caused to water and electricity installations and health care institutions.

Recommendation A

The political, social, environmental and health consequences of extreme weather events have increased in Europe in recent years. We recognize that climate is already changing, and that the intensity and frequency of extreme weather events, such as floods, heat-waves, and cold-waves, may change with a changing climate. These events will continue to pose challenges to current and future generations, in terms of health risk management and the reliability of infrastructure, including health services, power supply, and others.

Extreme weather and climate events pose a risk to human health in the European Region

6. A heat-wave struck France in early August 2003 after warmer than average temperatures in June and July. The period from 4 to 12 August broke all historical records since 1873 for Paris in terms of minimum, maximum and average temperatures, and duration. This unprecedented heatwave, associated with high levels of air pollution, was accompanied by excess mortality that began early and rose quickly: 300 excess deaths on 4 August, 1200 on 8 August and 2200 on 12 August. Total excess mortality between 1 and 20 August, compared with average daily mortality for the period 2000–2002, was 14 802. This represents an increase of 60 % in mortality from all causes. The observed excess mortality particularly affected the elderly (70 % for those aged 75 years and over), but was also severe for those aged between 45 and 74 years (30 %). In all age groups, female mortality was 15 % to 20 % higher than male mortality. Almost the whole country was affected by the excess mortality, although its intensity varied significantly from one region to another: 20 % in Languedoc-Roussillon (south), but 130 % in Île-de-France (Paris and suburbs). The excess mortality clearly increased with the duration of extreme temperatures. The mortality rate was highest in nursing homes where the number of deaths observed was twice that expected.[1]
7. High excess mortality rates were also observed in other European countries, such as Italy, where general mortality for all ages in the 21 capitals of the Italian regions was 15 % higher than during the same period in 2002; in particular, the figure for those aged 75 years and over was 21 %. It may be noted that deaths of elderly people represent more than 90 % of the overall excess mortality. Portugal recorded a

[1] Impact sanitaire de la vague de chaleur d'août 2003. Bulletin épidémiologique hebdomadaire, 2003, 45–46.

26 % increase in mortality in August 2003 compared to the average of the previous five years. Information on the full extent of the impacts of the 2003 heatwave will be available soon.

8. Studies from several countries indicate that mortality risk increases every winter in all European countries. During cold-waves, increased risk of injuries and frostbite has been observed in northern central Europe, northern Europe and the mountainous regions of Europe. Cold-waves may affect cardiovascular and peripheral diseases, cerebrovascular diseases and respiratory diseases, and may contribute to communicable diseases. Cold-waves are likely to affect poorer and marginalized population groups more severely.
9. Health impacts of floods can occur during or after flooding events. Relatively low numbers of flood-related deaths are recorded in Europe in comparison with other regions. Between 1980 and 2002, 260 flooding events resulted in 2500 victims.[2] The number of deaths associated with flooding is closely related to the life-threatening characteristics of floods (rapid rising of water, deep flood water and objects carried by the rapid flow of water) and the behaviour of victims. Injuries (such as sprains, strains, lacerations and contusions) may occur during flooding, but are more frequent in the aftermath of a flood disaster as residents return to their homes to clean up damage and debris. Infectious diseases are not common and are normally confined to illnesses endemic to the flooded region. Most of these illnesses are attributable to reduced sanitation or to overcrowding among displaced people. Some studies have shown an increased incidence of common mental health disorders for long periods after a flooding event. Anxiety and depression may last for months and possibly even years after the flood event and so the true health burden is rarely appreciated. During the 2002 floods in Dresden, two public health issues needed immediate attention: 1) the maintenance of public hygiene; and 2) the problems involved in evacuating complete hospitals.

Recommendation B

There is a need for ministries of health and other ministries to take actions to reduce the current and future burden of disease due to extreme weather and climate events and to include the prevention of health effects due to weather and climate extremes among national health priorities.

Knowledge gaps in risk assessment and management

10. Assessment of the environmental and health consequences of heat-waves has highlighted a number of knowledge gaps and problems in public health responses. In the past, heat-waves were not considered a serious risk to human health with “epidemic” potential in the European Region. In order to reduce the health impacts of future heat-waves, fundamental questions need to be addressed, such as whether a heat-wave can be predicted, detected or prevented, and how this may be done. Knowledge gaps exist: in characterizing the relationship between heat exposure and a range of health outcomes; in understanding interactions between harmful air pollutants and extreme weather and climate events; in harmonizing episode analyses; and in evaluating the effectiveness of heat-related public health interventions. There is ongoing debate on whether and how to develop heat health warning systems, provide space cooling in particular locations and develop public advice and community-based activities that support the social and medical welfare of the elderly and other high-risk groups in order to reduce their vulnerability to temperature extremes. Cost-effectiveness analyses will be needed.

[2] EM-DAT: The OFDA/CRED International Disaster Database. Brussels, Université Catholique de Louvain (http://www.cred.be/emdat, accessed 27 February 2004).

11. The vulnerability of European populations to extreme weather and climate events depends on the type of natural hazard, the geographical location, the capacity to anticipate the risk, the capacity to intervene and resist, and the ability to recover from the impact of the events. For example, although risk of flooding is clearly confined to coastal and mountain and riverbed areas, very little systematic knowledge is available on certain groups within communities at risk (e.g. the elderly, the disabled, children, women, ethnic minorities, those with low incomes and those living alone) or on public and health care infrastructure at risk. This is aggravated by the increasing populations in flood-prone areas and major economic and industrial activities carried out there. Flood protection is based on structural and non-structural measures. The level of available structural measures depends on the flood return periods. There is a need for more, better quality, epidemiological data before vulnerability indices can be used operationally to minimize the effects of flooding. With better information, the emphasis in disaster management could move from post-disaster improvisation to pre-disaster planning. A comprehensive, riskbased emergency management programme of preparedness, response and recovery would have the potential to reduce the adverse health effects of floods.
12. While public health authorities have begun to respond to heat-waves, cold-waves and floods, and in some cases have initiated programmes to prepare their populations and increase their capacity to tolerate extreme weather events, more work needs to be done to describe these responses and to evaluate their effectiveness.

Recommendation C

We urge ministries of health and other ministries, as well as research institutions, to improve the understanding of the regional and national burden of disease due to weather and climate extremes and the identification of effective and efficient interventions, such as early warning systems, surveillance mechanisms and crisis management.

Risk management and communication

13. The health crisis in France caused by the heat-wave in 2003 was unforeseen and was only detected belatedly. Health authorities were overwhelmed by the influx of patients; crematoria and cemeteries were unable to deal with the influx of bodies; and retirement homes were underequipped with air-conditioning or space cooling environments and manpower. The crisis was compounded by the fact that many elderly people were living alone without a support system and without proper guidelines to protect themselves from the heat. The heat-wave highlighted several problems in public health systems, including the limited number of experts working in the area of environment and health and the need for a significant improvement in the exchange of information between several public organizations and agencies, as well as a clear definition of responsibilities in this area.
14. Fortunately, other countries were affected to a much lesser extent. We therefore do not know how other countries or federal health systems would have managed with this level of crisis. However, there are some commonalities that are important to consider. First, the population of Europe is ageing and more elderly people are living alone. Second, the health and environment surveillance mechanisms needed to rapidly detect an "epidemic" due to a heat-wave do not exist. Finally, there is a lack of definition of responsibilities.
15. In 2002, the Council of Europe assessed the level of implementation of the Yokohama Strategy for the EUR-OPA Major Hazards Agreement.[3] The Yokohama Strategy stressed that more attention should be

[3] Member States signatory to the EUR-OPA Major Hazards Agreement: Albania, Algeria, Armenia, Azerbaijan, Belgium, Bulgaria, Croatia, Cyprus, France, Georgia, Greece, Lebanon, Luxembourg, Malta, Republic of Moldova, Monaco, Morocco, Portugal, Romania, Russian Federation, San Marino, Spain, The former Yugoslav Republic of Madeconia, Turkey, Ukraine.

paid to prevention. The results of the 26-country assessment showed that there is a complex distribution of competences in the management of natural and man-made hazards. There is also a lack of coordination; poor rehabilitation mechanisms; and poor sanction and control mechanisms. Most attention is focused on crisis management and relief interventions in case of a disaster; these are designed to address the most urgent needs of the affected population. It is certainly positive that much been invested in improving the speed and efficiency of relief intervention, but it should not be forgotten that many interventions could be avoided through better risk prevention and much human suffering could be reduced with structural compensation mechanisms.[4]

16. Flood risk management has been carried out for many years by a variety of organizations, commissions and agencies at international, national, river-basin and local levels. However, with larger vulnerable populations and more complex infrastructures, consideration must be given to including health aspects in (national) flood prevention plans. The possibility of climate change in decades to come further emphasizes the need for early warning and flood forecasting. A major risk in operating early warning systems lies in the possibility of false alarms, due to either under or over prediction of the hazard. An effective early warning and forecasting system that extends reaction time should be supported by meteorological information and the earliest possible warning of extreme weather conditions. A European Flood Alert System (EFAS) and a European Flood Forecasting System are being developed for a large part of Europe.[5] These forecasting systems should be used by the major health care providers and emergency managers.
17. So far, little information has been provided to the general public on how best to protect themselves from extreme weather and climate events. For this to be done, consensus needs to be built throughout Europe on the most effective measures and educational tools. Public participation in decision-making is a cornerstone of the successful implementation of integrated intervention plans. Important public information includes guidance on public health measures before, during and after floods, and on behavioural measures before, during and after heat-waves and cold-waves.

Recommendation D

We urge effective and timely coordination and collaboration among public health authorities, meteorological agencies (national and international), emergency response agencies and civil society in developing local, regional, and national monitoring and surveillance systems for rapid detection; developing civil emergency and intervention plans, including activities to prevent morbidity and mortality due to weather and climate extremes; and in improving public awareness of extreme weather and climate events, including actions that can be taken at individual, local, national and international levels to reduce impacts.

International cooperation

18. Weather and climate do not have borders, and so the entire European Region is at risk from the health impacts of extreme weather and climate events. Preventive measures, policies and strategies must therefore be part of a coordinated international effort to enhance and protect human well-being today and tomorrow.

[4] Comparative study of the regulations concerning major risk management in the 26 Member States of the Council of Europe's EUR-OPA Major Hazards Agreement. Strasbourg, Council of Europe, 2003 (document AP/CAT (2003) 39).

[5] Best practices on flood prevention, protection and mitigation. Discussion paper for review by the Water Directors of the European Union (EU), Norway, Switzerland and Candidate Countries at their meeting in Athens on 17 and 18 June 2003 (http://www.floods.org/PDF/Intl_BestPractices_EU_2004.pdf, accessed 1 March 2004).

19. There are many ongoing international activities, such as the UNECE guidelines on sustainable flood prevention and those for flood follow-up described in the European Water Directors' best practices guide. The World Meteorological Organization (WMO) and its national meteorological offices have a primary mandate to provide quick and reliable weather and climate information. Currently they are developing weather "indices" for human well-being and are involved in flood forecasting and weather forecasting activities, as well as in the development of showcases on "heat health warning systems". EEA is developing indicators, including some to express the impacts of heat-waves and floods on the environment and human health.
20. As indicated earlier, there is a need to develop guidelines to estimate the burden of disease due to weather and climate extremes, as well as to develop indicators for intra- and intercountry comparison and monitoring of progress. WHO and other organizations should provide guidance in this and set up international scientific teams. With the help of WMO, it may be possible in the future to provide timely information on the health impacts of weather and climate extremes at European level. It is very important to learn from lessons experienced at national and local levels. International cooperation might help to make this information available more widely and assist other countries in the development of similar measures and strategies. Civil protection agencies, the International Committee of the Red Cross and other organizations have long experience in risk preparedness and response, and could share that experience with national and local authorities. There is a particular need to develop optimal techniques for monitoring the health of populations during extreme weather events; monitoring sentinel health conditions may be one approach to the early recognition of effects on population health.

Recommendation E

We call on the World Health Organization, through its European Centre for Environment and Health, in collaboration with the World Meteorological Organization, the European Commission, the European Environment Agency, the International Committee of the Red Cross and other relevant organizations, to support these commitments and to coordinate international activities to this end. In particular, there is a need to develop guidelines for estimating the burden of disease due to weather and climate extremes; to develop indicators for intercountry comparison and monitoring of progress; to coordinate the development of new methods, including sentinel systems for monitoring and surveillance, to provide timely information on the health impacts of weather and climate extremes at European level; to develop and evaluate more effective and efficient interventions, such as early warning systems, to reduce negative impacts; and to harmonize interventions across regions and countries to facilitate the sharing of data and lessons learnt.

Conclusion

21. Based on this document, it is recommended that ministries of health and other ministries commit themselves to taking action to reduce the current burden of disease due to extreme weather and climate events. We call on the World Health Organization, through its European Centre for Environment and Health, in collaboration with the World Meteorological Organization and other relevant organizations, to support these commitments and to coordinate international activities to this end. We also recommend that progress should be reported on at the intergovernmental meeting in 2007.

Follow-up Programme on the Influence of Meteorological Changes Upon Cardiac Patients

Inge Heim

Introduction

It is well-known from experience that weather and climatic changes influence the health of people in general (1, 5) and particularly health of the already frail populations (2, 3, 4, 6, 7), but it is not known whether one factor or a combination of factors play the decisive role. It is also not yet clear whether it is the absolute level of key meteorological parameters or the speed at which they change which influences people's health.

Meteorological changes primarily affect the human organism through the autonomous nervous system, resulting in differing effects on psychological functions (9), the immune system (9) and the function of the respiratory and cardiovascular system. However, the exact mechanism of this influence is unclear.

The purpose of our study is to identify the relevant meteorological factors and to assess their influence on the health of cardiac patients (6, 7, 8).

Aim of the study

Main aims of the investigation are as follows:

- to collect data measuring absolute values of meteorological parameters (air temperature, humidity, winds and other relevant parameters) by defined time intervals and to analyse their changes;
- to determine the frequency of the most important disorders of the cardiovascular system in cardiac patients during the same time intervals;
- to search for and analyse connections between meteorological factors and human health disorders.

Materials and methods

The investigation has started in February 1999 as a 5-year programme. This is a prospective epidemiological study. Data has been collected continuously through these 5 years. The target population was the population of Zagreb, capital of Croatia (779,145 inhabitants – census 2001). The patients covered by the study all came to our Polyclinic for a check-up. They included coronary heart patients, patients with hypertension, patients with arrhythmias and patients with multiple risk factors for atherosclerosis. The sample size is expected to be about 10,000.

While they were sitting in the waiting room a nurse gave them a questionnaire and asked them to fill it in after she had explained the way to do it. This was done in two locations: in the center of Zagreb where mostly older people live and in the newly-built part of the city where mostly younger people live, so all age groups were included. These are the two locations where the Polyclinic has its premises. This is the only institution in Zagreb, apart from hospitals or private surgeries, where cardiac patients can come for a check-up.

The questionnaire has two pages and consists of 18 questions (date of filling in, gender, age, height, weight, reason for a check-up, risk factors: hypertension, diabetes, hypercholesterolemia, hypertriglyceridemia, smoking, and physical activity. Furthermore, there is a question about previous diseases such as: myocardial infarction, stroke, chronic heart failure, angina pectoris, arrhythmias, coronary artery bypass grafting or PTCA, and current therapy. This is followed by questions on their physical fitness, sense of fatigue, chest pain at rest or exertion, atypical chest pain, palpitations, dizziness, loss of consciousness, headache, nervousness, anxiety, sadness, sense of hopelessness, etc (subjective parameters). These answers are graduated on a 5-point scale from very weak, rather weak, average, quite strong to much stronger if they have had these symptoms over a certain period of time. Finally they have to answer 'yes' or 'no' to whether they are constantly exhausted and if something has recently shaken them (death of a relative or a close friend, car accident, financial loss, loss of a job, etc.). After the patients have filled in the questionnaire the nurse fills in the heart rate and arterial blood pressure. The Cardiologist describes the ECG changes (objective parameters).

Groups of patients

There are two main groups of patients:

- patients who come to our Polyclinic for a check-up because they do not feel well or for diagnostic reasons;
- patients who attend our out-patient rehabilitation program – continuous observation over one to three months.

We also register the number of patients per day admitted to Zagreb hospitals due to acute myocardial infarction, unstable angina pectoris or chronic heart failure and sudden death. The source of information is hospital protocol and the Register of acute myocardial infarction for Zagreb area.

Separately, data on the following meteorological parameters for the Zagreb area are collected:

- air temperature in °C
- relative humidity in %
- atmospheric pressure in h Pa
- winds (direction, strength in m/s), etc.

The biometeorologist decides which other parameters should be included. These parameters are measured at least once every 24 hours or, if necessary, in even shorter time periods if the biometeorologist thinks it necessary.

The analyses will include the correlation of subjective symptoms and objective health status parameters with meteorological parameters. So, this is a team effort. The team consists of different experts: epidemiologist (Inge Heim), cardiologists (Mirjana Jembrek-Gostovic, Vladimir Jonke), psychologist (Dubravka Kruhek-Leontic), nurses and biometeorologist (Ksenija Zaninovic). The study has been carried out in two institutions: Polyclinic for Cardiovascular Prevention and Rehabilitation and Croatian Institute of Hydrometeorology, and financed by the City of Zagreb – Office for health, occupation and social care.

Public health benefits

We expect that this study will cast some light on the influence of meteorological conditions on cardiac patients. It could help us in introducing some additional public health measures which could help the patients to overcome unfavourable weather conditions with fewer consequences. More lives could be saved and less money spent on health care. In addition, employers could achieve better working results by improving working conditions.

References

1. Aylin P, Morris S, Wakefield J, Grossinho A, Jarup L, Elliot P (2001) Temperature, housing, deprivation and their relationship to excess winter mortality in Great Britain, 1986–1996. Int J Epidemiol 30:1100–8
2. Danet S, Richard F, Montaye M, Beauchant S, Lemaire B, Graux C, Cottel D, Marecaux N, Amouel P (1999) Unhealthy effects of atmospheric temperature and pressure on the occurrence of myocardial infarction and coronary deaths. A 10-year survey: the Lille-World Health Organization MONICA PROJECT (Monitoring trends and determinants in cardiovascular disease). Circulation 100:E1-7
3. Kloner RA, Poole WK, Perritt RL (1999) When throughout the year is coronary death most likely to occur? A 12-year population-based analysis of more than 220,000 cases. Circulation 100(15):1630–4
4. Koken PJ, Piver WT, Ye F, Elixhauser A, Olsen LM, Portier CJ (2003) Temperature, air pollution, and hospitalization for cardiovascular diseases among elderly people in Denver. Environ Health Perspect 111(10):1312–7
5. Lawlor DA, Davey Smith G, Mitchell R, Ebrahim S (2004) Temperature at birth, coronary heart disease, and insulin resistance: cross sectional analyses of the British women's heart and healthy study. Heart 90:381–8
6. Nayha S (2002) Cold and the risk of cardiovascular diseases. A review. Int J Circumpolar Health 61:373–80
7. Pell JP, Cobbe SM (1999) Seasonal variations in coronary heart disease. QJM 91:689–96
8. Pleško N, Goldner V, Rezakovic Dz, Zaninovic K, Zecevic D (1983) Karakteristike vremenskih prilika u sedmodnevnim razdobljima s velikim brojem infarkta miokarda u Zagrebu. Acta Med Jug 37:3–17
9. Sher L (2001) Effects of seasonal mood changes on seasonal variations in coronary heart disease: role of immune system, infection, and inflammation. Med Hypotheses 56:104–6

[illegible] the unfavourable weather conditions will [illegible]
[illegible]
[illegible]

[illegible]

[illegible]

Subject Index

P

Y

Z

MIX
Papier aus verantwortungsvollen Quellen
Paper from responsible sources
FSC® C105338

If you have any concerns about our products,
you can contact us on
ProductSafety@springernature.com

In case Publisher is established outside the EU,
the EU authorized representative is:
Springer Nature Customer Service Center GmbH
Europaplatz 3, 69115 Heidelberg, Germany

Printed by Libri Plureos GmbH
in Hamburg, Germany